INTRODUCTION TO MARINE BIOLOGY

INTRODUCTION TO MARINE BIOLOGY

Barry Fell
Museum of Comparative Zoology
Harvard University

HARPER & ROW, PUBLISHERS
New York, Evanston, San Francisco, London

Sponsoring Editor: Joe Ingram
Project Editor: Holly Detgen
Designer: T. R. Funderburk
Production Supervisor: Will C. Jomarrón

INTRODUCTION TO MARINE BIOLOGY

Library of Congress Cataloging in Publication Data

Fell, Barry.
Introduction to marine biology.

Includes index.
1. Marine biology. I. Title.
QH91.F4 574.92 74-31119
ISBN 0-06-042034-0

The Prayer of Maui

For us mariners, far from the shores of our motherland, for our six ships, and for our Admiral Rata. For the protection of these my brothers upon the waters I pray to Rono to calm the wind and to reveal to us the homeward track. Let not my brothers perish, I ask of thee, O Kopu.

Inscribed in the month of November, 15th regnal year of Pharaoh Ptolemy III (232 B.C.), on the eve of the Polynesians' epic voyage to America. The prayer was found with the navigator's calculations in Sosorra Cavern (see page 12), and deciphered by the author. Rono is the Polynesian god of fair weather, Kopu is the planet Mercury personified, the patron god of travelers.

CONTENTS

PREFACE

This text is based on courses in marine biology offered at Harvard University. Most of the illustrations are from the laboratory notebooks of students and former students, and they were deliberately selected so as to exemplify the level of systematic competence that can be achieved without undue demand in courses of one or two semesters' length. The stress on acquiring a familiarity with the organisms most commonly encountered in the various regions of the oceans is also deliberate. Former students have repeatedly confirmed the value of this approach, finding it easier to grasp the fundamentals of new ecological situations, or unfamiliar waters, when they had a prior training in how to recognize the major groups at sight and to give immediate names to the commonest genera of the world's seas. However, for courses in which systematics is of lesser import, the material on classification can be omitted at the instructor's discretion.

Attention has been given to the historical context in which discoveries were made in the field of oceanography. As the last pages of the book were set in type, the important discovery of the Caves of the Navigators occurred, too late for inclusion in Chapters 1 and 10. Preliminary study of the numerous inscriptions in these New Guinea caves shows that the discovery and settlement of Polynesia occurred shortly after November, 232 B.C., when a Libyan Maori naval squadron sent out by Ptolemy III attempted to circumnavigate the earth, in accordance with the newly propounded doctrine of Eratosthenes that the world is a sphere of 250,000 stades (ca. 28,000 miles) circumference. The inscriptions, now in preparation for publication, disclose an astonishing knowledge of navigation, astronomy, and the use of sophisticated instruments for reducing zenith-angle observations to yield latitude. The fishing gear used by these early explorers, and illustrated in the cave inscriptions, corresponds closely to the inferences given in Chapter 1.

Many of the species of fishes illustrated in *Introduction to Marine Biology* were collected by Peter Garfall, who also helped in other ways and taught me much natural history. Haris Lessios and Joseph D. Germano made collections in the Mediterranean.

It is a pleasure to record the generous assistance of members and former members of my classes in bringing together the photographs and drawings in this book. Each illustration carries the name of the contributor. The majority of the drawings of fishes are the work of Gerald Heslinga, and most of the photographs of marine invertebrates were made by Joseph D. Germano and by George Putnam. Professor John H. Dearborn, who participated in teaching the courses for several years, provided photographs of marine birds for Chapter 16. James F. Clark, with the cooperation of the U.S. Naval Academy, contributed illustrations in Chapters 3, 12, and 28. David F. Moynahan is represented by both drawings and photographs. Other contributors of photographs and drawings are Veronica Wilcox, Alan Baker, Peter Garfall, Isidoro Zarco, F. Julian Fell, John Bullivant and the New Zealand Oceanographic Institute, Mervyn D. King, Patricia M. Ralph, Bernard Zimmermann, Stephen H. Dart, Eric Barham and the U.S. Navy Electronics Laboratory. I offer my grateful thanks to these collaborators. Like all marine biologists, my work has been aided on numberless occasions by good friends and colleagues, both at home and overseas, on ships and boats, and many remote islands. To them all aloha nui.

To the late Sir Charles Cotton I am indebted for some of the ideas used in Chapters 7 and 8. The section on geoelectric fields is due to Dr. Sentiel Rommel, who also gave generously of his time and knowledge in discussing other sections of the book.

Photographs of ancient Polynesian tablets were supplied by the U.S. National Museum, and some of the information deciphered from them has been incorporated in the historical sections. For advice on special topics I am grateful to Dr. David L. Pawson of the U.S. National Museum (Holothuroidea); Dr. Ruth Turner, Museum of Comparative Zoology (thermal pollution); Professor Kenneth J. Boss, Museum of Comparative Zoology (Bivalvia); Charlene Long (Annelida); and the anonymous scribes and artists of ancient Egypt are our authorities for most of what we know of the origins of marine biology.

Joe Ingram and latterly Holly Detgen gave of their editorial skills and other special knowledge in smoothing out many rough spots in the original manuscript and in coordinating the variety of elements into a more cohesive pattern; I am grateful for their judgments and advice.

H.B.F.

Museum of Comparative Zoology
Harvard University
November, 1974

INTRODUCTION TO MARINE BIOLOGY

1

THE MARINE SCIENCES

Scope of marine biology / Importance of the sea for the welfare of all life on earth / Importance to man / Subsidiary disciplines / Fisheries / Mariculture / The marine sciences that constitute oceanography / Biological and physical oceanography / The marine biotope / Marine biota / Marine ecosystem

Marine biology is the study of life in the sea. It is, therefore, one of the broadest subdivisions of the natural sciences, for it encompasses all the subsidiary disciplines of biology as a whole, merely restricting the scope of the enquiry to the marine habitat. The marine habitat, together with the plants and animals that occupy the sea, forms part of the entire ecosystem of the earth—so the study of marine biology offers a wide-ranging view of life on our planet and the conditions that make such life possible. For this reason, perhaps, many people whose calling does not particularly involve the sea find, nonetheless, a peculiar intellectual or philosophical interest in watching or reading about the life of the sea. This is well illustrated by the numerous films shown on television in which a public at large can gain insights into a world beneath the waves and share vicariously the activities of marine biologists. However, apart from the charm or strictly intellectual exercise of such experiences, marine biology is, of course, a major science with practical aspects of direct concern to the maintenance of human welfare. The greater part of the food chain of animals important to man, and the terrestrial plant life depend in one way or another upon the oceans. Most living tissue is located in the surface waters of the ocean, and much more than half of the sum total of plants on earth comprises the floating marine organisms called **diatoms.** It is by the activity of diatoms that valuable mineral nutrients derived from the ocean floor are made available to fishes and, ultimately, to man in the shape of fish proteins; and most scientists now agree that nearly all the oxygen we breathe is generated by the marine diatoms. As the world's food resources are strained, man will be obliged to pay more attention to the careful management of other food resources yielded by the sea. Thus, the practical importance of marine biology is obvious.

Like the parent science, marine biology has many subsidiary disciplines, so many different specialists are to be found in the membership of a marine biological expedition or on the staff of a marine biological research station. Besides this, many marine biologists are to be found in colleges and universities, often working on topics requiring only an occasional visit to the sea, for example, the anatomical structure of marine animals or plants (**morphology**), their classification (**taxonomy**), or their manner of respiration or excretion in a watery medium, and other functional aspects (**physiology**). Of marine biologists whose work requires a more intimate acquaintance with the sea and longer periods of time spent at, on, or beneath the sea, ecologists perhaps come first to mind. Theirs is the task of observing the reaction of organisms to one another and to the environment; and, in recent years, it has also become their important and often unhappy task to report the effects of marine pollution. There can be little doubt that, in the decades ahead, such forms of marine pollution as develop from oil spills and the disposal of industrial and urban waste will inevitably place great strains upon the capacity of the surface waters of the ocean to continue to support a healthy population of plants and animals. Thus, the role of the marine ecologist is bound to increase in importance. Such a man or woman will have the satisfaction of knowing that the investigations performed will not only advance the total of human knowledge of nature, but also will benefit humanity directly.

As Fig. 1.1 shows, marine biology is only one of a group of **marine sciences,** each one of which is a self-contained discipline, yet every one of which is in some way dependent upon or interrelated to the others. Study the chart and notice that all the major physical sciences, such as geology, physics, and chemistry, have their marine branches too. Each one of them yields data of concern to the marine biologist insofar as they explain more fully the manner of environment in which a marine organism must survive. Notice, too, how each of these disciplines contributes to the applied technology of the sea, that is, to human activities not necessarily scientific in origin, but nonetheless requiring scientific methods for their efficient performance. If you are planning to major in the area of marine biology, the chart will guide you in selecting other subjects for study because a marine biologist would do well to have an understanding of the disciplines of those other scientists with whom he works. Do not think of the technological subjects as merely peripheral; as Chapters 2 and 3 point out, marine biology is largely dependent upon the skills and perseverence of mariners, marine engineers, geographers, and others for the successful recovery of important biological samples from the seafloor and from other inaccessible places, such as the Antarctic Ocean. All these disciplines—both the pure and the applied sciences—together comprise the great field of endeavour called **oceanography**; and, indeed, marine biology is often called **biological oceanography.** Oceanography is one of the widest of all the categories of scientific research. It commands a literature of more than a hundred regular periodicals reporting research in many different languages. It offers a meeting ground, often on ships at sea, equally often on land-based institutions, where men of science from different lands may work together and ponder common problems in an environment that is not subject to the laws or customs of any one nation.

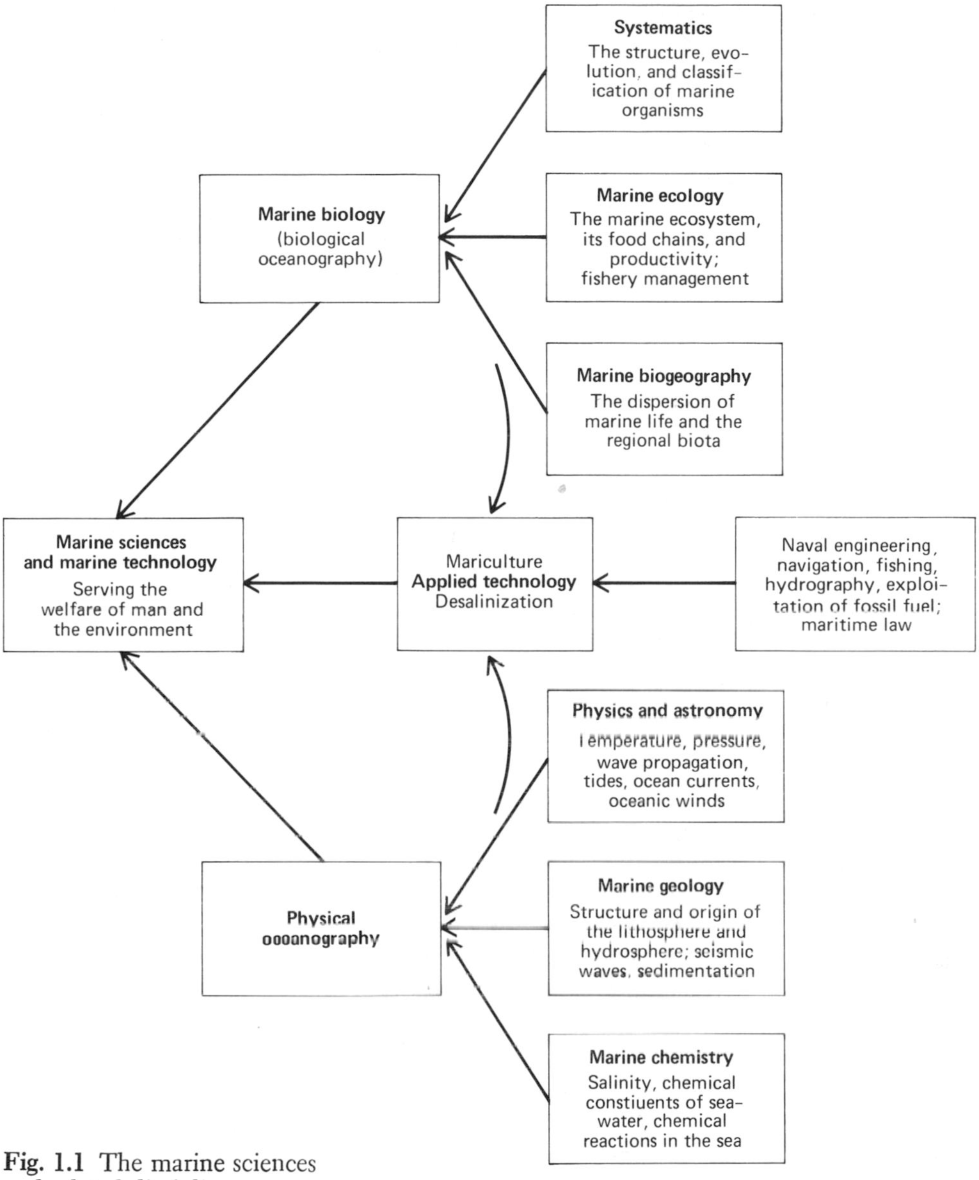

Fig. 1.1 The marine sciences and related disciplines.

Perhaps that is part of the attraction of oceanography and why notable advances have been made in fields where the sheer physical problems posed by the ocean itself might seem to be forbidding.

Fisheries and Mariculture

The importance of sea fisheries as a source food has been recognized since prehistoric times, and primitive societies understood the need for conservation. Particular lengths of strand would be assigned to particular tribes whose responsibility it was to protect the young stages of fish. Polynesian methods of fishing, for example, are generally selective and permit young fishes to escape.

In more recent times it has become clear that the continental shelf is really part of the landmass it surrounds and serves as a breeding ground for many species of plants and animals. Thus, international agreements have become necessary to prevent overfishing on those parts of the shelf that are accessible to all nations. Sometimes, because of climatic change or other factors, the natural range of schooling fishes, such as herring, may undergo change. The fishermen of a land where herring or cod fishery has long been an established activity may find themselves obliged to sail ever further from home in order to take their traditional catch; and, in due course, the fishery may be entirely transferred from one continental shelf to another. For example, in recent years fishermen of Britain have been encroaching upon areas considered by the Icelanders and Greenlanders to be their exclusive resources; and the so-called cod war has come about. Marine biology can throw light on these otherwise mysterious movements of commercially significant fishes.

The United States is now taking a harvest of nearly 2 billion kg of fish annually from its coastal waters, and the demand is rising. There is an ever increasing need for a fuller understanding of the effective management of sea fisheries. **Mariculture** is a newly developing marine application of an ancient art, **aquaculture,** by which food species of fish and shellfish are kept in holding tanks or ponds until, by artificial enrichment of the fishes' food supply, their body size reaches commercially acceptable values. These new techniques require research to bring them to successful fruition.

If you glance over the contents of this book you will notice that the topics are arranged in three general sections, namely the **biotope** (Chapters 6 to 12), the **biota** (Chapters 13 to 22), and the **marine ecosystem** (Chapters 23 to 29). By biotope is meant the physical environment in which living organisms form natural communities. Many different communities occur in the sea: floating ones (collectively called **plankton**); bottom-dwelling ones (**benthos**); swimming assemblages, such as schools of fishes (**nekton**); communities comprised of burrowing animals on the seafloor and wind-drifted floating forms at the surface of the sea; and communities that live around the shores of islets. These very different community require different biotopes, so in speaking of the biotope we are using the word in a collective sense. Therefore, Chapters 6 to 12 relate to physical conditions or processes, such as coasts, waves, substrates, currents, and so on, that are the aspects of marine biology more particularly related to the sciences of marine geology, marine physics, and other areas of **physical oceanography.** They are important to the marine biologist, but not of core importance, because they deal with inorganic phenomena. They are treated here in a nonmathematical way; but should you want to know more about them, refer to a text on physical oceanography or marine geology.

The second category, comprising Chapters 13 to 22, describes the biota. This term is a collective noun embracing all the living plants and animals of a given flora or fauna, in this case the marine organisms. These chapters do not deal with phyla that are not conspicuous in oceanographic samples (for in such case this book would cease to be an introduction to marine biology). Instead, they concentrate on those organisms that are most conspicuous in oceanographic samples—that is to say, the organisms you are most likely to see if you are on deck when a trawl is emptied, or if you are a member of a shore expedition somewhere in the world. These chapters refer to the actual genera most frequently encountered on the various coasts of the world. If you are able to spend enough time in the laboratory examining specimens of these conspicuous organisms so that you can recognize them and name them when you go into the field, then you can be sure you will soon become proficient at recognizing the important features of the ecology of whatever region you may be called upon to study.

The last group (Chapters 23 to 29) describes the ecosystem, in this case the marine ecosystem. It deals with the various assemblages of biota you can expect to find in the principal environments of the sea. Stress has, naturally, been placed on environments most likely to be seen in North America, although the material is presented in a manner that should make other similar environments meaningful. Thus, Chapter 24 on North Pacific shores, although written mainly with west-American shores in mind, nonetheless gives a general picture of marine life found in, say, Japan or northern China or Siberia. Similarly, Chapter 23 on tropical seas tends to stress Caribbean forms; but similarities and differences in other tropical seas are noted.

The book closes with examples of some unresolved problems where marine biology and the physical sciences seem to be in conflict. These are intended to make you think about possible solutions.

2

THE ORIGINS OF MARINE BIOLOGY

Polynesian fishing techniques / The first ships / Researches in early times in ancient Egypt / Development of seamanship / The Egyptian trawlers of the Eleventh Dynasty / Nautical literature of Egypt / The first Suez canal / Hatshepsut's expeditions to the Indian Ocean / The Egyptian trawl / The Roman tragulum / The Egyptian Ih / The Greek and Roman sagene / Origin of the modern net types / Evolution of early types of sailing ships / Early anchors / Aristotle and marine biology

When Captain Cook sailed the New Zealand coasts in 1769, his ship, *Endeavour,* evoked the admiration of his Maori hosts, although they were surprised at her slow speed. (On other islands the Polynesian mariners amused themselves by sailing in loops around *Endeavour* as she pressed forward at her best pace.) However, when Cook's men brought out their official British Navy seine net to catch fish, the Maori were reduced to laughter and inquired why grown men should use children's playthings. Cook asked to see their nets and was led to the net shed. "It was five fathoms deep," wrote Cook afterward, "and by the room it took up could not be less than three or four hundred fathoms long."

With such effective sampling equipment at their disposal, it is not surprising that the Polynesians were well-informed on the denizens of their seas. The Maori of New Zealand had no difficulty in recognizing and naming more than 30 species of edible fish that appeared regularly in his nets; some of these have no English name, so the Maori names continue in use by the European settlers. But in addition, many special names were applied to growth stages of fish and to different kinds of schooling behavior. The Hawaiians have a large vocabulary devoted to special modes of catching fish, and all Polynesians have a large vocabulary of words applied to special types of fishing gear. For example, the Hawaiian recognizes about 60 categories of fish hook, all for particular purposes, made of particular materials, and for attracting particular species. The New Zealand Maori has a generic term for a net (**kupenga**) but find it needful to have special names for at least 24 types, ranging from small shrimp nets (**whakapuru**) to enormous seines (**kaharoa**) requiring many men to haul. The Hawaiians distinguish about 60 kinds of net, although some of these are probably local names. In addition, there is a special vocabulary for the tools by which nets were made, the manner of placing them, the manner of towing them, and so forth. About 30 names were applied to various kinds of marine mollusks; other marine invertebrates also have their particular names, irrespective of whether they are edible or not.

As noted in Chapter 1, an effective sampling technology is essential for the pursuit of marine biology, and this means above all, the successful design and deployment of nets for trapping fish and of dredges or trawl nets for securing specimens of animals that live on the seabed. These in turn predicate the use of boats or ships to transport the gear to and from the region to be studied and to tow the gear while it is in use. If, now, the supposedly neolithic Polynesians had techniques so far in advance of those in use in eighteenth-century Europe, perhaps the proper place to search for the origins of marine biology may well be in the archeological remains of Neolithic and Bronze Age peoples. The oldest remains of a maritime culture are those of the Middle Eastern peoples of the fourth millenium before Christ, and they occur in Babylonia and Egypt. However, the records of Egypt far surpass in detail those of all other ancient peoples. So it is to Egypt that we may direct our attention.

The equipment used by the ancient Egyptians in the harvest of the sea proves to be beyond all doubt the prototype of that used in modern times, and we have no difficulty in recognizing basic items that form part of a present-day oceanographic survey. This means, of course, that we can apply the names that are used today for corresponding equipment. But the precise form of the Egyptian gear comes closest to that used by the Polynesians, and even the names of some items correspond. This is not the place to explore that interesting trail, leading as it does into the field of cultural anthropology. But this certainly is the place to examine the origins of marine biology.

From remains of pre-Dynastic Egypt, known as the Nagadah II culture and dating from about 3500 B.C., we have a series of paintings on linen and pottery showing longboats manned by 30 rowers in two tiers of 15 (a tier to each beam) and having a helmsman at the stern to control the large steering oar (forerunner of the rudder). The rowers face the bow and evidently paddled in the Polynesian manner. Some of the paintings depict a voluptuous female figure, larger than the men, standing mysteriously amidships. The pictures antedate writing, so we can only look and wonder.

By the Third Dynasty, around 2680 B.C., permanently incised rock hieroglyphic inscriptions were cut under the direction of the Egyptian chancellor Imhotep. It appears that, on Imhotep's advice, the pharaoh Djoser established what in effect was an academy of sciences, and among the surviving records are engraved illustrations of the plants and animals of the Nile Valley. Apparently, then, freshwater ecology preceded marine research. By the Fourth Dynasty, 2613–2492 B.C., written records were more extensive. We learn from them that the basin system of water conservation had been instituted, the greatest pyramids were built, the formula for calculating the area of a circle was discovered, water clocks and sundials were invented, and the length of the solar year was determined to be 365 days. Marine biology was now being pursued, and the species were illustrated so accurately that they can be identified.

How were the marine specimens obtained? For a specific answer we have to turn to Eleventh Dynasty records (2133–1991 B.C.). Here the techniques are documented by series of scale models of boats, ships, and fishing gear buried in the tombs of the Egyptian nobles of the period. And for a further thousand years after that epoch the records continue, providing us with the mechanical details and dimensions of their vessels, the size of the crews, the mode of propulsion, steering, and anchoring, the sounds for determining the depth of the bottom, and the nature of the collecting devices. As the records come nearer to our time there is increasing evidence of exploration overseas, beyond the confines of the waters that lap the coast of Egypt. Creating graceful vessels, square-sailed on two yards hung upon a mast amidships, a bank of oarsmen on either beam, these ancient sailors carried a battery of nets, dredges, floats, sinkers, hooks, harpoons, and cordage, setting their squadrons down the searoads of the pharaohs in search of rarities from afar. The records are sometimes complicated by breaks in the sequence, as well as by difficulties in interpreting some of the hieroglyphs. But the fundamental truth is plain enough: Egypt was bent upon the peaceful harvest of the seas. To build her ships she needed great trees, of which none were to be found in Egypt; so the earliest voyages were in papyrus-boats, to the coasts of Phoenicia, where cedars of Lebanon might be obtained. The lumber was rafted home, and there delivered to the shipyards, whose artisans we see depicted in contemporary paintings.

The Egyptian Trawlers

During the Eleventh Dynasty, so the chronicles report, the pharaohs Amenemhet and Sesostri (ca. 2100 B.C.) cut a canal by way of Ismailia from the Nile to the Red Sea, thus connecting the Mediterranean with the Indian Ocean. From the tombs of the nobles of that era we have many model ships depicting the *khenp enti em mu* (the drawing out of that which is in the waters). One of the nobles named Meket-re seems to have owned and operated a trawling fleet.

Nautical Literature of Egypt

Like seafarers of all ages, the Egyptian mariners spun tall yarns on their return from a voyage. What may be the oldest sea story is a novelette from the twenty-first century B.C., *The Tale of the Shipwrecked Sailor.* Its hero would appear to have been the same as he who later figured under the name of Sinbad; apparently old stories in Egypt survived the onslaught of Islam and found their way to Bagdad to beguile the thousand-and-one nights of a sleepless Kalif.

Voyages were long and dangers were many, but homecomings were joyful events. "The mooring post has been driven in, praise has been rendered, God has been thanked, and every man embraces his companion. Our crew has returned safe without loss." So wrote the unknown scribe of the Ermitage papyrus, and across the space of 4000 years we sense the bonds that unite all those who go down to the sea in ships. The voyagers brought back cedarwood and precious stones from Arabia and spices and balsams (for the embalmers) from Punt, a land believed to be Somalia. One bas-relief (Fig. 2.1) of the time of Queen Hatshepsut (1500–

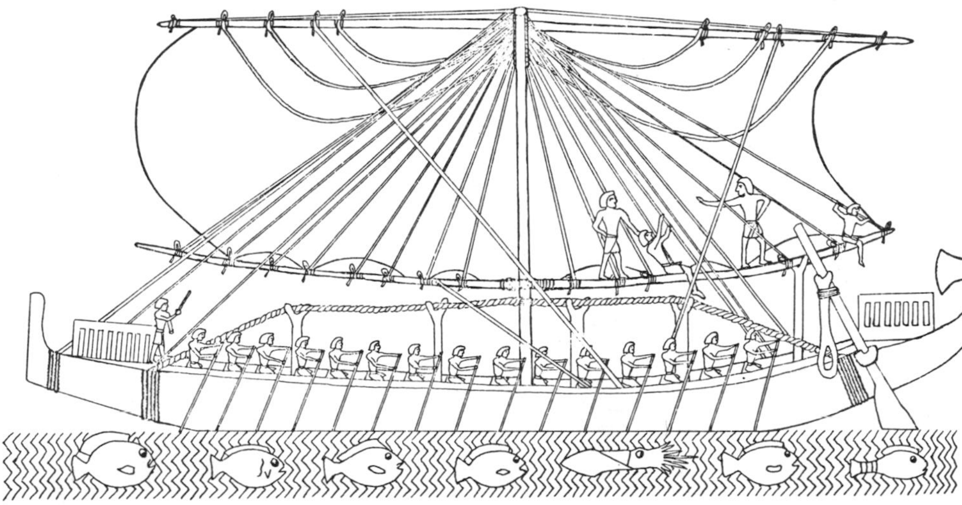

GERALD HESLINGA

Fig. 2.1 One of Queen Hatshepsut's five ships represented on her temple at Deir el Bahri, ca. 1475 B.C.

1475 B.C.) shows botanical specimens being unloaded and young trees in pots; the reliefs also depict a variety of tropical fishes and squid (Fig. 2.2). One can imagine that the latter were brought home salted in barrels and dispatched to the queen's sculptors to be perpetuated in stone on temple pylons where we can yet admire them.

The Egyptian Trawl

From these ancient records we can see that the Egyptian sailors in the time of Meket-re used a conical bag net as a trawl, apparently about 2 m across the mouth to judge by the models. The lower edge of the mouth was weighted to keep it in contact with the seabed, the upper edge was supported by a hoop-shaped piece of wood—we would call the device a beam trawl. It was towed by two vessels, a rope to each running from the lateral corners of the trawl. The Polynesians use such a device and call it a **kete**, a word meaning either a bag net or a bag or basket. We cannot be sure what the Egyptians called the device, but it is interesting to note that the ancient Egyptian word for a basket is **ket.**

The Roman Tragulum

From Latin sources 2000 years after Meket-re, we learn that a similar device was employed in Italy by the fishermen on the continental shelf and that their word for the trawl was **tragulum**, meaning that which is dragged. Some lexicographers believe that the English word **trawl** is derived from this Latin root. The trawl was brought to New England by the colonists. By the 1670s, we find Governor John Winthrop II writing to the Royal Society in London to report "a curiously contrived fish lately found in Massachusetts Bay"; it was, undoubtedly, trawled. In the nineteenth century, fishermen in both England and New England adopted the then newly invented **otter trawl,** a device still widely used both for commercial fishing and for scientific research. It is described in Chapter 3.

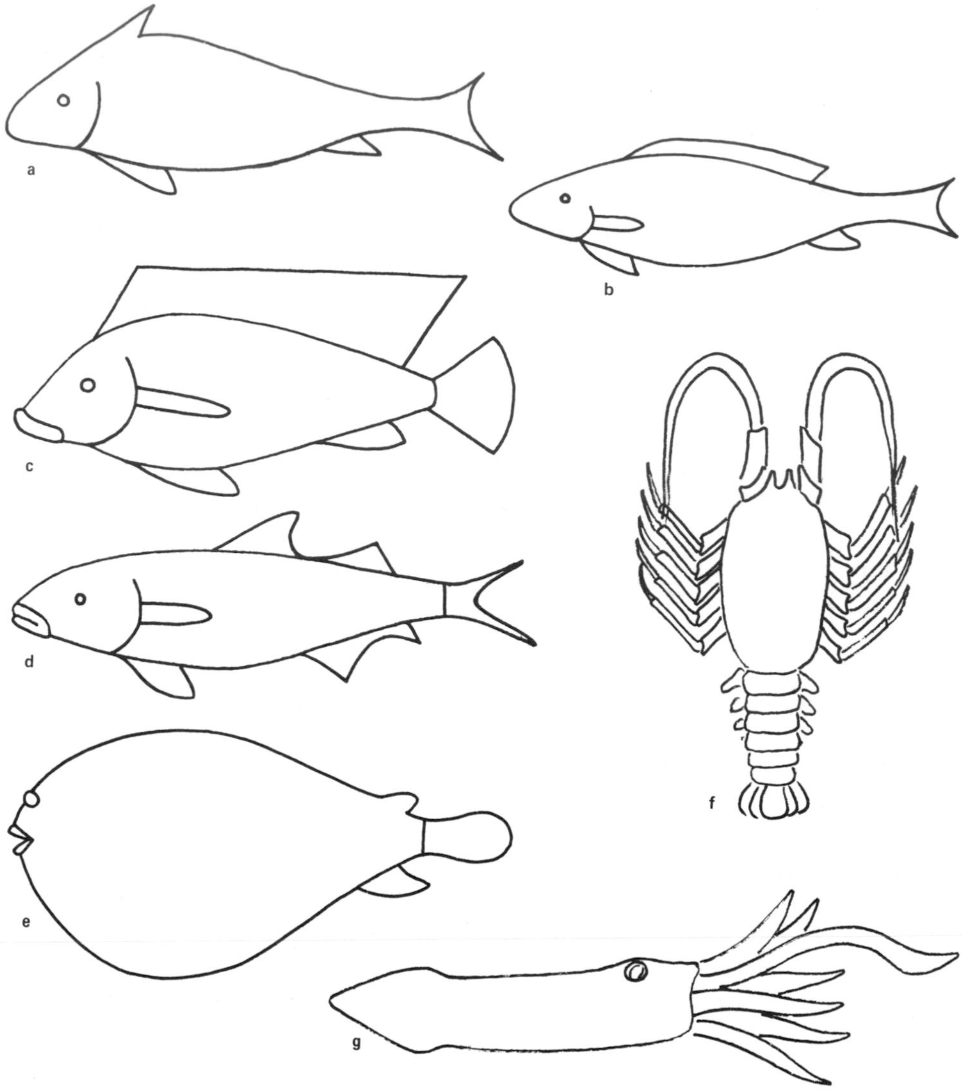

VERONICA WILCOX

Fig. 2.2 Mediterranean and Red Sea fishes and marine invertebrates as depicted on ancient Egyptian monuments and tomb paintings, Eleventh to Eighteenth Dynasties (2100–1420 B.C.). (a) *Barbus*. (b) *Petrocephalus*. (c) *Tilapia*. (d) *Mugil*. (e) *Tetradon*. (f) *Panulirus*. (g) *Loligo*.

The Egyptian Ih Net

The **ih** was the direct ancestor of the modern seine; that is to say, it had a series of rectangular pieces of netting sewn together to make a long strip, similar to a tennis court net. The lower edge was weighted to make it hang down, and the upper edge was provided with wooden floats to keep the net vertical. Ropes from either end permitted the ih to be set across a mouth of a bay and then drawn to land by men pulling on either end of the ropes. Small dragnets of this type are called **lau** or **hukilau** in Hawaii and **kaka** in New Zealand. The act of fishing was called **ikan** in ancient Egyptian—in Polynesia the fish themselves are called **ika.**

The Greek Sagene

In the *Iliad* the Greeks are described as using the ih net, but they called it a **sagene.** The word also appears in the vernacular Greek of the New Testament and evidently corresponds to the late Latin **sagena** and to the modern English **seine** (or sean). Today seines are constructed on the same principle as they were in ancient Egypt. As already noted, the Polynesian seine is the largest known example of this gear. In New Zealand, where I have observed its effective operation by a Maori tribe, the length formerly reached 1 mile from end to end; such a net was loaded on a platform constructed across two **waka** (large canoes). The vessels sailed in catamaran fashion across a bay from headland to headland, laying out the net. Later the whole tribe would combine to draw it in. As the last 100 yards are left, the men run into the sea to take the catch, which is so concentrated by that time as to make the water appear to be boiling.

The earliest Egyptian and Mesopotamian boats were river craft made of reeds tied into a boat shape. When seagoing vessels were built, the Egyptians constructed the hull in carvel style; that is, the timbers were laid edge to edge, not in clinker fashion as used in northern lands. The remains of these boats show how the timbers were fastened by dowels. The Polynesians used a similar construction, the fastenings being however of twisted fiber. The sail of the Egyptian ships differed from all others except those of Polynesia—a square sail with an upper and a lower yard. The Phoenicians discarded the lower yard, producing the ballooning **lug-sail,** which was inherited by the Greeks and Romans and also used by the Viking ships. In Polynesia, the Egyptian sail was later converted to a triangular lateen sail, still retaining the Egyptian predilection for two yards to support the sail. In Egypt during later centuries, the Arabs introduced the lateen sail minus its lower yard, and that is the form now to be seen in the Middle East. The vessels called **luggers** were similar northern variants used by fishermen, and old engravings of Boston show that luggers were to be seen in the harbor there as late as 1830.

Anchors were initially stones with a hole bored through one end to take the rope, alternatively, a mesh bag filled with stones served as an anchor. Flukes of wood were added in Phoenician times. Scuba divers have recently recovered cast anchors of many ancient ships in the Mediterranean. The Polynesians used all the varieties just mentioned. The Polynesian lobster trap is shaped like a pot with the entrance at the top and one "parlor" within. Its design is effective, and it is employed by the European fishermen who now live in New Zealand. In America, on the other hand, the lobster trap is usually a rectangular wood device with entrances at either end that lead to two parlors within. The origin of both designs is probably ancient. In marine biology lobster traps are used to take any large crustacean.

Around 325 B.C. Aristotle in Greece wrote *Peri ta Zoa Historia* (an account of animal life) in which he described numerous marine organisms. The names he used—such as *Echinus, Asterias,* and many others—are still employed in scientific contexts. Aristotle is believed to have operated marine biological research stations on the island of Lesbos and later in Euboea on the mainland. He is generally regarded as the father of marine biology; but it is probable that he had access to earlier works, now lost, composed by Babylonian or Egyptian writers. He paid much attention to what Greek fishermen told him. The *Historia* is thought to be a collection of lecture notes assembled by his students after his death, for the prose style is not polished; it seems likely that his original work was either lost or deliberately destroyed by political enemies.

The voyages of mariners in ancient times opened up the seas to the people of the west, even so far as the north Pacific, and doubtless increased man's knowledge of life in the sea. Some of the voyages are referred to in other chapters. Man's knowledge of the deep sea is much more recent; historical aspects of that topic are touched upon in the chapter on life in the deep sea.

A major archeological discovery of early navigation records is in progress as this book goes to press. Preliminary reports (Fell, 1974) show that six Libyan ships were dispatched by Ptolemy III in 239 B.C. and entered the Pacific by way of New Guinea. The navigator, named Maui, employed detailed star charts and sophisticated equipment, including two types of analog computers invented by the Libyan astronomer Eratosthenes. The records include advanced mathematical calculations relating to the curvature of the earth's surface and the size of the earth. The fishing gear illustrated includes examples of all Egyptian and Polynesian equipment discussed in this chapter. Detailed observation of an eclipse permits one record to be dated exactly to November 19, 232 B.C. The new data, presently under study at Harvard University and at the U.S. Naval Academy, include evidence of the earliest-known use of the astrolabe as a navigational aid by ancient mariners.

3

BASIC SAMPLING PROCEDURES

Sampling and field observations / Shore collecting / Small aquariums / Larger aquariums and cold rooms / Observations and photography using aquariums / Collecting in offshore waters / Sampling procedures for benthos / Skin diving / Baited lines / Scuba diving and photography / Photography by remote control / Deep submersibles / Lobster traps / Pipe dredge / Rectangular dredge / Triangular dredge / Trawl net / Otter trawl / Beam trawl / Grabs / Epibenthic sledge / Core samplers / Simple tubes / Augers / Explosive-driven core samplers / Drill cores / Preservation of samples / Storage / Labeling / Collecting plankton

This chapter outlines some of the basic sampling procedures in marine biology. It is assumed that opportunities exist for visiting the coast and that the equipment described is either available in the laboratory or can be inspected by going aboard a research vessel operated by an oceanographic institution. Colleges located far from the seaboard can generally come to an agreement with another college that has a marine station or operates a research vessel, enabling a class to spend a day at sea observing or participating in offshore sampling. Modern laboratory aquariums also permit the study of many interesting marine animals at locations far from the sea.

SHORE COLLECTING

Much can be learned about marine organisms and their ecology merely by the systematic observation of rock pools or wet sands left exposed when the tide is low. If you are a skin diver, you are less dependent upon the low tide and can plan your visits with a more flexible timetable. Although skin diving is a powerful asset, it is not absolutely essential because other methods can be used. A good camera will yield valuable information for subsequent study. Notebooks, collecting jars and buckets, waders, a strong knife (for removing cemented organisms from the substrate), and plastic bags for algae are other basic items for field work. In the cooler latitudes, spare, warm clothing is advisable as well as any other items that common sense may dictate. Collect examples of the commonest animals and plants, a few at a time, and consult a naturalists' field handbook so that you can identify what you have found. Memorize the names of the most common species and make lists of all you recognize. (It aids the memory if you write the names out from time to time.) Soon you will become proficient at using scientific names for animals and plants. Also, you will become aware that the marine environment is both complex and fragile; animals and plants are all too easily destroyed if their natural environment is altered in any way. Therefore, make it a practice *always* to follow these rules:

1. Don't disturb animals needlessly. Investigate by all means, but only as many times as are needed to satisfy your inquiries.
2. If animals are found sheltering under stones, replace the stones as you found them. If animals live attached to the upper surface of a stone, take care you have not left it overturned.
3. If you collect samples to keep as permanent specimens, do not remove too many from one place; in any case, don't collect unreasonably large samples.
4. Don't convert collections of shells or other marine creatures into knickknacks.

If you are a skin diver you will have more extended opportunities to compare and contrast the organisms on the bottom with those in the overlying waters. On the bottom, it is apparent that particular species seem to prefer particular areas and shun others—not all species have the same preferences. Now is the time to find the relationship between these. Make these observations:

1. Date and time of day of your dive.
2. Water temperature at the surface (use Celsius scale).
3. Water temperature at the bottom.
4. Substrate type, that is, kind of bottom. Examples are soft bottom (sand, sandy silt, gray mud, brown mud), hard bottom (cobble, gravel, shells), mixed bottom.
5. Depth of bottom.
6. Predominant organisms on the bottom. The bottom may be covered by seaweed (if so, what kind of seaweed?), or it may be covered by starfishes, and so on. It may be lifeless and polluted (if so, what pollutant?), or it may be covered by the bodies of dead animals (what animals?).

SMALL AQUARIUMS

Marine aquariums offer a means of studying live organisms in the laboratory. If the following suggestions are adopted, their use should be practicable, even in noncoastal areas.

1. In the continental United States, air temperatures fluctuate widely during the course of the year. Sea temperatures, especially near the bottom, do not. Marine animals are very sensitive to temperature changes. You must plan your aquarium to minimize these fluctuations. If you propose to operate the tank throughout the year inside a house or in a classroom where central heating is normally in use in the cold months, then you must use animals that tolerate ambient room temperatures. The solution is to collect your specimens on some occasion when you are visiting Florida or the Gulf States where the water is always warm; such animals are able to live in higher latitude interiors without harm from the high summer climatic temperatures or from the high winter interior temperatures.

2. Obtain an all-glass or all-plastic tank, such as pet shops supply, having a 5-gallon or 10-gallon capacity. A 10-gallon tank is illustrated in Fig. 3.1. Set it in a cool, shaded place, out of reach of direct sunlight, and if possible, near a draught—a north window is suitable. Do not move it once it has been filled. If it has a metal frame, be sure that the metal joints are absolutely sealed by a thick layer of rubber putty so that they cannot come into contact with the seawater.

3. Arrange a visit to the coast, taking sufficient empty glass containers to collect fresh, clean seawater at low tide. Collect also some *small* animals for stocking the tank and place them in a plastic bucket of seawater, which is to be kept out of sunlight. If a long journey follows, surround the containers with ice cubes.

Select, for example, one or two small sea stars, about 2 inches across, or a small sea urchin and a sea snail; to these could be added a small hermit crab, a small limpet or periwinkle, none to exceed 1 inch across. Select some rounded, clean

DAVID F. MOYNAHAN

Fig. 3.1 Lionfish in a simulated reef environment in an aquarium.

pebbles and clean, dead shells and a few stones carrying coralline seaweed (Fig. 3.1) and place all in the bucket. These animals will survive a journey of several hours but should be transferred to the tank as soon as it is set up, and the original water in the bucket is to be discarded. An alternative to visiting the coast is to place an order for a suitable collection of aquarium species with one of the biological supply houses.

4. Mark a water line on the side of the tank, at the level of the surface of the water. As water is lost by evaporation, replace it with distilled or fresh rainwater, to keep the level constant; this should be done each week. Keep dust from entering using a glass cover with a chink for air.

5. If any animals die, remove them at once. Do not overstock the aquarium. After a few weeks, the sides of the tank will become greenish as a result of the growth of minute algae. These will be cleaned off and eaten by the starfishes. Other microscopic organisms in the water will support the animals for the cool months of the year. Small pieces of chopped fish or shellfish can be fed to the animals; but if they are not eaten, they should be taken away again. In general, the animals will find food for several months in the minute growth that forms in the tank.

A tank set up in this way should be self-supporting. The algae and the wide surface exposed to air will provide oxygen, so long as the air temperature is not above 55° F. If any of the animals die soon after they are introduced, they may be unsuitable species. Species in rock pools around the midtide zone will generally be found hardy enough to live for at least some months under artificial conditions. Schools near the sea will be able, of course, to change the animals at frequent intervals and thus exhibit a much wider variety of forms. If a scum should form on the top, there is no remedy but to start the tank again with fresh seawater, after a thorough cleansing. This, however, should not occur if overcrowding is avoided. Never attempt to keep large specimens—they will die and so will all the small animals.

Use of Simple Aquariums

The main value of a simple aquarium lies in the opportunities it affords for watching the behavior, feeding habits, reactions to light, response to other organisms, and so forth of animals brought in from collecting expeditions. A second important use is for still or motion picture photography of the organisms under easily controlled conditions and with adequate illumination.

LARGER AQUARIUMS AND COLD ROOMS

Simple aquariums as described above cost little, but they suffer from the disadvantage that animals from the cooler parts of the United States cannot survive the relative high indoor temperatures encountered in buildings where central heating is installed. To overcome this limitation more expensive aquariums with a cooling apparatus built into the tank must be employed. Biological supply houses can provide tanks of this type in sizes ranging from 25 gallons upward. An alternative is to operate a cool room, in which a constant temperature is maintained at about 5°C. In such a room a large number of simple aquariums can be housed, each holding various type of organism until they are needed for laboratory study.

COLLECTING IN OFFSHORE WATERS

Sampling Procedures for Benthos

Sampling procedures for benthos depend upon the type of bottom. A sampling device that works well on soft bottom may be totally destroyed if it strikes hard bottom, and a device that gathers sparse harvests from stony ground may fill up too rapidly if it passes over soft sediment. In addition, the sudden change in strain may break the tow line. Thus, successful seafloor sampling requires careful reconnoitering with expendable devices first.

Procedures for All Types of Bottom

1. **Skin diving,** with collecting bag or basket carried by handles, or slung around the waist, wrist, or over the shoulder. This technique was probably in use in Europe in 6000 B.C. by the Neolithic tribes that left the great shell middens. Naked diving was practiced in Greece in Homeric times, as noted in the *Iliad,* and it is still used today by Polynesian peoples in the south Pacific.

2. **Hooked and baited fishing lines.** These procedures date from Neolithic times; they are of greater value in midwater sampling for nekton.

3. **Scuba diving,** with sampling containers and/or with camera (Fig. 3.2). This is a modern technique developed during World War II.

JAMES F. CLARK

Fig. 3.2 Scuba diver with still camera and strobe.

4. **Photography** by remote control devices.
5. Use of **deep submersibles.**

Methods 4 and 5 are discussed in Chapter 28.

Procedure for Hard Bottoms

1. **Lobster traps.** These have been known since Neolithie period.

Procedure for Rough Rock Bottoms

1. **Pipe dredge.** This is a dredge made by attaching a bridle to one end of a strong metal pipe about 30 cm in diameter and closing off the other end of the pipe. This crude device can be dragged along the bottom. It is not very efficient, but, on the other hand, it does not break or tear. When the bottom is quite rugged (or unknown) it may be the only device that succeeds when others have failed.

Procedures for Smooth Bottom (Hard or Soft)

1. **Naturalist's dredge.** This technique has been known since Neolithic times (Fig. 3.3). It consists of a rectangular wooden frame with a bridle on the leading edge attached to the tow rope that trails a woven mesh bag. A variety of this dredge was still in use among the Polynesians a century ago, and, therefore, the Neolithic fishing procedures could be seen and recorded. In Neolithic times a successful model would receive a given name, like Viking swords, and karakia (prayers and incantations) to Neptune were regarded as essential when the dredge was lowered (Fig. 3.4). The modern naturalist's dredge is a metal version of the prehistoric pattern.

2. **Triangular dredge.** This is either wholly of metal or has a triangular metal frame trailing a mesh bag. A three-part bridle is attached to each corner. The dredge operates in any of the three possible positions it will adopt when it hits the bottom.

3. **Trawl net.** This was used in medieval Europe. It consists of a large net towed by two lines each leading to one of two ships (the two ships sailing on a parallel course). It operates equally well on the bottom or in midwater (for nekton).

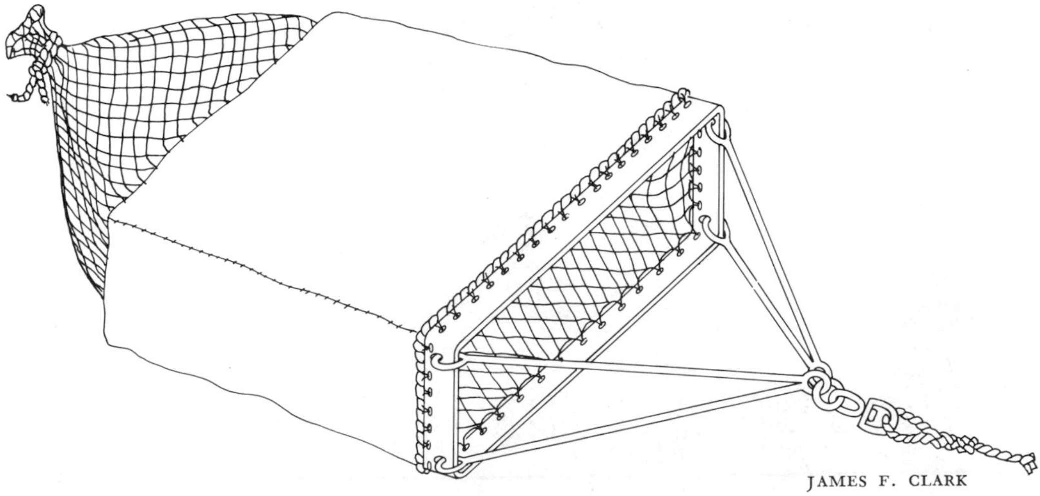

Fig. 3.3 Naturalist's dredge.

JAMES F. CLARK

Fig. 3.4 Preparing to lower a naturalist's dredge over the side.

4. **Otter trawl** (Fig. 3.5). The otter trawl was developed around 1800. It requires only one towing vessel. The type usually employed for commercial fishing has a large conical net that is towed with the wide end leading. The mouth of the net is strengthened by a rope that attaches the net around the circumference of the mouth. From the left and right sides, a cable leads an "otter board," which is a rectangular wooden element. Each otter board is held by adjustable chains or rope shackles on the front face by which the trawl is towed. The otter board responds like a kite when it is drawn through the water, moving outward in accordance with physical laws. As each board moves outward, to left and right, it causes the mouth of the net to open to its maximum extent. The head rope over the mouth and between the other boards carries floats which keep the mouth open; the foot rope is weighted with weights or a continuous chain, to glide over the surface of the seabed. The foot rope is about 1.5 times as long as the head rope, causing the lower leading edge of the bag to lag behind the upper leading edge (head rope). Thus bottom animals are not disturbed by the foot rope until the head rope has already passed overhead, so when the benthonic animals rise from the seabed they are already engulfed within the net. The trawl is towed at a speed fast enough to ensure that the captured animals are swept back into the so-called cod end of the net, a cylindrical bag in which the catch is held. Such

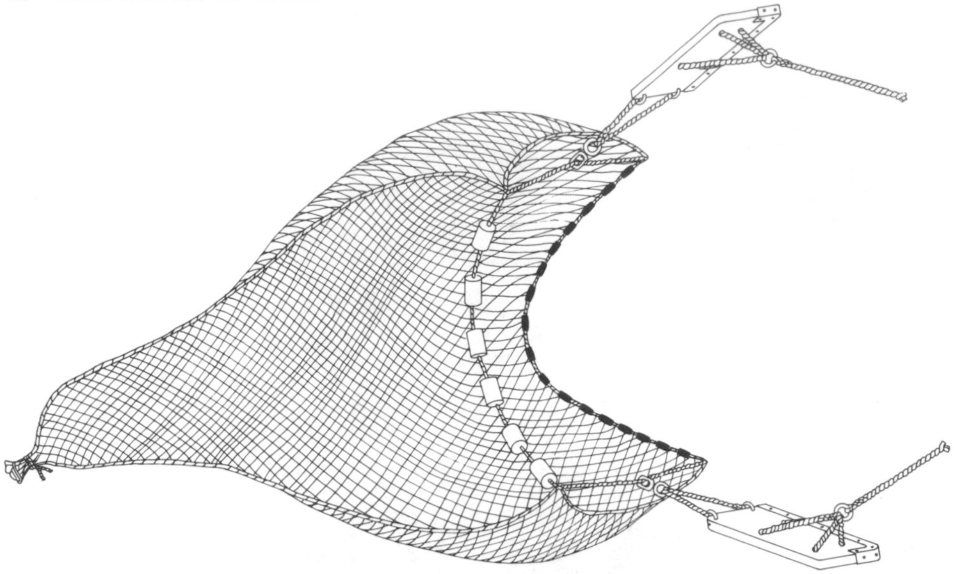

JAMES F. CLARK

Fig. 3.5 Otter trawl (chain drag).

DAVID F. MOYNAHAN

Fig. 3.6 Clamshell grab.

trawls are made up to about 80 feet across. In smaller versions the otter boards may be set closer to the actual net. An otter trawl may be adjusted to operate in either midwater or on the bottom.

5. **The beam trawl.** This is really an enlarged version of the rectangular dredge, with the mouth of the trawl net being held open by a rectangular metal frame or by a metal beam that replaces the foot rope of the otter trawl. The beam trawl is easier to control and is used more in research on very deep bottom, where the otter boards could become entangled.

Procedures for Soft Bottoms

1. **Grabs.** These are devices that collect samples of seafloor sediment together with any organisms that happen to be in the sediment. A **clamshell grab** (Fig. 3.6) has two scoops that can be triggered to close like the valves of a clam. An **orange-peel grab** (Fig. 3.7) has four valves that appear to form the four segments of an orange when cut along its meridians. The four valves bite into the seabed and close around the bite. The disadvantage of grabs is that they collect very little, and most of that is mud.

2. **Sledge.** This is a device designed to slip along the surfaces of the seabed while a cutting edge, placed along the leading edge, peels off a specified thickness

JAMES F. CLARK

Fig. 3.7 Orange-peel grab.

of silt and feeds it into a collecting bag. In the best models there are two knives, each operating when the surface of the sledge to which it is attached is the one on which the sledge comes to rest when it first strikes the bottom. Such a sledge may be designed as either a quantitative or a qualitative sampler. If the mesh has wide spaces, most of the mud escapes, leaving behind only those organisms of the infauna that are too large to escape. If the mesh is made very fine, the entire sample is retained and is similar to that of a grab sample, although much larger. In the latter configuration the sledge yields valuable information about the young forms and, hence, about the structure of seafloor populations of a given species, which can be studied by size and age groups.

Core Samplers

Originally restricted to soft substrates, core samplers are now so sophisticated that they can be used on any substrate, including the hardest known, such as sheets of lava deposited on the ocean floor and chert layers produced by the solution and redeposition of silica derived from the skeletons of marine plankton. The principal use of a core sampler (Fig. 3.8) is to obtain vertical cylindrical cores from the seabed. Inside the core is the sample that, when extruded, provides a chronologically stratified record of the successive sheets of sediment. The longer and

JAMES F. CLARK

Fig. 3.8 Pfleger corer (see also Chapter 28).

deeper penetrations provide proportionately larger continuous chronologies. These contain embedded fossils that appear in the layers that were the interfaces between the water and the seabed at the time of their entombment. Following types of core samplers are used.

1. **Simple metal tubes.** These are dropped on the end of a light line so that they fall on the seabed, hopefully in a vertical position, and are driven some distance into the sediment by a weight attached to the upper end of the tube. They collect only short cores.

2. **Augers.** These are screwed into the seabed and then withdrawn by upward pulling. The sample emerges as a helix coiled around the archimedian screw. Augers require a diver to perform the sampling, and they are good for selective sampling in relatively shallow waters.

3. **Explosive-driven core samplers.** Developed by Swedish oceanographers, these were the first devices to collect long cores with samples ranging back into the early Pleistocene (i.e., about 4 million years of continuous recording).

4. **Drill cores.** These are obtained using the same principle as in oil rigs but adapted by highly sophisticated computer-controlled techniques to sample the seabed in very deep water and to penetrate earth's crust to great depths. At the present time, such cores have penetrated to seafloor layers deposited down about 135 million years ago, proving that the oceans have existed continuously, at least since the Jurassic period. These cores have not yet reached any layers that could be recognized as antecedent to the formation of an ocean.

Underwater Habitats

The recent progress of scuba techniques has led to the development of **habitats**—living chambers of varying sophistication within which a diver may find temporary refuge while working on the bottom. One of the simplest of these is the **Clark habitat,** which is held in place by weighted anchor ropes and is inflated and kept habitable by a steady flow of air pumped into a plastic envelope forming the sides and roof of the habitat.

PRESERVATION OF SAMPLES

One task that usually faces a biologist on board any vessel that is operating a benthic research program is the sorting, labeling, and preservation of trawl or dredge samples. An efficient program would allow for this work to be done on board large vessels or immediately after return to base on smaller vessels (Fig. 3.9) so that the materials can be submitted to appropriate investigators with minimal delay. However, in cases where there are too few biologists to service an ambitious sampling program, the sorting has to be postponed and carried out at a later date.

The mixed sample should first be sorted roughly so that the larger animals are separated from the more delicate ones. This work can be done on deck if the weather is good (Fig. 3.10). The whole sample should be photographed on color film to record the natural colors before they are bleached by preservatives. Most animals can be preserved in either ethyl alcohol or formaldehyde (4 percent aqueous solution). Isopropyl alcohol can be used also, although not for echinoderms.

In addition animals should not be preserved in a contracted state because subsequent identification will require that all external organs be readily visible. Contracted specimens must therefore be relaxed and anesthetized prior to preservation. The exact procedures vary somewhat according to the kind of animal being collected. The following discussion explains the treatment of starfishes; it is rather more complicated than treatment required for animals that do not usually have to be relaxed, such as crabs or mollusks.

Bring the specimens into the laboratory alive in seawater and in a covered plastic bucket or similar container. Before preserving them, they should first be killed and relaxed. To do this, place them in a flat-bottomed dish, such as a photographic developing tray. Do not put more in at a time than will form a single layer. Then run some warm, fresh water into the dish, about as hot as the hand can bear. No matter how twisted the specimens may be, they will die at once and become quite flexible. They can then be arranged neatly into the position you wish them to keep. Omit the warm water in the treatment of fishes; in that case the fins should be spread out with the aid of clips or pegs so that the finrays can be counted when they are being identified. Next add to the dish of water

DAVID F. MOYNAHAN

Fig. 3.9 Oceanographic research can be carried out on board small vessels. Here an otter trawl is hoisted on board a 44-foot fishing rig off the New England coast.

some 40 percent formaldehyde, about 4 tablespoons to each pint of water (you will soon learn to judge this without having to measure it), taking care not to disarrange the specimens. Put the dish away for two or three days. Formaldehyde is mildly poisonous. The fumes should not be breathed because they are irritating, otherwise the substance is quite safe to use.

After a few days, the starfishes, which by now will be quite stiff and strong, can be removed from the dish and dried out in some well-ventilated place. Other specimens, such as crustaceans and fishes, should be transferred to large containers of fresh formaldehyde or alcohol after the pegs are removed from the fins. Small brittlestars, mollusks, sea urchins, and corals can be placed directly into the formaldehyde without relaxing in warm water, if this is inconvenient. If this procedure does not give a good result, consult with the specialists at the nearest museum.

STORAGE

Dried specimens and shells can be kept in strong cardboard boxes. Line the boxes with tissue paper not with cotton–wool, which will adhere to the spines and will be unsightly and annoying in microscopic study. Soft-bodied animals, of course, cannot be dried and must be stored in fluid preservatives.

Small animals are best kept in 1-ounce, 2-ounce, or 4-ounce jars, either dried or, alternatively, immersed in ethyl alcohol (not in formaldehyde, because it slowly dissolves the lime in the skeleton). Specimen labels should be placed inside the jar or vial, not gummed onto the outside because they easily fall off. It is best if labels are written on strong paper or card with india ink, however, black pencil can be used also. Do not use ballpoint because this fades or dissolves in preservatives.

DAVID F. MOYNAHAN

Fig. 3.10 A sample from the flounder grounds off New England is taken by otter trawl.

PLANKTON

Plankton is usually collected by means of a **plankton net** (Fig. 13.1). The planktonic plants of the marine plankton (**phytoplankton**) are normally concentrated within the first few fathoms of the surface of the sea, where the sunlight is strongest. The animals of the plankton (**zooplankton**), on the other hand, are usually provided with some type of muscular swimming organ, and they can occupy different depths. It has been observed that zooplankton occupy the surface waters during the night, but it sinks into deeper zones during the day. Thus, in the day, the best hauls for upper-level zooplankton are usually obtained at depths of 20 or 30 fathoms (roughly 35 to 50 m).

The plankton net is normally towed at about 1 knot (at speeds over 1.5 knots the water is not filtered fast enough and spills out of the net at the leading edge instead of passing through). The tow wire or rope is held at an angle by the tension exerted by the 3-kg weight and the bridle holds a 50-cm diameter hoop. The conical net, made of bolting cloth (nylon), has a collecting jar attached to the narrow opening by means of a collar, which can be loosened for removal of the jar.

Collecting at greater depths requires successively heavier weights on the tow line. For example, although 3 kg is suitable for a surface tow with a small net, about 10 kg is the weight needed to tow the net at a depth of 20 or 30 fathoms. When the net is hauled up after the tow, it may continue to collect samples from the overlying water, or the contents may escape on the way up. To prevent such mishaps, the net may be provided with a lid that can be closed by sending a metal weight called a **messenger** down the line before it is hauled up. When the net reaches the deck, the sampling bottle is removed from it. Organisms adhering to the net are washed free, through the opening at the narrow end, and are caught in another container. After sorting (if this is necessary) all parts of the sample are preserved, generally in 5 percent formaldehyde. Labels enclosed in the jars, giving the station data, that is, the date, time of day, latitude and longitude, depth at which the sample was taken, weather and sea state, name of the vessel, and number of the station. The sample can then be stored for later study.

Collecting Plankton from Greater Depths

A larger vessel is required for collecting plankton from greater depths than for sampling shallow waters. Plankton becomes progressively scarcer with increasing depth, therefore much larger plankton nets must be used in order to filter larger volumes of water. A large plankton net, such as that employed on an ocean-going research vessel, may measure up to 5 m in diameter. The techniques of sampling deep-water plankton were developed in the closing years of the nineteenth century by the German *Valdivia* expedition (see Chapter 13).

4

THE LITERATURE OF MARINE BIOLOGY

Basic library resources / The news media of marine biology / The laboratory bookshelf / Consulting original research papers

An important part of the work of a marine biologist consists in keeping abreast of current research as well as tracing significant older research. To do this effectively, a minimum of library materials is required.

The smallest library must have certain minimum resources, or documents essential for the study of marine biology. The *Zoological Record*, an annual publication of the Zoological Society of London, is one essential resource. It is an annual list of the authors of all known papers on zoological topics published during the year noted on each volume (the volumes are usually compiled several years after the publication of the papers). With this source you or your library can trace almost any paper you are likely to need.

Your library probably subscribes to some regular periodical that functions as a scientific news medium. Examples of periodicals that report briefly on recent advances in marine biology are: *Science, Nature, Science News,* and *Marine Science Contents Tables.* In addition some of the larger oceanographic institutions issue newsletters.

THE LABORATORY BOOKSHELF

Some of the most useful types of books to have readily available in the laboratory when specimens are being examined and identified are suggested below. The list also includes examples of general works for background reading or for expanding upon topics discussed in the previous chapters. (E, elementary. A, advanced.)

General Books

Berril, N. J. 1966. *The Life of the Ocean* (McGraw-Hill). E.
Carson, Rachel L. 1961. *The Sea Around Us* (Oxford Univ. Press).
Ekman, Sven. 1967. *Zoogeography of the Sea* (Sidgwick & Jackson). A.
Fairbridge, Rhodes W. 1966. *Encyclopedia of Oceanography* (Reinhold). A.
Hunt, C. A., and Garrels, R. M. 1972. *Water, the Web of Life* (Norton).
McConnaughey, B. H. 1970. *Marine Biology* (Mosby).
Reish, D. J. 1969. *Biology of the Oceans* (Dickenson).
Sverdrup, H. U.; Johnson, M. W.; and Fleming, R. H. 1942. *The Oceans* (Prentice-Hall). A.
Turekian, K. K. 1968. *Oceans* (Yale Univ. Press, New Haven). E.
Weyl, Peter K. 1870. *Oceanography* (Wiley).

Scuba and Marine Archeology

De la Varande, J. 1952. *La Navigation sentimentale* (Flammarion). A.
Dodd, Edward. 1972. *Polynesian Seafaring* (Dodd, Mead).
Dugan, James. 1956. *Man Explores the Sea* (Hamish Hamilton).
Taylor, J. du P. 1966. *Marine Archeology* (Crowell).

General Marine Ecology

Hedgpeth, Joel. 1957. *Treatise on Marine Ecology (Geol. Soc. Amer.)*. A.
Nybakken, J. W. 1971. *Readings in Marine Ecology* (Harper & Row). A.
Russell, F. S., and Yonge, C. M. 1966. *The Seas* (Warne). E.
Thorson, Gunnar. 1971. *Life in the Sea* (McGraw-Hill). E.

Fishes

Frank, S. 1971. *Pictorial Encyclopedia of Fishes* (Hamlyn).
Perlmutter, A. 1961. *Guide to Marine Fishes* (Bramhill). E (West Atlantic).
Roughley, T. C. 1966. *Fish and Fisheries of Australia* (Angus and Robertson).
Smith, J. L. B. 1949. *Sea Fishes of South Africa* (Central News Agency, Cape Town).
Thomson, D. A., and Eger, W. H. 1966. *Fishes of the Gulf of California* (Univ. Arizona Press).
Zim, H. S., et al. 1970. *Fishes* (Golden Press). E (Pac. and Atl. coasts of America).

North Atlantic Shore Life

Amos, W. H. 1966. *Life of the Seashore* (McGraw-Hill). E.
Barrett, J., and Yonge, C. W. 1958. *Guide to the Sea Shore* (Collins). E (Europe).
Collins, H. H. 1959. *Field Guide to American Wildlife* (Harper & Row).
Gosner, K. L. 1971. *Guide to Estuarine and Marine Invertebrates* (Wiley).
Miner, R. W. 1950. *Field Book of Seashore Life* (Putnam).
Morris, P. A. 1951. *Field Guide to the Shells* (Houghton Mifflin).
Zim, H. S., and Ingles, L. 1963. *Seashores*. (Golden Press).

North Pacific Shore Life

Hokoryukan Encyclopedia of the Fauna of Japan., n.d. (5000 ill.; text in Japanese.)
Publications of the Bishop Museum, Hawaii.
Hedgpeth, Joel, et al., publications of Univ. California Press, Berkeley.

Southern Oceans

Dakin, W. J. 1952. *Australian Seashores* (Angus and Robertson).

Deep Sea

Burton, Maurice. 1958. *Life in the Deep* (Phoenix, London).
Murray, John, and Hjort, Johan. 1912. *Depths of the Ocean* (Macmillan).

Tropical Reefs

Allan, Joyce. 1959. *Australian Shells* (Angus and Robertson). (Indo-Pacific.)
Abbott, R. T., and Zim, H. S. 1962. *Sea Shells of the World* (Golden Press).
Clark, Hubert H. 1921. *Echinoderm Fauna of Torres Strait* (Carnegie Institute). (Best general source on tropical echinoderms.)
Darwin, Charles. 1962. *Coral Reefs* (Univ. California Press, Berkeley).
Smith, F. G. W. 1971. *Atlantic Reef Corals* (Univ. Miami Press, Miami). (See also Hokoryukan.)

CONSULTING ORIGINAL RESEARCH REPORTS

In this text facts may be stated without citing an authority if they have been known for many years and are widely accepted. More recent contributions, however, are likely to be cited with the name of the author and usually the year in which the research was published. You should know how to obtain a copy of the cited report, if you want to explore a topic more fully. The following examples are illustrative.

Author's Name and Date Are Known. On page 90 of this book it is stated that a Floridian ophiuroid *Ophiophragmus filograneus* was found by Thomas (1961) to tolerate low salinities. You now wish to consult the original report. Go to your college or city library and locate the series of volumes called *Zoological Record.* Take the volume for the year 1961, and find the section headed "Echinodermata." Here you will see an alphabetic list of all papers on echinoderms published in the year 1961, showing that the paper you need is: Thomas, L. P., "Distribution and salinity tolerance in the amphiurid brittlestar *Ophiophragmus filograneus,*" *Bull. Mar. Sci. Gulf Caribbean,* **11,** 158–160. Copy this reference and give it to the librarian, who will either direct you to the shelf where the *Bulletin of Marine Sciences of the Gulf and Caribbean* is stored or, if necessary redirect you to another library that holds the journal. Alternatively you or your librarian can obtain a copy or microfilm of the paper from a library that has the journal. The *Union List* gives the latter information.

Author's Name and Date Are Known, but the Date Is Recent. If a paper is quite recent, not enough time will have elapsed for it to have been abstracted in the *Zoological Record.* For such cases, your library should possess one or another of the *Current Contents* journals where the source you seek may be located. For example, on page 90 of this book Mohsen et al. (1972) are cited as reporting a salinity dependence of the alga *Ulva* in regard to its metabolism. Reference to the *Marine Science Contents Tables,* issued monthly by the United Nations Food and Agriculture Organization, shows that the paper in question is: A. F. Mohsen; A. H. Nasr; and A. M. Metwalli, "Effect of Different Salinities on Growth Re production . . . of *Ulva fasciata,*" *Botanica Marina,* **15,** 177–181.

These two examples show that it is now no longer necessary to print detailed bibliographies in textbooks; all that is required is the author's name and the date of publication. Similarly the long lists of synonymies of the scientific names of particular organisms, formerly considered to be a mark of scholarship in technical papers, are in fact only burdensome to the institutions that have to finance the publication of scientific reports. The best researcher is the one who knows the literature so thoroughly that he or she can use the resources provided by abstracting journals to find his way effectively to the desired papers. Practice these sleuth-hunts till you are efficient and confident, and at the same time you will learn a great deal about how science is reported and its results stored. In the course of tracking a paper one normally encounters many other interesting references en route, so the task is neither sterile nor unrewarding.

Topic Known, but No Author Cited. You may want to know whether anyone has published anything at all on a particular topic. Or you may want to review the literature on some topic. In this case you merely go to each annual volume of the *Zoological Record,* locating the appropriate subject head, and then scan the list of authors and topics under that head for each year. In short order you can compose an important list of references to consult for the research data wanted. For example, suppose you wish to write a review of work done on the effects of pollution on echinoderms, and you happen to take down the 1968 volume of the *Zoological Record.* Listed alphabetically under "Echinodermata, Ecology and Habits" you will find the entry "Pollution." In 1968 three authors wrote papers on that topic; Beyer in Oslo, Norway; Leppakoski in Sweden; and Smyth in Scotland. Turning now to the alphabetic listing of "Authors" under "Echinodermata," you will find each of the papers cited in full under the name of the author, and you can then ask your librarian for the journals concerned, or request microfilm or xerox copies if the journal is not held by your library.

Review Journals. From time to time certain journals, such as *Oceanography and Marine Biology Annual Review,* publish articles summarizing work in a particular topic over the past several years. This is, of course, the easiest and most effective way to grasp the present status of research or to establish a base upon which you can build by consulting all papers published *after* the period covered by the review.

There are now more than 100 periodicals publishing reports on marine biology and allied disciplines, each averaging perhaps 100 articles a year. The annual increment is about 10,000 papers. The publication rate is increasing exponentially, doubling approximately every 8 years. It is clear that the traditional methods of circulating scientific data will have to be changed to meet the information glut and to prevent scientists from drowning in their own output. One way to prepare for the changes that are sure to come is to follow the procedures here recommended and to make yourself familiar with the effective use of available resources for identifying and locating papers.

The use of copying machines has greatly relieved the pressure upon smaller libraries to subscribe to numerous journals. In the last resort, the only essential journals are those such as *Zoological Record* and the current content series, for they are the ones that identify the required papers. The current content series can always be obtained as a microfilm or xerox copy from a larger center that has the original journals. Thus it is now possible for even a solitary scientist in a remote location to keep abreast of current work without paying regular visits to a large library.

In following these recommendations, take care not to infringe the law of copyright, which stipulates that a xerox or other copy of a copyrighted work must be for an individual's personal use. Thus you are free to copy whatever you need, but you cannot legally distribute copies to your friends. Tell them to make their own arrangements!

Recognizing the need for significant historical papers to remain accessible to students and scientists, some publishers have now begun to issue individual reprints of classic papers. These can be obtained either separately or in sets, and in some cases publishers have issued volumes composed of collections of important papers all dealing with a stated topic. Your instructor will be able to guide you to sources of these if you let him or her know that you would like to start a basic library.

5

THE MARINE ECOSYSTEM

Ecology and the ocean / Biota / Autotrophs, primary and secondary heterotrophs / Food webs and food chains / Concept of the ecosystem / Ward's experiment / The balanced aquarium / Tansley coins the term ecosystem / Thynne develops the saltwater aquarium / Systematics / Ecosystems change through time / Structure of the marine ecosystem / Biomass / Standing crop / Productivity / Primary productivity / Autotrophs of the sea / Thallophytes and embryophytes / Heterotrophs / Phyla of marine invertebrates / Marine vertebrates / Dispersion of marine life

ECOLOGY AND THE OCEAN

The organisms that inhabit any particular region or environment are collectively termed the **biota.** The biota comprises plants and animals. Most plants can capture the energy radiated by the sun as light and heat (electromagnetic energy) and can use it to convert inorganic constituents of the environment into organic materials. Organisms with this power are termed **autotrophs.** Animals, on the other hand, need preexisting organic materials for their nutrition; in this role animals are collectively termed **heterotrophs.** Animals that feed upon plants are termed **primary heterotrophs,** and those that feed upon other animals are called **secondary heterotrophs.** The interlocking nexus of autotrophs and primary and secondary heterotrophs is termed a **food web,** made up of many distinct **food chains.** The ultimate products of any food web are feces, carrion, gases, and other lifeless materials of organic origin. These are converted by bacteria into simpler inorganic substances, are thus returned to the environment, and then become available to the autotrophs again. Any such self-sustaining food web, together with the input of solar energy and the environment containing the food web, constitutes an **ecosystem.** Ecosystems may be large or small; the entire ocean conforms to the definition just given and therefore may be viewed as a single ecosystem—but so also does a forest pond or a desert oasis. The study of organisms in relation to their environment is named **ecology.** A primary task of the ecologist is to identify the various plants and animals in an ecosystem and to determine the roles they play in it.

CONCEPT OF THE ECOSYSTEM

In 1837 Ward predicated the theoretical properties of a self-maintaining natural community, thus in effect defining an ecosystem, although the word did not come into use until Tansley coined it a century later. Ward's inferences were based on prior work of Priestley and his contemporary chemists, who had shown that plants require sunlight and that they liberate oxygen in the presence of sunlight. Ward reasoned that it should be possible to assemble a community of plants and animals and provide sunlight and soil, with the result that the plants would grow and the animals would feed on the plants or on one another. If reproduction occurred, the community would continue to exist. Four years later, in 1841, Ward successfully constructed a freshwater 20-gallon aquarium, stocked with plants and fishes, and placed it in the sunlight. For several years thereafter the aquarium continued in a healthy state without any further attention. This **balanced aquarium,** as he called it, was soon afterward paralleled by a saltwater aquarium made by Thynne, operating on the same principle; by this means the people of London in 1847 saw marine life for the first time. The immediate result was that the cultured amateur now began to take a close interest in the seashore and its life. Philip Henry Gosse brought out his book *The Aquarium, An Unveiling of the Wonders of the Deep Sea* in 1854 (actually the book dealt with shallow-water biota), and the vogue for aquariums continued through the next 50 years. These balanced aquariums were simplified artificial ecosystems in which only two **trophic levels** occurred, namely autotrophic and primary heterotrophic (animals were chosen that would not attack one another).

Systematics

An ecosystem can be adequately described only if the biota comprising it are correctly named. The identification of organisms is the task of **systematic biology,** which also seeks to determine the evolutionary relationships of organisms. Most ecosystems can be subdivided into two or more parts, some of which may themselves qualify as independent ecosystems; usually ecosystems are, at least in part, interdependent. The entire biosphere may be viewed as comprehending a single **planetary ecosystem,** itself made up of numerous subsidiary ecosystems.

Ecosystems Change Through Time

Organisms have evolved through geological time, from which it follows that ecosystems must have undergone change—the constituent biota cannot always have comprised the same species of animals or plants. Thus temporal as well as spatial elements enter into the study of the ecosystem. The overwhelming importance of the minute floating plants called **diatoms** is discussed in Chapter 13, where they are noted as comprising the basic elements of all marine food chains. The fossil record of diatoms shows them to have been common marine plants only after the Jurassic period (ca. 120 million years ago). Thus the marine ecosystem must formerly have been somewhat different.

Productivity

As has already been stated, any ecosystem is a self-sustaining community of living organisms comprising autotrophs, which produce organic substances from inorganic environmental materials in the presence of sunlight, and heterotrophs, which consume the autotrophs or one another, and yield lifeless organic products in the form of carrion, feces, carbon dioxide, and other excreted substances. These organic products are degraded by bacteria, are converted back to inorganic constituents of the habitat, and so are recycled for the renewed use of the autotrophs. So long as solar energy continues to be supplied the ecosystem is permanent.

A few more definitions will be introduced at this juncture. One useful term is **biomass,** which can be defined roughly as the wet weight of living tissue, including unassimilated organic materials ingested by animals that are not yet digested or defecated. In recent years the world's oceanographers have devoted much time to measuring the amounts of biomass in various parts of the sea. The amount of biomass existing at any given instant under a stated area of the sea surface is called the **standing crop.** It can be measured simply as so many pounds per square yard; in scientific contexts, however, the amount is given as kilograms per square meter (kgm^{-2}). The wet weight varies with the water content, which is not constant for various kinds of organisms, therefore a more useful measure is the dry weight, or still better the ashfree dry weight, which is the difference between the dry weight and the residual inorganic salts after total combustion of the dry biomass. Chemists can measure the energy stored in the biomass by a calorimeter and so give its calorific value in calories or kilocalories per square meter. If the area under investigation is studied at fixed intervals of time, the increase in the standing crop per day can be calculated. This value is termed the **productivity,** which can be reported as kilocalories per square meter per day ($kcal\ m^{-2}\ day^{-1}$). If the biomass of autotrophs can be determined alone, then we have a measure of what is called **primary productivity.** Such measures are obviously of great importance in determining the relative richness of communities in different regions of the sea and their ability to yield crops of fish for the use of mankind. The measurements have shown that the most productive areas of the sea lie in the cool northern oceans and also occur where upwelling of deep water takes place, bringing dissolved mineral nutrients to the surface. The warm tropical oceans are relatively low in productivity. This rather surprising fact can be explained if we take into account certain properties of phytoplankton; this is discussed in detail in Chapter 13.

THE AUTOTROPHS

The primary producers, or autotrophs, of the oceans are the marine plants, nearly all of which are called algae (Fig. 5.1). The largest and best-known algae are the

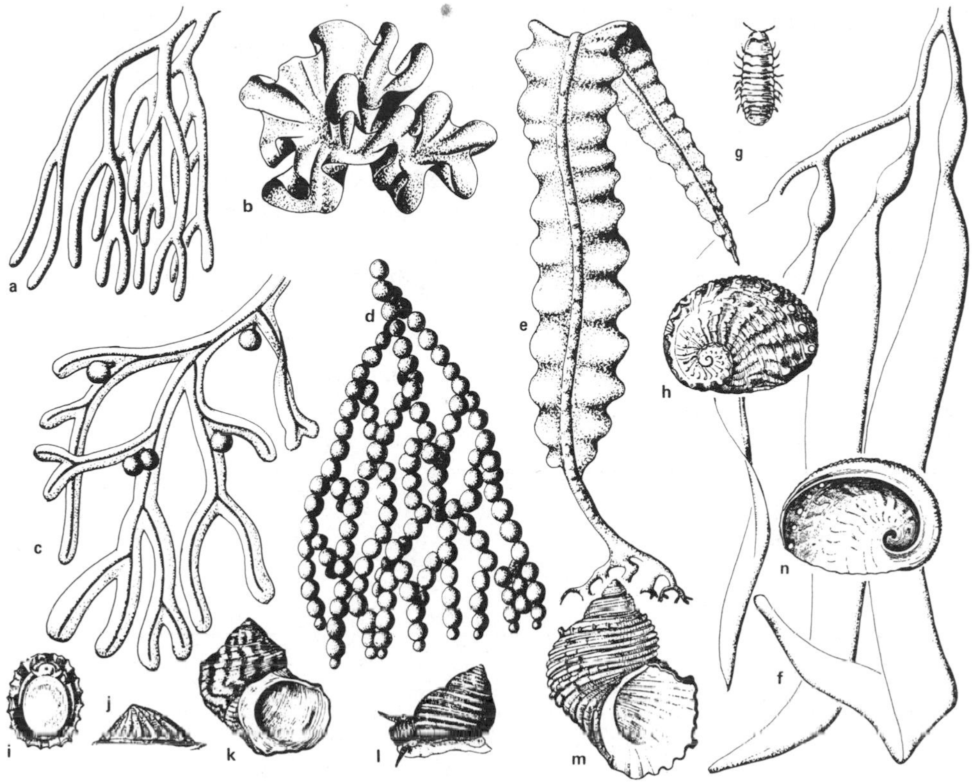

Fig. 5.1 Representative autotrophs of the seashore with associated primary heterotrophs. (a) *Codium* ×0.5, subtropical and tropical coasts, Chlorophyta. (b) *Ulva* or sea lettuce, ×0.5, all seas, Chlorophyta. (c) *Fucus* or sea wrack, rocky coasts of Northern Hemisphere seas, Phaeophyta. (d) *Hormosira* or mermaid's necklace, rocky wave-beaten coasts of southern Pacific where it replaces *Fucus*, Phaeophyta. (e) *Laminaria* or kelp, ×0.25, North Atlantic rocky coasts. (f) *Macrocystis* or kelp, rocky coasts of Southern Hemisphere and northern Pacific, Phaeophyta. (g) *Limnoria* or gribble, ×1, boring into drift wood, Isopoda. (h) *Haliotis* or abalone, ×0.1, Indo-Pacific and Mediterranean, Gastropoda. (i, j) *Patella* or limpet, ×0.5, most temperate seas but lacking from America, rocky coasts, Gastropoda. (k, l) *Turbo* or turban shell, ×0.5. (m) *Littorina* or periwinkle, rocky coasts of most seas, Gastropoda.

seaweeds, which are common on all coasts. However, the most important in their contribution to overall biomass (and hence to primary productivity) are the microscopic single-celled algae called **diatoms.** These tiny plants float in the surface waters of all oceans, unnoticed by the casual observer, although they are the most abundant living things on the earth at the present epoch. The term **alga** is no longer (as it once was) a part of the formal classification system of plants, but it is so widely used that its meaning should be made clear.

In ordinary scientific contexts the terms autotrophs and plants may be taken as synonymous. These organisms are grouped together in one large assemblage called the kingdom Plantae. This kingdom is further subdivided by botanists into two sections:

1. **Subkingdom Thallophyta:** Plants without leaves, roots, or vascular tissue, that release the sex cells when they are formed. (Seaweeds and phytoplankton.)

2. **Subkingdom Embryophyta**: Plants with leaves (or leaflike organs of nutrition), that retain the female sex cells in the sex organ until after fertilization and development of the embryo. (Mangrove, seagrass, and land plants.)

The first of these groups, the Thallophyta, comprises a number of subsidiary groupings called phyla; each phylum possesses some distinctive character. Some of the phyla include plants that contain chlorophyll, the pigment that produces the green color of plants and that has a complex molecule capable of causing sugars to be synthesized in the presence of sunlight. All Thallophyta possessing chlorophyll are termed algae. Some supposedly degenerate Thallophyta lack chlorophyll; if these are multicellular, they are called fungi, if unicellular, they are called bacteria. Thus the fungi are organisms that have the structure of plants but behave like animals (heterotrophs) because they obtain their energy from carrion or dead plant materials. The bacteria, as already noted, break down lifeless organic residues into inorganic components, thereby obtaining energy. In addition some bacteria obtain energy from inorganic sources. A summary of the main phyla of algae is given in Table 5.1.

Table 5.1 The Phyla of Marine Algae

PHYLUM	CHARACTERISTICS	EXAMPLES
Cyanophyta	Minute floating one-celled plants that lack a distinct cell nucleus (the genetic material being dispersed in the cell); often colored blue-green on account of the presence of the pigment phycocyanin	Blue-green algae
Chrysophyta	Minute floating or attached one-celled plants, with a distinct nucleus, and having the cell body enclosed in a 2-valved secreted silica capsule; usually brown or yellow pigments present, plus chlorophyll	Diatoms
Pyrrophyta	Extremely minute floating and swimming one-celled plants (requiring a magnification of at least $\times 500$ to reveal detail); having 1 or 2 threadlike flagella used as swimming organs, and usually with several cellulose plates forming a jointed capsule	Dinoflagellates
Phaeophyta	Conspicuous brown seaweeds attached to rocks on coasts and to hard substrates of the continental shelf; some have hollow flotation bladders enabling them to float if their attachment is lost; contain brown pigments plus chlorophyll	Kelp Sargasso weed
Chlorophyta	Green seaweeds, some minute and one-celled or of a few cells, others larger and resembling moss or lettuce, the larger kinds attached to the seabed; chlorophyll the main or only pigment	Sea lettuce
Rhodophyta	Red or purple seaweeds, usually a few inches long, never massive, containing red pigments as well as chlorophyll; sometimes also containing lime (in which case they resemble encrusting coral or paint spills on rocks); usually attached to substrate	Dulse Corallines

The subkingdom Embryophyta consists almost entirely of terrestrial and freshwater plants. One group, the phylum Tracheophyta, or vascular plants with roots and leaves, includes a few marine members. The best-known of these are the **eelgrasses,** which grow in shallow seas on soft bottom, and the **mangroves,** which are low tropical trees found in estuarine environments and along muddy shores subject to tidal waters.

THE HETEROTROPHS

The consumers, or heterotrophs, of the oceans are animals (Fig. 5.1). Unlike plants, the marine animals are varied in complexity of structure and are, accordingly, classified in nearly all the various phyla into which the animal kingdom is divided. Many marine animals have distinct preferences for particular habitats. (This is discussed in the following chapters with a detailed analysis of each classification.) The term Invertebrata covers all the phyla of animals that lack a backbone; they are mostly rather small animals. The Vertebrata are the larger and better-known animals, all members of the phylum Chordata and (except for some transitional forms called protochordates) distinguished by having a vertebral column.

Table 5.2 summarizes the characters of the more conspicuous phyla of marine invertebrates. Although space limitations prohibit an extensive discussion of the classification of organisms, a few comments about general characters may be helpful here. Botanists refer both unicellular and multicellular plants to various phyla because their rationale for classification is not based on cellular complexity. Zoologists, on the other hand, draw a sharp line between unicellular animals (all of which are referred to a single phylum called Protozoa), and multicellular animals, which are referred to other phyla. Thus any one-celled animal is immediately recognizable as a member of the phylum Protozoa. Animals may lack obvious symmetry, have a radially repeated symmetry (e.g., starfishes or jellyfishes), or have a bilateral symmetry in which the left side is the mirror image of the right (e.g., lobsters, bristleworms, or man). These contrasted patterns of symmetry provide valuable diagnostic characters in recognizing the phyla, as shown in Table 5.2. Sponges (phylum Porifera) are peculiar in having no single mouth opening and no single internal digestive chamber. Of the phyla that have a mouth, some have only a single internal digestive chamber, whereas others have a general body cavity (coelom) within which the digestive chamber (enteron) is separately placed (these are called coelomates). In some phyla the skeleton may be lacking, as in flatworms (phylum Platyhelminthes), take the form of an external limy shell (phylum Mollusca), comprise an exoskeleton investing the whole animal (phylum Arthropoda), or take the form of an internal bone or cartilaginous endoskeleton (phylum Chordata). Table 5.3 lists various combinations of these contrasted diagnostic characteristics that serve to define the different phyla.

The phylum Chordata, of which the best-known members are the vertebrates, is so well represented in terrestrial environments that nearly everyone already knows the principal subdivisions by their everyday vernacular names. The phylum is subdivided into a number of classes, corresponding for the most part to groupings known by venacular names. The marine representatives, therefore, are included in Table 5.3. Begin now to memorize the technical names of the phyla and classes; they will frequently be cited without further definition.

Table 5.2 The Most Conspicuous Phyla of Marine Invertebrates

PHYLUM	CHARACTERISTICS	EXAMPLES
Protozoa	Minute one-celled animals, marine forms often with a limy or silica capsule; differing from minute algae by lacking chlorophyll; floating or benthic	Forams Radiolarians
Porifera	Multicellular anchored forms with no mouth or central digestive cavity; grow on seabed	Sponges
Coelenterata	Multicellular anchored or floating forms with a radially symmetrical body and a mouth leading into a central digestive chamber, but no coelom; solitary or colonial	Jellyfishes Sea anemones Corals
Echinodermata	Radially symmetrical coelomates, mostly free roaming on the seabed, sometimes attached or floating; surface of body commonly spiny; solitary	Sea urchins Starfishes
Platyhelminthes	Bilaterally symmetrical flattened wormlike forms, with mouth and digestive cavity present in free-living forms, but lost in some parasitic forms	Turbellarians Fish flukes Tapeworms
Aschelminthes	Body cylindrical, wormlike, tapering at tips, with mouth and digestive cavity; coelom imperfectly developed, no obvious segmentation; often parasitic	Roundworms
Annelida	Body elongated, wormlike, flattened or cylindrical, with well-developed segmentation, mouth, and digestive tract, and a segmented coelom; mainly benthic	Bristle worms
Arthropoda	Body bilaterally symmetrical, covered by an external chitinous skeleton, and having well-developed jointed paired limbs, mouth with jaws, and complex gut	Shrimps Crabs Lobsters
Mollusca	Body bilaterally symmetrical (though sometimes coiled in a helix), usually secreting a protective limy shell of 1, 2, or several valves; mouth and gut well developed, but coelom secondarily reduced	Sea snails Clams Octopuses

DISPERSION OF MARINE BIOTA

Marine organisms differ conspicuously according to their means of dispersion. Organisms that float in the sea or that swim too feebly to be able to overcome motions imparted by ocean currents are termed the **plankton,** the corresponding adjective being **planktonic.** Those that live on the bottom, either as fixed (**sessile**) forms or creeping about, are spoken of as the **benthos,** and are described as **benthic.** Those which swim actively, mainly fishes, and can overcome some of the effects of ocean currents, are termed **nekton,** and described as **nektonic.** The differing habitats and degrees of mobility of these three categories influence the nature of the distribution achieved by marine organisms; they also each require special collecting apparatus and techniques. Thus in the study of marine biology, plankton, benthos, and nekton are important concepts. It is generally believed that the earliest forms of life on earth were planktonic organisms.

Table 5.3 The Classes of Marine Vertebrates (Phylum Chordata)

<table>
<tr><th>CLASS</th><th>VERNACULAR NAMES</th><th>SKELETON</th><th>BODY CHARACTERS</th><th>REMARKS</th></tr>
<tr><td>Cyclostòmata</td><td>Lampreys and hagfishes</td><td>Cartilage only</td><td>Body eellike, no paired fins, mouth suctorial, without jaws</td><td rowspan="3">Aquatic animals swimming by tail motion, fins acting as stabilizers; and breathing by means of gills in water</td></tr>
<tr><td>Chondrichthyes</td><td>Sharks and rays</td><td>Cartilage only</td><td rowspan="2">True fishes with paired fins and paired jaws (one or both pairs of paired fins occasionally lacking)</td></tr>
<tr><td>Osteichthyes</td><td>Bone fishes, codfishes, flounders, herrings, eels, perches, and so on</td><td>Bone and cartilage</td></tr>
<tr><td>Reptilia</td><td>Marine crocodiles, marine turtles, and sea snakes</td><td>Bone and cartilage</td><td>Cold-blooded egg-laying, scaled tetrapods</td><td rowspan="3">Originally terrestrial tetrapods, readapted to life in the sea; using limbs as paddles, and breathing by means of lungs in air</td></tr>
<tr><td>Aves</td><td>Oceanic and shore birds, penguins, gulls, terns, and so on</td><td>Bone and cartilage</td><td>Warm-blooded egg-laying, feathered tetrapods</td></tr>
<tr><td>Mammalia</td><td>Dolphins and whales, seals, manatees, and so on</td><td>Bone and cartilage</td><td>Warm-blooded viviparous, haired, milk-secreting tetrapods</td></tr>
</table>

6

THE HYDROSPHERE

Hydrosphere / Lithosphere / Atmosphere / Biosphere / The world ocean / Continental shelf / Continental slope / Abyssal zone / Hadal zone / Neritic, pelagic, bathyal, and abyssal habitats / Zonation of sea surface / Age of the earth / Age and origin of the ocean / Geological time scale

THE HYDROSPHERE

Oceans cover 71 percent of the earth's surface and are collectively termed the **hydrosphere.** The average depth of the oceans is 3.8 km, or about 2½ miles; the ocean floor has a thickness of between 5 and 10 km. See Fig. 6.1.

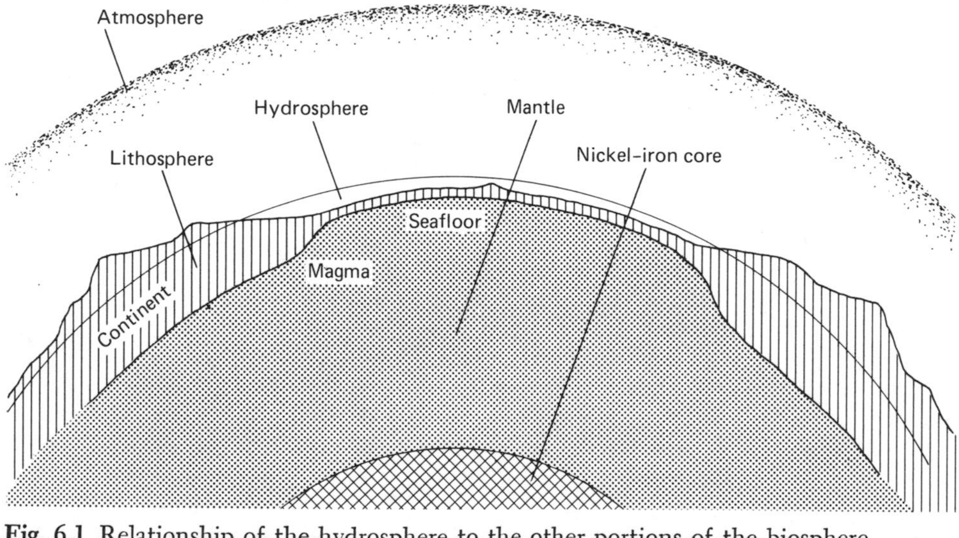

Fig. 6.1 Relationship of the hydrosphere to the other portions of the biosphere.

The rocky portion of the earth's crust, including the emergent continents and the submerged seafloor, is called the **lithosphere.** Beneath the lithosphere lies magma molten, semimolten material; the continents, with an average thickness of about 30 km (about 20 miles), seem to float like rafts on the magma.

Surrounding the lithosphere and hydrosphere is the **atmosphere.** Living organisms occupy the lowest part of the atmosphere, the hydrosphere, and the uppermost part of the lithosphere; this habitable shell of the planet, about 20 miles deep, is termed the **biosphere.**

THE WORLD OCEAN

Since all the oceans are interconnected, geographers speak of a world ocean or, more simply, the ocean as a collective entity of all the saline waters of the hydrosphere. Its total area amounts to 360 million km^2, or 70.8 percent of the earth's surface. If we plot the entire ocean as a circle representing an area of 360 million km^2, then the areas of the separate oceans are yielded in proportion by sectors subtending angles of 82° each for the North Pacific and the Indian oceans, 66° for the South Pacific, 47° for the North Atlantic, 37° for the South Atlantic, 32° for the Antarctic Ocean, and 14° for the Arctic Ocean. The average depth of the entire ocean is 3.8 km. The deepest point is in the Indian Ocean, with an average depth of 4.3 km, the shallowest point is in the Arctic Ocean, where it is only 1.2 km deep.

The continents are bordered by the **continental shelf,** a zone of relatively shallow seabed that slopes gradually downward until it reaches a depth of about 200 m. Beyond that the slopes become steeper. The steeper parts of the seabed are called the **continental slope,** and they continue to a depth of about 4 km. At that depth the level of the seafloor flattens to form the **abyssal plain** (in fact the abyss is undulated rather than forming a flat plain). In a few regions (defined in Chapter 28) there are **trenches,** reaching a depth of 11 km.

The waters and the associated biota of the continental shelf region are called **neritic;** thus one speaks of neritic faunas, neritic habitats, neritic seas. The corresponding terms used for the other regions of the ocean are as follows: the open sea, beyond the shelf, is **pelagic;** the continental slope is **bathyal** or **archibenthic** (both terms in current use); the abyss is **abyssal;** and the trenches are **hadal.** Biota and habitats on the margin of the sea and the land are called **littoral.** Biota and habitats between the upper and lower limits of the tides are called **intertidal.** These features are illustrated diagrammatically in Fig. 6.2.

ZONATION OF THE SEA'S SURFACE

Examples of zonation of the biosphere are given in Table 6.1 for the oceanic east–west zones and corresponding climatic zones of the land and lower atmosphere. To visualize the zones more clearly, compare Table 6.1 with a globe of the earth. Notice that the Southern Hemisphere is approximately, but not exactly, a mirror image of the Northern Hemisphere. The equator defines the axis of symmetry on a Mercator map or the actual plane of symmetry, as shown on a globe.

AGE OF THE EARTH

Astrophysics presently incorporates a general theory of the evolution of stars from which it is inferred that the earth, and probably also the other planets of the Solar System, arose as a consequence of a disruptive explosion of some large star in the transient phase called a **supernova.** As a result of this explosion, the internal heavy elements of the star were discharged into space and subsequently were condensed as planets and captured by our local star, the sun. Study of the radioactive decay of elements in the earth's crust, supposed to have originated in definite proportions in the interior of the supernova, yields a calculated age for the earth of 4.5 billion years. The oldest known fossil remains of recognizable organisms on the earth date from about 3 billion years ago. So it would appear that the planetary ecosystem began to evolve sometime between 3 and 4.5 billion years ago.

AGE OF THE OCEAN

The study of volcanic eruptions shows that when magma reaches the earth's surface the loss of pressure causes it to differentiate into steam and a residual material that solidifies as rock (lava). The annual production of water vapor and lava by the world's active volcanos seems to be in the correct quantity and proportions to account for the creation of the oceans and the continents over a timespan of about 4 billion years, and that is the same order of time as that yielded by the radiometric data. Theory implies, therefore, that the oceans and the continents began to form as soon as the earth itself was formed, and that magma has been continuously rising to the surface and differentiating into oceans and continents ever since. The oldest rocks so far found on the moon also indicate an age of about 4.5 billion years—a fact that suggests that the earth and the moon are sister planets formed by some common event.

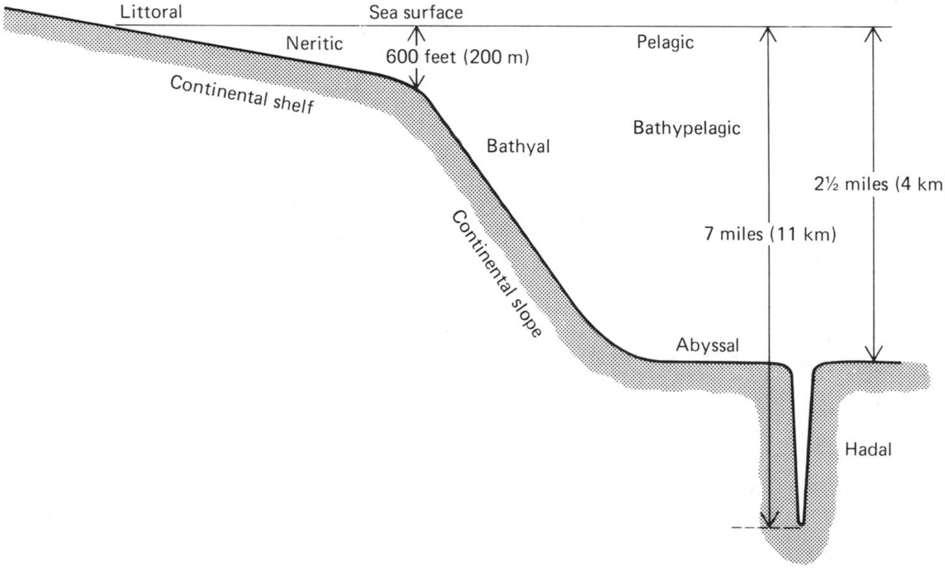

Fig. 6.2 Diagrammatic section through an ocean in the vicinity of a continent showing the principal life zones.

Table 6.1 Zonation of the Surface of the Earth

LATITUDE	NATURE OF LAND SURFACE	CONDITION OF SEA SURFACE	ATMOSPHERIC FEATURES	MEAN ANNUAL SURFACE TEMPERATURE
90°–70° N	Arctic ice desert	Floating sea ice	Arctic high pressure	Below zero
70°–50° N	Arctic tundra	Currents flow from west	Sea winds blow from west	0° C
50°–40° N	Temperate forests and grasslands			6° C
40°–30° N				12° C
35° N	Savannas, deserts	Seasonally reversing currents Northern limit of coral reefs	Seasonally reversing sea winds	18° C
30° N–0°	Tropical forests	Currents flow from east	Sea winds blow from northeast (trade winds)	26° C
0°–30° S	Tropical forests	Currents flow from east Southern limit of coral reefs	Sea winds blow from southeast (trade winds)	26° C
35° S	Savannas, deserts	Seasonally reversing currents	Seasonally reversing sea winds	18° C
30°–40° S	Temperate forests and grasslands	Currents flow from west	Sea winds blow from west	12° C
40°–50° S				6° C
50°–60° S	Island tundras			3° C
60°–90° S	Antarctic ice desert	(No ocean)	Antarctic high pressure	Below zero

NOTE: The Northern and Southern hemispheres are mirror images of each other.

THE GEOLOGICAL TIME SCALE

Occasional reference is made to the various periods of the past into which geologists divide the history of our planet. To follow such discussion, you should know the most commonly employed divisions of the geological time scale. These are presented in Table 6.2.

Table 6.2 The Geological Chronology

ERA		PERIODS AND EPOCHS	MILLIONS OF YEARS BEFORE PRESENT	ENVIRONMENT AND LIFE
CENOZOIC	QUATERNARY PERIOD	Recent (Holocene) epoch		Modern climates; first agriculture
		Pleistocene epoch	2	Ice-age tundras; modern man
	TERTIARY PERIOD	Pliocene epoch	10	Apemen appear
		Miocene epoch	25	First extensive grasslands
		Oligocene epoch	40	Mammals become dominant
		Eocene epoch	60	Mammals become conspicuous
		Paleocene epoch		Extinction of dinosaurs
			70	
MESOZOIC	Cretaceous period		135	First flowering plants
	Jurassic period		180	First podocarp forests
	Triassic period			First araucarian forests
			225	
PALEOZOIC	Permian period		270	Conifers become forest trees
	Pennsylvanian period		310	First reptiles
	Mississippian period		350	First swamp forests
	Devonian period		400	Lungfishes first appear
	Silurian period			First bog plants
	Ordovician period		440	First land animals (scorpions)
	Cambrian period		500	Marine invertebrates conspicuous
	(Precambrian time)		600	Marine invertebrates conspicuous
				Marine invertebrates; probably lichens on land
	Creation of the earth		4500	

7

COASTS

Types of coastline / Coasts of submergence / Submerged castles of Lebanon / Aristotle's observations of sea level change / Omar Khayyam detects change in level of the Caspian Sea / Submergence of the Godwin sands / Disappearance of Dunwich in eighteenth century / Truncated rivers / Submarine canyons / Coral atolls / Coasts of emergence / Piedmonts / Leonardo da Vinci's inferences / Raised beach platforms / Stepped coastal profiles / Complex coastlines / Greek temples that rose out of the sea / Sea-level changes / Stable coasts / Origin of raised beaches / Stillstands / Wisconsin advance / Post-Wisconsin glacial retreat / Carbon dating / Melting polar caps / Submerged migration routes of ancient man and contemporary fauna / Submarine wave-cut platforms

The coasts of the world are so intricately embayed that their total length is estimated to be about a million miles. They provide the land base for countless fishing communities; a habitat for those animals and plants that frequent the shallower seas; marginal beaches, cliffs, and wetlands; and the international frontiers of maritime states. They are also the resort of many city dwellers on vacation. So their broad, coastal features as elements of landscape are known to nearly everyone. The geological processes that shape a coast are less obvious and are the subject of this chapter.

TYPES OF COASTLINE

The main types of coast are coasts of **submergence** and coasts of **emergence.** The distinction, of course, seems to imply that coasts have either been sunken beneath the sea, or raised above it—and, in general, it seems that such changes have in fact occurred very widely in recent geological time as well as throughout most of former periods of earth history. A few coasts seem to be stable, but most show evidence of the heaving that periodically afflicts parts of the earth's crust.

Coasts of Submergence

Coasts of submergence are often recognizable by the extreme degree of indentation in the land–sea margin, sometimes with more or less complexly shaped peninsulas separated by intervening arms of the sea. Shores in this category have evolved from earlier land surfaces on which rivers and streams carved out a dendritic pattern of confluent valleys, lying between equally complex ranges of hills. Then, at some later geological date, the whole area suffered down-warping, or uniform depression without warping, so that the sea was able to flood the lower portions of the original valley systems. In some parts of the world, submergence of a coastline has occurred within historic times, with the result that buildings or other works of man and even whole ports or cities have been swallowed up by the sea and now lie wholly or partly beneath the waves.

The coast of Lebanon near the ancient Phoenician city of Sidon has undergone a partial submergence during the past one thousand years. One result has been that fortresses built by the marauding crusaders in the middle ages now have their lower floors awash by the sea; we may be sure that the architects of these forbidding castles would not have placed them in situations subject to daily inundation by the tides. Submarine archeologists have found still older remains of the classical port of Tyre, now completely submerged; the same is true of Sidon and other ancient ports.

It is a matter of historical record that so long ago as the fourth century B.C. it had become clear to thinking men that the sea and land may change their relative levels. The transient nature of coasts and of river valleys is noted by Aristotle, for example, in his *Meteoritica.* The Roman poet Ovid also commented on such matters, which evidently were thought of as quite reasonable interpretations by the ancients. In medieval times such ideas became heretical. The Persian astronomer Omar Khayyam in the twelfth century A.D. observed that the Caspian Sea had changed its level. His deductions, however, were found in conflict with Koran, and he was exiled for the rest of his life to central Asia, where he found time to write the *Rubaiyat.*

In Europe the course of events finally forced upon men the realization that land and sea interchange. Most startling of the evidence for observers in Britain was the disappearance beneath the sea of the Godwin pasture lands (belonging to the Saxon earl of that name) in the time of Edward the Confessor and of William the Conqueror (eleventh century), which was recorded in the *Domesday Book.* (Indeed, the very inferences that cost Omar his liberty were simultaneously recorded as observed facts by contemporary thinkers some thousand miles to the west.) More alarming still for England was the total loss of her greatest port, Dunwich, as described in Chapter 8. A new port, London, has taken the place of Dunwich, but, to judge by the increasingly severe tidal floods now periodically lapping the lower reaches of the Thames, it would appear that the days of London are numbered.

When a terrain that is dissected by a river system is submerged, the lower part of each river valley is now transferred to the continental shelf, where it may form a canyon on the seabed. (Not all **submarine canyons** are formed in this way, however.) The rivers are said to be **truncated,** and their tributary streams now emerge directly into the sea (Fig. 7.1). The truncated rivers give rise to sounds or arms of the sea bounded by peninsulas and form most of the world's major harbors. The **coral atolls** of tropical seas are a particular form of submerged volcanic island, described in Chapter 23.

Coasts of Emergence

In marked contrast to coasts of submergence are shorelines that have suffered upward warping and have therefore recently emerged from the sea, or shores that have undergone gradual accretion by the action of the tides or current. In such cases sand is piled up along the beach, and it advances in a seaward direction. Coasts of emergence generally have relatively simple contours, in the form of gentle curving strandline or almost straight strandlines. It should be noted, however, that in desert regions without rivers, the coasts tend to have simple contours, irrespective of whether there has been any change in the relative elevation of the sea and the land.

The typical feature of an emergent coastline is a long sandy beach that stretches seemingly without limit into the distant horizon. Hills, if present, do not make direct contact with the sea. Instead, there is an intervening plain or **piedmont** between the strandline and the foot of the hills, with sand dunes defining the seaward margin of the plain. As one travels inland from the coast itself older dunes can be observed, now covered over by a veneer of grass or other low vegetation; farther inland still one encounters scrub and woodland covering the land. If the soil is examined, it may often be found to consist largely of marine sands. This

Fig. 7.1 Coast of submergence (Picton, New Zealand) with drowned and truncated river valleys continuing out on to the continental shelf.

can be demonstrated by the presence of subfossil seashells, particularly clams' shells and remains of other organisms that inhabit the sandy seafloor. Observations of this kind, and inferences which we now see to have been correct, were made by Fracastori in Italy in the seventeenth century. His ideas, however, were considered to be contrary to the biblical revelation of the Deluge, so they were dismissed and forgotten. Leonardo da Vinci's drawings show that he, too, understood these matters. Since he did not publish his thoughts, they went to the grave with him, only to surface when the English monarchs began to publish the archives of Windsor including the notebooks and drawings of the Italian polymath.

If actual uplift of the land has occurred through a series of sudden jolts with accompanying earthquake, distinct terraces may be observed. These become progressively higher as one passes inland. Such **stepped profiles** are common on the restless Pacific margin. Usually the lowermost terraces or **raised beaches** still carry wave-cut boulders, dry rock pool basins, and even the remains of seashells left by the former denizens.

If human settlement has extended over a sufficient space of time, historical records are often available from which one can relate a particular raised terrace to a particular earthquake. Often the raised terraces are seen to be tilted; in such cases the uplift has been differential. In the 1855 earthquake of Wellington, New Zealand (the first case to be studied by the British geologist Charles Lyell), a differential uplift took place such that tilted wave-cut platforms now stand above the present strandline. These are 3 m above mean sea level where the uplift was most severe, falling gradually to zero where the fulcrum of the tilting was located.

Complex Coastlines

Peculiarly complicated coastal formations occur in some parts of the world where earthquake and volcanic activity is common. An example is the central and northern margin of the Mediterranean Sea. On the island of Sicily we find archeological evidence of alternating uplift and submergence. Striking evidence is seen in the columns of Greek and Roman temples, now standing above the sea, yet carrying the unmistakable borings made by marine mollusks into the limestone of the temple columns. The borings for any given temple are found to extend to the same height on the various columns, showing that at some unrecorded epoch *after* the temple was built the land must have sunk beneath the sea, and then, at some later epoch, rose again. The lack of historical records of such events is a reflection, probably, of the decay and collapse of learning during and after the dark ages when barbarians overran much of Europe. Evidence of the retreat of the sea since the eleventh century is seen in the present location of the bronze rings to which the Saracens anchored their ships in Sicily. The rings (cemented into bedrock) are now found at distances up to 3 miles inland from the coast. Evidence of similar changes of this kind can be found at points scattered along hundreds of miles of the Sicilian and Italian coasts.

SEA-LEVEL CHANGES

The foregoing examples are from regions of the earth's surface where earthquakes and other seismic disturbances are common and where the earth's crust periodically heaves and warps, so that the surface tends to be elevated or depressed. So consequences of the kind noted are not particularly surprising. But what do we make of cases where we can recognize what appear to be definite wave-cut platforms or former beaches, elevated well above the present level of the sea, and occurring in South Africa and Southeast United States where seismic disturbances are rare and only of minimal character, without measurable effects on the level of the land? In regions of this kind, believed to be extremely stable, it is nonetheless quite common to find well-preserved wave-cut platforms lying at a height of around 30 m, or roughly 100 feet, above the surface of the sea.

Origin of the Raised Beaches

About 100,000 years ago the climate of the earth seems to have been decidedly warmer than it is now. At that time the cypress swamps of the present-day southern coasts probably extended further northward, and there were probably mangrove sloughs on the Massachusetts coast. But that coast was not located where we now find it, instead it lay further inland. This was because the higher temperature had caused the polar ice to retreat, in fact, probably the whole Arctic ice cap and some part of the greater Antarctic ice cap had also melted. So water was added to the world's oceans from the melted ice that the general sea level increased. Therefore, wherever stable crust exists, we may expect to find remnants of the old beaches cut by the waves of that ancient high-level ocean. In restless regions of the earth's crust, however, the old beaches no longer necessarily remain at a height of 100 feet above the sea, for some of them have since been warped upward, others downward, and still others tilted at various angles, or arched or troughed. Our best estimates of their original heights have to come from the shores of older and more placid regions, such as have been mentioned.

The sea must have remained at this elevated level for a long period, for there was time enough for cutting quite extensive wave-cut platforms. The term **stillstand** is applied to any such prolonged episode of constant sea level in earth history, when sets of corresponding platforms are cut by the sea at matching levels in different parts of the world. In the case noted, the sea seems to have remained at about the same level until about 80,000 years ago. Then something happened to the earth's climates, and our planet began to grow colder. We can infer from this and other evidence that ice re-formed at the North Pole, and the ice sheet gradually spread outward until it crossed the northern coasts of Europe, Asia, and North America. This spread of the ice 80,000 years ago is known as the Wisconsin advance (because the glacial deposits associated with it are best known from Wisconsin examples; the corresponding term for European deposits is Würm). From various data elucidated by geologists, we now believe that the rate of southward advance of the ice front averaged about 160 feet per annum, roughly 50 m per year. As the ice crept southward, it overwhelmed the tundra region, then the taiga or boreal evergreen belt, overturning the spruce trees and burying the remains of forests beneath it. Meanwhile a new and more southerly tundra zone became established south of the southern margin of the ice, and polar animals ranged the lands of a desolate belt that stretched westward from the latitude of New York (as well as through much of France and Germany and other lands in similar latitudes).

Our ability to date the event, and to measure the rates of advance of the ice, stem from the circumstance that fragments of the old forest woods were preserved beneath the ice. Their age can be estimated by **carbon dating.** The carbon atoms bound up in the organic substances in the wood, such as lignin, were formed (at the time the tree was actually growing) from the carbon dioxide of the earth's atmosphere of that epoch. The carbon dioxide of the earth's atmosphere is of a dual nature: One part of it constructed from a stable molecule of carbon called ^{12}C. A second part, of known original percentage in the atmosphere, consists of an unstable isotope called ^{14}C, which derived from atmospheric nitrogen atoms that have been struck by cosmic rays. The ^{14}C slowly disintegrates into ^{12}C at a rate such that half of the ^{14}C decays to ^{12}C in 5000 years. If we assume that cosmic ray bombardment of the earth's atmosphere has occurred always at a stable rate, then the proportion of ^{14}C to ^{12}C in the air will always have been the same. In that case the initial ratios of the two carbons in the wood must have matched that of living wood today. Therefore any difference in the ratios in the fossil wood from that of living wood must be due to the disintegration of the original ^{14}C, at the known rate. From these facts derive the equations that permit the determination of the age of fossil wood.

The Wisconsin advance proved to be the last of a series of ice ages that we know to have occurred at various times over the past 3 million years. This last glaciation episode lasted from about 80,000 years ago until about 16,000 B.C. Then climates became warmer.

Ever since, the polar ice caps have been melting—sometimes slowly, sometimes rapidly—with occasional reversals of the warming trend. On the average during the twentieth century, the ice-melt has caused the sea level to rise at an annual rate of 1.5 mm; and this is probably about the rate for the last few hundred years. If this is so, then Boston Harbor today has a sea level about 3 feet, or 1 m higher than when the Pilgrims landed three-and-a-half centuries ago.

Over the past 25 years the rate of sea level rise has averaged 2.5 mm per year, or about 1 inch in 10 years. So, during the 20 years since the great North Sea floods, when a spring tide, a barometric low (which causes the sea level to rise), and an onshore storm wind all combined to make the sea rise to an unusual height, the general level of the sea has risen by another 2 inches all over the world. But 2 inches was that narrow margin of safety by which London escaped by the skin of its teeth during the last abnormal tides, so the margin of safety of the subways of that city is now exceedingly narrow.

The sea has still another hundred feet to rise before it reaches the high levels of the period before the Wisconsin glaciation. At present no one can predict how fast this rise will continue, or if the sea will return to its former height. However, should the polar ice melt more completely, all the ports of the world will be submerged.

It follows, of course, from what has been noted about the higher sea levels during warm or **interglacial** periods, that there must have been former lower sea levels during the cold or **glacial** periods. So there must exist somewhere beneath the present sea level some invisible wave-cut platforms that were formed during the last ice age when the sea level was lower. This is a topic on which scuba divers can throw light, by watching the seabed for signs of submarine wave-cut platforms, former storm beaches, which are marked by horizontal bands of boulders or pebbles, or for other proof of the submergence of coastal strips of land that once lay above the sea when the ocean was at a lower general level.

Evidences of such things come in the shape of bones and teeth of ice-age animals, such as mastodons, terrestrial by habit, yet their bones are dredged in fishermen's nets from shelf seafloors which were once dry land. Similarly, from the submerged Dogger Banks of the North Sea, fishermen have, for generations, fished up stone axes and other signs of material culture, which ancient Teutons left behind, men who once hunted the coastal lands before these were swallowed up by the rising level of the world ocean.

Explorer divers working on the sides of coral reefs have discovered features resembling wave-cut platforms beneath the sea. These are now interpreted as former wave-cut platforms, dating from various standstills during the general progress of the return of milder climates after the end of the Wisconsin glaciation.

8

SHORES, WAVES, AND SUBSTRATES

Seabeds and shorelines / Surface waves / Descriptive parameters of a wave / Ground swell and rollers / Breakers / Progradation / Emergent coast / Cliff erosion / Raised beaches / Longshore current / Sandspits and sandbars / Hooks / Tombolos / Lagoons / Barrier islands / Zonation of shorelines / Substrates / Hard substrates / Reefs, boulders, and slabs / Cobble, gravel, sand, and silt / Temporary hard-bottom shellbanks / Angle of slope / Ratio of organic to inorganic constituents of bottom / Glacial shores / Fjords / Ooze / Berg-cast hard bottom

The **continental shelf** is a fringe of relatively shallow seabed surrounding the major landmasses to a depth of about 100 fathoms (600 feet or roughly 200 m). Beyond the edge of the shelf the seafloor plunges more steeply to a very much greater depth. The relatively steeper slope, or **continental slope** to give it its full name, penetrates downward to a depth of about 3 miles where the seabed evens out into the abyssal plain that underlies most of the world oceans. The average depth of the continental shelf throughout most of the world is about 132 m or 70 fathoms (420 feet). The upper part of the continental shelf is accessible to divers using standard types of scuba equipment.

A coast or **shoreline** is formed when the continental shelf intersects the surface of the sea. Coasts owe their characteristic features to various physical processes, including waves, to be considered in this chapter. The position of a shoreline is variable from one epoch to another. During cold or glacial stades in the earth's history, the sea level falls (Chapter 7), and the shoreline moves seaward; the reverse occurs in warm stades. At the present epoch the shorelines lie about midway between former upper and lower extremes. So the beaches of the world were previously submarine features of warm stades, and conversely the shallow marginal seabeds of today were formerly beaches in the cold glacial stade that ended about 12,000 years ago.

When strong winds pass across the surface of the sea, some of the energy imparted by the atmosphere to the water is dissipated by waves and some gives rise to a wind-driven surface current. In the latter case the particles of water move continuously in one direction; in the former case the particles oscillate. By observing the motion of floating objects on the surface of a sea across which waves are passing, one can readily see that the particles of water merely oscillate, for the floating objects do the same, moving to and fro and up and down so as to describe a circle oriented vertically, with the plane of the circle being normal to the wavefront and parallel to the direction of propagation of the wave.

Characteristics of a Wave

A surface wave has **amplitude,** that is to say, magnitude of oscillation of the water particles. It has **wavelength,** the distance from one crest to the next or from one trough to the next, and it has **frequency**, that is, the number of oscillations that occur in a stated time interval. The highest point on a wave is called the **crest,** the lowest point the **trough;** the vertical distance from crest to trough equals the amplitude. When the wind tears away the crest of a wave, the wave is termed a **whitecap;** the flying water droplets are the **scud.**

Some waves originate in areas that are remote from the point at which they are observed. Such waves, having a long wavelength and a low frequency, are termed **rollers,** and the sea is said to have a **ground swell.** A ground swell may therefore occur on a calm sunny day, because the storm that generated it may have been thousands of miles away. When a roller passes across the continental shelf toward the coast, there comes a time when its trough begins to reach to the shallowing bottom. This results in a severe braking action on the lower part of the wave, while the crest, free from such obstruction, continues to travel forward at an unaltered rate. The result is that the crest ultimately passes over and in front of its antecedent trough, and so collapses in a seething mass of foam; the wave is then said to be a **breaker.**

When a breaker forms, a great deal of wave energy is suddenly released as energy of motion in the forward direction. The wave rushes up beach as its energy is dissipated, and the rapidly moving water carries stones or even boulders, hurling them against the land. If the land is low lying, sand is thrown high on the beach with the cascading water, and some of the sand remains there when the water runs back into the sea. Such a coast will undergo **progradation**, that is, growth in the seaward direction as more and more sand is cast upon it. Thus one type of **emergent coast** develops (see Chapter 7). If the coast is hilly, as is commonly the case on a coast of submergence (Chapter 7), the stones, boulders, and sand hurled against the land cut it into a **cliff.** Material abraded from the cliff falls into the sea to add to the battery of material thrown against the cliff, until the debris is ground to sand and swept off to be deposited in some more sheltered place. Meanwhile, the cliff is cut progressively back in the landward direction, with the sea advancing behind it.

Major changes in a coastline may occur quite rapidly (Fig. 8.1). A great earthquake on February 2, 1931, raised 20 square miles of seafloor in New Zealand and destroyed two cities. Over a span of a thousand years from the seventh to the seventeenth centuries England's greatest medieval port, Dunwich, rose to eminence, became the seat of a bishopric, and could claim 52 churches built within its four walls as well as a palace for the king when in residence. Yet within the space of several centuries it was swallowed up by the advancing sea, with as many as 250 houses disappearing at a time during severe storms in the eighteenth century. Portions of one church still remained in 1910, the others all now sunk beneath the waves or shattered by the breakers. A line of cliffs now marks the former inland boundary of the port, the rest of which forms the floor of the North Sea shelf (H. Walker, 1970).

The energy of waves does not extend very many fathoms below the surface, so the erosion of a cliff is limited to the part that lies within a few meters below sea level (the part above sea level of course falls off as it is undercut). These processes produce what is called a **wave-cut platform.** If the sea level should rise, wave-cut platforms will be submerged and preserved in deeper water. If the sea level falls, or if the land is elevated by earthquakes or other seismic change, elevated horizontal platforms occur in the topography of the country near the shore. Several such **raised beaches** or **raised wave-cut platforms** may be observed on the sides of hills adjacent to the sea, especially on Pacific coasts, where seismic activity is always high. Wave-cut platforms may subsequently be tilted by buckling of the earth's crust.

Longshore Current

If incoming ocean waves strike a coast at an oblique angle, the breaking wave is reflected in accordance with the laws of physics (such that the angle of incidence equals the angle of reflection). In practice this means that the water that runs back into the sea as the wave is dissipated always is deflected a little to the lee

Fig. 8.1 On May 29, 1937, a violent eruption of the volcano Matupi, in New Britain, was followed by a sudden collapse of the adjacent land to a new level below that of the sea. Here is seen the remains of a former forest, decaying on the new seafloor.

side of the place where the wave broke; the particles of sand carried along by seawater in motion are therefore also carried a little to the lee side each time a wave breaks. The sum total of these effects for constantly breaking surf is the effective generation of a **longshore current**, for particles of sand are observed constantly to move in one direction along the coasts.

Sandspits and Sandbars

When such a longshore current encounters the mouth of a river, the collision of the outward stream of the river with the longshore current results in dissipation of energy of motion and loss of velocity of the waters. The power of water to carry suspended material, such as sand, is a function of its velocity. Thus the suspended sand grains are eventually dropped. Their accumulation over a period of time produces a longshore **bar**, closing the mouth of the river, except at one place where the pent-up river forces its way out to sea. If a longshore current encounters a bay, where the seawater is more sheltered and hence nearly stationary, sand is deposited across the bay mouth as a bar or spit. A **spit** is an exposed bar; it forms if wind-blown materials are added to the water-borne materials. If two mutually opposed longshore currents meet, each cancels out the current of the other, and all their joint carrying capacity is lost. Sediment rapidly accumulates at the junction point, growing out to sea as a spit. If the spit grows so far as to encounter incoming rollers the tip of the spit becomes deflected toward the land, producing a **hook.** Cape Cod and Sandy Hook are examples. If an offshore island lies in the path of an incoming pattern of rollers, the shallow margin of the island breaks the waves on either side as they pass the island on the way to shore. But farther away no such breaking occurs. Hence the wavefront is refracted by the island, as if it were a prism in the path of light waves. The wavefronts on either side of the island are now forced to advance upon each other behind the island. Being so refracted, they collide, lose their carrying power, and drop the sediment on the lee side of the island. In this way the island becomes tied to the land by a bar; this is termed a **tombolo.** The former island of Nahant, near Boston, is now a tombolo.

Lagoons or regions of quiet water are present behind offshore bars if these form in such a manner as to lie parallel to the coast. Subsequent erosion may break such bars into longshore **barrier islands,** as occur along and south of the New Jersey coast. The processes may be complex with alternate shore building (progradation) and shore erosion.

From the foregoing it can be seen that:

1. Coasts of submergence are characterized by peninsulas, headlands, drowned valleys, and, in glacial terrain, the steep-sided drowned valleys called fjords.
2. Coasts of emergence are characterized by bars, spits, hooks, and tombolos—sometimes with offshore barrier islands.
3. Neutral coasts, namely deltas, volcanic islands, and coral reefs are not formed by the processes mentioned and have special features peculiar to themselves.

THE ZONATION OF SHORELINES

The organisms that inhabit shorelines are found to be zoned according to the length and type of immersion in seawater that they require for survival. The physical cause of the zonation is the rhythmic rise and fall of the sea called the **tide.** We may recognize the following simplified classification of shore zones:

Upper littoral habitats
Salt marsh, littoral ecotones
Rocky substrates
a. Biota of the midtidal zone
b. Biota of the lower tidal zone
c. Shelf biota
Soft substrates
a. Biota of the midtidal zone
b. Biota of the lower tidal zone
c. Shelf biota

The term **littoral** is applied to phenomena connected with the sea margin. An **ecotone** is a transitional region between one type of habitat and a different type. In the above example this is a transition from land to sea, as in the dunes or wetlands (Chapter 26), or from forest to sea, as in mangrove sloughs. The tidal zone is the band of shoreline between mean high tide and mean low tide; that is, a zone which is under water for some time each day. The term **substrate** refers to the type of bottom upon which a given community is located; it is classified as hard and soft.

SUBSTRATES ON THE CONTINENTAL SHELF

We may now glance at the various kinds of bottom, or substrate, which a diver will encounter as he explores the continental shelf. These different types of bottom provide various kinds of habitat for the animals and plants that inhabit the shelf. Some animals and plants are better adapted to live on one particular kind of substrate, which they prefer or upon which they have better chance for survival.

Substrates may be sampled by mechanical or photographic devices. For the purpose of this chapter it will be assumed that a scuba diver is moving across the continental shelf, and observing, collecting, and photographing the habitats as he encounters them. This, in practice, is now becoming a standard procedure among younger marine biologists. For the deeper parts of the continental shelf we must still, of course, rely upon the older methods. The remote-controlled cameras of the oceanographer are still the most important technique in studying the ecology of deep-water faunas beyond the edge of the shelf. However, knowledge of man's physiology under great pressures is increasing rapidly, and the probability now appears great that before very long the entire continental shelf will become the range of the professional scuba diver.

Hard Substrates

This type of substrate commonly occurs on coasts where there are wave-cut cliffs overlooking wave-cut platforms that carry masses of emergent bedrock. They also occur on more gently profiled coasts where a hard bedrock yields large rounded boulders or cobbles. Examples of the first case are very frequent around the margins of the Pacific Oceans, and examples of the second are found on most ancient igneous or metamorphic foreshores of the world or in places where hard ancient sandstones encounter the sea. If a diver explorer enters the sea from such a coast, he will first encounter an intertidal zone of hard slabs, rock reefs, or boulders covered by such plants as the large brown seaweeds and by such animals as barnacles and mussels—all able to withstand the heavy surf commonly generated in these localities. Then, as he enters the water and crosses boulder bottom, he passes gradually into masses of smaller rounded stones called cobble and, finally, into coarse gravel. Foreshores and inshore bottoms of this type are common in cold climates, but they can also occur in the tropics if the coral is not able to gain a permanent footing. On parts of the Virgin Islands, for example, cobble bottoms occur near coral reefs and are invaded and occupied by the same kind of hard-bottom denizens as also frequent coral reefs. This circumstance also illustrates the fact that organic structures such as **coral reefs** are themselves a type of hard bottom. A logical extension is the realization that a mass of shellfish, or the dead shells of shellfish, can also be considered as hard bottom because, even though the shellfish rest upon mud or sand, they themselves offer a firm substrate to any organisms that live upon them, such as sea urchins or seaweeds. Such **shellbanks** may often constitute a **temporary hard bottom.** A casual disturbance may cause silt to accumulate over a shellbank, or the shellbank may be overwhelmed by a submarine silt flow or turbidity current, converting the region back to soft substrate, and also stressing the temporary nature of some hard bottom.

Angle of Slope

If you enter the sea from a coast, such as that of New Brunswick or Nova Scotia, where the intertidal zone is of a kilometer or so in breadth, you are obviously going to encounter similarly gentle-sloped conditions for a considerable way out under the water. These slopes have a great deal of sand, fine gravel, pebbles, and shells. **Sunstars,** such as *Crossaster*, favor these substrates. In addition fossil starfishes that are quite similar are known from sedimentary rocks of ancient seabeds that are up to 300 million years old. This is just one of many hints we have that the underwater seascapes seen by divers are probably among the most ancient scenery to be found on earth. A diver glancing about himself on the seabed may often, in effect, be peering down the corridors of time and seeing the world as it was and as it has remained during vast intervals of elapsed prehistory. Farther out beyond the pebbly region is a finer bottom of sand. With the decrease in the size of the inorganic constituents of the bottom there has an increase in the proportion of organic material, such as shells of dead animals, the less visible decaying soft parts, and the bacteria that feed upon them.

On some coasts the slope of the bottom may be so steep as to be almost vertical, and perhaps the entire distance from the seaward margin of the forest, across the splash zone, the barnacle zone, the green and brown alga zones, down to low-tide level may occupy no more than a distance of 3 or 4 m. A shelf margin of this kind occurs on the seaward edge of mountain regions that were heavily glaciated during the ice ages, as, for example, the fjords of Norway, of Alaska, or of the southwestern part of New Zealand. A **fjord** is a very deep inlet of the sea, formerly occupied by the lower part of a great glacier and capable of being eroded below sea level (on account of the weight of the ice), unlike a river. Generally in fjord country, the depth of water inshore is of the same order of magnitude as the height of the mountains that rise directly out of the sea. In the case of southern New Zealand this means a water depth of 1 mile within a few hundred yards of land. In such localities the continental shelf may be said to be nonexistent.

As one proceeds out across the shelf, only soft sediments are encountered for the most part, because the finest particles of silt brought down by rivers are able to float in suspension far out from the coast before sinking to the bottom. However, hard cobble may occur where melting icebergs have dropped moraine rubble or where submarine volcanoes have erupted lava. As we cross the edge of the shelf and pass down the continental slope on to the abyssal plain, only very fine soft sediments or **oozes** and the seastars and sea cucumbers that live under such conditions are found.

9

PLANETARY RHYTHMS AND THE OCEAN

Cycles of activity / Random environmental parameters / Scalar and vector influences / Scalar rhythms / The diurnal cycle / The tides / Cause of the tides / Tidal friction / Diurnal growth lines / The annual cycle / Annual ring counts / Rings of fossil corals / John Wells / Monthly or lunar sexuality / Scrutton's hypothesis / Absolute measurement of the lunar cycle / Kepler's laws / Sidereal and synodic months / Coral data yield distance of moon in Devonian times / Cause of the retreat of the moon from the earth

An important feature of life in the ocean is that the cycles of activity, feeding, growth, and reproduction tend to be controlled more by planetary influences rather than by local environmental effects. This leads to some marked differences between marine organisms and those that inhabit dry land. Events that may cause terrestrial organisms to take shelter for several days and to give up hunting and feeding very seldom have any effect upon marine organisms.

Random or seasonal variations in rainfall have virtually no effect upon bottom-dwelling organisms protected by an overlying water mass. Exposure and desiccation rarely occur, for mean sea level is determined by the mass of water in the entire ocean, and local fluctuations or irreversible changes only take place as a result of rare events, such as earthquakes or great hurricanes. The aqueous medium protects its denizens against the effects of violent short-term fluctuations in the temperature of the overlying air masses, such as commonly occur during a single day over land. Wind variations only affect substantial bodies of water if they are prolonged, for there is a considerable delay in the transfer of air motions to water.

Scalar and Vector Influences

Two main categories of influences may be distinguished: (a) **scalar,** that is, effects which can be represented by quantitative measurements, such as temperature, light intensity, and duration of light or darkness; and (b) **vector,** that is, effects which involve directional components, such as wind-flow and the course of ocean currents. Vector effects often require special notation, such as rotating arrows. The scalar factors will be discussed in this chapter.

SCALAR RHYTHMS

Three categories of scalar rhythms may be distinguished, each associated with a celestial body:

1. **Diurnal (or terrestrial) cycle:** the repetitive 24-hour cycle of feeding, growth, activity, and rest associated with day and night and with the tides that are imposed by the daily rotation of the planet.
2. **Annual (or solar) cycle:** the repetitive yearly cycle of growth, activity, migration, reproduction, and hibernation that are imposed by the annual orbital revolution of the earth about the sun and accentuated by the fact that the earth's axis is inclined to its orbital plane.
3. **Monthly (or lunar) cycle:** the 30-day cycle of sexual activities related to the phase of the moon and, hence, to the orbital motion of the moon in revolution about the earth.

The conspicuous alternation of day and night governs the behavior of many marine organisms, at least in the illuminated zone of the hydrosphere. This diurnal cycle, which has a present frequency of 365 days per year and a wavelength therefore of 24.0 hours, is so well known that it requires little comment. Organisms tend to conform to the geophysical frequency quite precisely, performing corresponding activities on consecutive days at the same time of day. The rhythms are obviously superimposed by the environment and, in the same species, will differ therefore according to the latitude.

In subpolar regions, where the winter nights are long and the winter days are short, the biological rhythms conform to the long night and short day. Nearer the equator, where seasonal variations in the length of day and night are very slight, the biological rhythms show a corresponding degree of near uniformity throughout the year.

But there are significant differences between terrestrial biota and marine biota. On land, a week of bad weather or a prolonged major storm may cause the biota to take shelter, to refrain from hunting or feeding activities, or, perhaps, to sleep; growth, of course, ceases during such periods of enforced inactivity. On the other hand, the relative immunity of the marine environment from atmospheric variations ensures that the biota carry on with their normal diurnal cycle activities, whether the weather above the air–sea interface is fair or foul. This means that growth continues much more regularly; most organisms have a daily increment of growth. Some marine organisms, such as mollusks, corals, and similar shell-secreting forms, add a marginal zone of calcium carbonate around the edge of the growing part of the shell every day.

Many littoral organisms, such as the shore crab *Carcinus*, pattern their activities on the semidiurnal cycle of the tides. Successive high tides occur at intervals of 12.4 hours, so that about 700 cycles occur in the space of a year. Such activities, however, are also moderated by light and darkness, and it appears to be a general rule that a single growth line forms each day.

THE TIDES

The level of the sea at any given place rises and falls twice each day, a **high tide** being followed by a **low tide** some 6 hours later, in turn to be followed by another high tide 6 hours after that, and another low tide 6 hours later still. Thus, there are two **tidal cycles** every day or, more precisely, every 24 hours 50 minutes. Because of the 50 (or 51) minutes required to complete the cycles over and above the 24-hour day, the times at which high tides and low tides occur become 51 minutes later on each successive day.

Ancient astronomers noticed that the moon rises about 51 minutes later each day, too, so it was natural to conclude (correctly) that the moon is in some way responsible for the tidal effect. Early astronomers also noticed that tides reach a maximum amplitude twice every month (**spring tides**) and that tides of minimal amplitude occur in the intervening weeks (**neap tides**). Furthermore, it was noticed that spring tides occur only at the time of the full moon and the new moon, whereas neap tides occur when the moon is at quarter phase. In other words, spring tides are the tides raised when the moon, the earth, and the sun lie along the same straight line in space and neap tides occur when the moon, earth, and sun form a right angle in space. So it was clear that the sun also raises a tide, and, when the **solar tide** is raised in the same direction as the **lunar tide**, the two tides reinforce each other producing the spring tide. The neap tide occurs when the solar and lunar tides are drawn at right angles to each other, with the neap tide as the resultant.

Cause of the Tides

The attractive power of a heavenly body upon the earth varies as the product of their masses and inversely as the square of the distance between them; this is Newton's well-known law of gravitation. The **tide-raising force** is the difference between the gravitational attraction of a heavenly body upon the solid sphere of the earth, and its attractive force upon the fluid hydrosphere. For example, the moon is 240,000 miles from the hydrosphere of the side of the earth turned toward it, and by (240,000 + the earth's radius) miles from the center of gravity of the semirigid earth-sphere. On account of this difference, the hydrosphere on the side of the earth toward the moon is more strongly attracted than the solid earth as a whole, so the sea appears to rise up toward the moon to a greater extent than does the lithosphere, thus producing a high tide. At the same time on the opposite side of the earth, the hydrosphere is separated from the moon by (240,000 + 1 earth radius) miles, whereas the earth as a whole is separated by only 240,000 miles. Inserting the squares of these distances into Newton's equation shows that the differences between the two attractions is such as to produce a tide on the far side of the earth, where the hydrosphere is less strongly attracted than the semirigid earth sphere as a whole. The theory of tides is complex; and, as will be noted later, a phenomenon known as **tidal friction** causes delays in the propagation of tides across an ocean so that the moon does not always stand overhead at the time of high lunar tide at any given place. The degree by which the tide is delayed with respect to the meridian passage of the moon is known for all major ports and is called the **establishment** of a port.

So far as marine organisms are concerned, those that live on the upper part of the continental shelf feel the effects of tides most markedly because in the beach zone parts of the seabed are left exposed to air every 12 hours when low tide is occurring. All marine organisms in the shallow-water zone are affected in some way by the diurnal cycle of light and darkness and by the tidal cycle. Thus, as noted above, growth lines tend to form in their skeletal material, reflecting the successive daily spurts of activity of tissue formation related to the cyclic occurrence of favorable feeding conditions.

DIURNAL GROWTH LINES

Growth lines may remain visible months or years after they were originally secreted, thereby providing a permanent record of the successive sizes and stages of growth of an individual organism. About 2 percent of the time a marine invertebrate may fail to perform its normal growth increment. This may be due to some internal factor (such as bacterial infection or other disease), to some external factor (such as temporary food storage), to an accidental injury, from predator attack or other natural cause, or an unusually severe storm that disturbs the seafloor. Whatever the case, there is a slight nonconformity between diurnal activities and the diurnal planetary cycle, but the discrepancy is only a matter of a few omitted days' growth in the course of a year. Panella (1971) has recently demonstrated that daily growth lines are readily observable in the bones of the inner earth (otolith) of certain fishes, thus bringing chordates into line with nonchordates in respect to diurnal growth cycles.

Growth lines of essentially the same kind as those in existing marine organisms, can be recognized in fossil skeletons of extinct mollusks, corals, and other biota dating back as far as the early Paleozoic, thus proving that diurnal cycles of growth and behavior are very ancient features of living organisms.

THE ANNUAL CYCLE

With the exception of the central belt of the tropics, where the elevation of the sun and the consequent mean annual temperature vary little in the course of a year, seasonal variations in the number of hours of daylight and darkness and the rise and fall of environmental temperatures are quite common. Metabolism (biological activity of tissues and cells) occurs faster at higher temperatures than at lower ones (within the viable range). Therefore, corals and other marine organisms that live on or near the equator grow at a nearly uniform rate throughout the year. On the other hand, corals that inhabit the northernmost or southernmost limits of the coral reef zone, in latitudes of about 30°N and 30°S, experience a pronounced seasonal variation in water temperature. In such cases the rate of growth in winter is much less than that in summer, when the water temperatures are higher. For example, several species of the genus *Fungia* have been found to grow in a sequence of about seven annual installments, the radius of the corallum increasing by about 10 or 11 mm each year. The periods of rapid growth (corresponding to summer) are marked by thicker growth lines, and the winter lines are discernibly thinner. Similar observations have been made on corals in the Caribbean and Floridian reefs.

Annual Ring Counts

Because of the differences in the thickness of the summer and winter growth lines, it becomes possible to make counts of the growth lines and to classify them into successive sets of about 360 rings, representing successive annual increments. This is not as easy as it sounds because the rings are produced in a thin layer of the coral skeleton called the **epitheca,** and unfortunately the epitheca is attacked by organisms such as boring worms and is eroded by the abrasive influences of the inorganic constituents of the environment. Accurate counts of growth rings are quite difficult to make. Genera for which detailed counts have been carried out with some success include *Manicina* and *Lophelia* (Fig. 9.1).

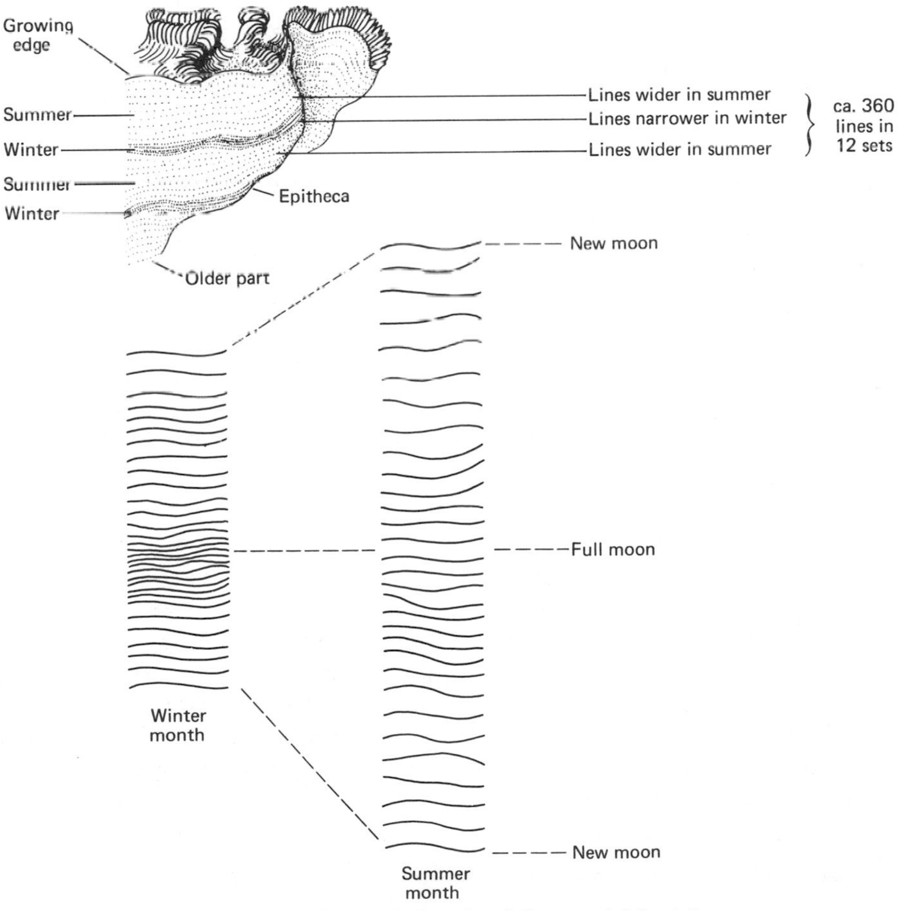

Fig. 9.1 Diurnal and seasonal growth bands of the coral *Manicina.*

Rings of Fossil Corals

The observations noted above seem relatively straightforward and predictable. However, in 1963 John W. Wells, an authority on fossil and living corals, discovered that annual sets of growth lines can also be recognized on some fossil corals and that the number of growth lines in an annual sequence **increases with the antiquity** of the coral fossils. He found that whereas living corals such as *Manicina* produce about 360 growth lines per year, a coral of the Pennsylvanian period (about 300 million years ago) will have between 385 and 390 rings in an annual sequence and a coral from the middle Devonian period (about 380 million years ago) will show between 385 and 410 diurnal rings.

Biologists, astronomers, and physicists have been studying Wells' findings ever since. Although some uncertainties remain (as Wells himself was careful to point out), particularly because of the difficulty in making precise counts of growth lines, most of all on worn fossils, nonetheless a general body of literature now exists on the subject. The following is a summary.

1. The year may once have lasted longer and, therefore, comprised a greater number of days than now is the case or
2. The day may have been shorter, so that more days elapsed in a year or
3. Both the length of the day and the length of the year may have varied.

Fortunately the first and the third of these possibilities may be dismissed. From a long series of classical investigations, beginning with those of Laplace on the properties of the earth's orbit, continuing with those of Leverrier (1843) on the mutual perturbations of the planets, leading in 1846 to the discovery of Neptune, and concluding with the further development of Leverrier's equations by later investigators, it has been established that the parameters of the earth's orbit are such that the eccentricity varies slowly between a maximum and a minimum value. The extremes are separated by intervals of hundreds of thousands of years, but the period of revolution about the sun is virtually stable. This means that the absolute length of the year is a fixed quantity. If, therefore, the number of days comprising an annual cycle in the Carboniferous or the Devonian periods was greater than is the case today, then the absolute length of the day must have been increasing through geological time. In other words, the earth is gradually rotating more and more slowly.

When Wells plotted his data on the number of days in a year against the elapsed time since the observed values for Devonian and Pennsylvanian periods based on radiometric determinations, he found that the points for the Recent, the Pennsylvanian, and the Devonian periods all lay on the same straight line. This suggests a steady increase in day length from a Cambrian value of 21.5 hours. Astronomers further noted that the slope of the line was that which might have been expected from observed variation in the earth's rate of rotation within historic times. Assuming the correctness of these inferences, we could interpolate and extrapolate from the points given by Wells and so obtain a set of expected values for other geological periods. Such interpolation and extrapolation must still be treated with reserve, but it is encouraging to think that these are reasonable deductions.

TIDAL FRICTION

Causes of the Lengthening of the Day

In 1897 George H. Darwin pointed out that the earth's rate of rotation is being retarded by tidal friction or the delay in the time at which high tide occurs. The mechanism operates in this way: Tidal friction causes delays in the transmission of the tidal crest across an ocean. Thus, the high-tide crest arrives **after** the moon has already crossed the meridian. This means that the tidal bulge lies always **to the east** of the sublunar point. The tidal bulge may be thought of as a protuberance of the earth. As such, it suffers gravitational attraction by the moon, which therefore tends to pull it toward the west. However, the earth is rotating from west to east, so the earth's motion of rotation is carrying the tidal bulge in the opposite direction from that in which the moon is drawing the tidal bulge. In other words, the moon is exerting a torque upon the earth itself in such a direction as to counteract the earth's rotation. Therefore a brake is being applied, and the day is gradually lengthening.

THE MONTHLY OR LUNAR SEX CYCLE

The sexual activities of many marine animals are mediated by the nervous system, especially by the photosensitive visual or light-detector organs. At particular thresholds of illumination or duration of illumination, hormones are released that cause the male and female sex cells to mature. This is followed by their liberation some hours or days later. Peak illumination at night by the full moon serves as an effective maturation stimulant in the cases of many marine organisms that are nocturnal. In such forms the sex cells are likely to be liberated several days after the night of the full moon. One possible advantage of this arrangement is that most or all members of a given species of a community will become sexually active on the same night; thus wastage of sperm or eggs is avoided, because large numbers of gametes of both types are simultaneously liberated. For any given species the time that elapses between the full moon and the mating is fixed within rather precise limits, which may vary according to the season.

The nuptial dances of the polychaete, *Platynereis dumerilii,* for example, are held on the French coasts at the first and last quarter of the lunar month. In *Ceratocephale,* the Japanese Palolo worm, the July and August meetings are held 6 to 9 days after new and full moon, whereas the September gatherings are delayed till the fifteenth night after the new and full moon (doubtless on account of slowing maturation as the water temperatures fall). In reef-building corals such as *Stylophora* the reproductive stages (planulae) are released with a lunar periodicity throughout the year. In the Palao islands, the planulae of *Fungia* are released at the time of the new moon every month from September to April. On the other hand, *Rhizopsammia,* a small Japanese littoral coral, liberates its planulae only between about August 10 and September 10, the period of maximum water temperatures. Thus corals differ, with lunar periodicity most marked in the cases of species living in the tropics, where water temperatures are high at all times. It is perhaps true to say that in corals the more conspicuous the annual cycle is, the less conspicuous is the lunar cycle. Water-temperature factors would explain this relationship.

Effects on Growth of the Lunar Sexuality

When corals are actively producing and liberating reproductive stages, there is a reduction in the fraction of metabolic activities directed to normal growth, including growth of the skeleton. This means that the growth lines laid down during the sexually active period are narrower than those produced during the nonsexual parts of the month. Consequently, when a long series of growth lines is studied, it is usually easy to distinguish the sets of consecutive lines laid down during the days or week of reproductive activity, and the alternating sets of broader growth lines laid down on the intervening days. Numerous data on these have now been obtained and published. A variety of other lunar-related factors probably also influence the rate of growth of marine organisms; whatever the factors may be, the net result is a lunar-related cyclic fluctuation in the growth rates. When a large number of such lunar growth lines is measured and analyzed statistically it becomes apparent that the existing marine organisms follow a lunar periodicity that matches the **synodic month.** (The synodic month is the time that elapses between two successive matching phases of the moon, from full moon to full moon, new moon to new moon, and so on.)

Soon after Wells detected annual periodicity in Devonian corals, Scrutton (1964) in England observed lunar periodicity in a number of genera of corals of the Devonian period. In 1964 he published his conclusion that the Middle Devonian year contained 13 lunar months each having 30½ days. Subsequent studies by other scientists have confirmed this conclusion and have further implied a gradual change in the length of the synodic lunar month from about 31½ days in the Middle Cambrian (about 550 million years ago) to 29.53 days, its recent value. It must, of course, be remembered that the days vary in absolute length according to the geological period to which they relate. Therefore, in order to obtain a true picture of the change in the length of the lunar cycle, we have to convert the units of measurement into uniform standard absolute values.

Absolute Measurement of the Lunar Cycle

This measurement is achieved very simply by converting the units of time into fractions of the year, for we know that the absolute length of the year has not varied significantly. Since there are 29.53 days in the present (recent) synodic month, we can say that the length, or **period**, of the synodic month at the present time is

$$1/\frac{365.24 \text{ days of 24 hours}}{29.53 \text{ days of 24 hours}} = \frac{1}{12.4} \text{ year}$$

This value can now be compared directly with Scrutton's value for the Middle Devonian synodic month of 1/13 year. From this comparison it is now clear that the synodic month has increased in length (13 — 12.4)/13, or 4.6 percent during the past 380 million years.

Kepler's Laws

The previous section implies a change in the period of revolution of the moon about the earth. The statement is of special interest because Kepler's third law of planetary motion tells us that "the square of the period of a planet (or satellite) is proportional to the cube of its distance from the primary body." In other words, for the moon to have orbited the earth in Devonian times in the observed period of 1/13 year, its distance from the earth cannot have been the same as is now the case. The difference can be computed by inserting the known values into Kepler's equation. Before this calculation can be performed, however, one adjustment is needed, for the synodic month is not the true period of revolution of the moon about the earth.

Sidereal and Synodic Month

The synodic month is the time that elapses between one full moon (or new moon, or any other recognizable phase) and the next corresponding phase. As already noted, at the present epoch 12.4 such synodic intervals occur in a year. However, it has to be remembered that while the moon is completing one revolution about the earth, the earth itself is traversing its own orbit about the sun. In the space of one month the earth moves through about 30° of its own orbit, with a consequent alteration of the direction in space in which the sun now lies. The change in direction is such that the moon is required to traverse an extra 30° of its own orbit before it returns to a position that appears, when viewed from the earth, to match its former position relative to the sun. In the course of one year, therefore, the moon actually completes one extra orbit around the earth. Thus, for the present epoch, 12.4 synodic months correspond to 12.4 + 1 = 13.4 revolutions of the moon about the earth. The period 1/13.4 year, or 27.32 days, is called the **sidereal month,** and this is the true period of revolution of the moon. Similarly for the Middle Devonian the sidereal month would be 1/13 + 1, or 1/14 year.

Distance of the Moon in Devonian Times

This value can now be calculated from the growth-line data of corals, given that the present distance of the moon from the earth is 238,857 miles, or 384,400 km. By substitution in Kepler's equation:

$$\text{Devonian distance of the moon} = \text{Present distance}\left[\frac{(12.4+1)^2}{(13+1)^2}\right]^{1/3}\text{km}$$

$$= 384{,}400\left(\frac{13.4^2}{14^2}\right)^{1/3}\text{km}$$

$$= 372{,}800\text{ km}$$

from which it appears that the moon has retreated over 11,000 km during the past 380 million years. If certain modified values for the number of days in the year be used, as recommended by Newton (1969), then the distance the moon has retreated since Devonian times becomes about 15,000 km.

Cause of the Retreat of the Moon from the Earth

Although the growth-line data from the marine fossils have yielded the first measures of the rate of retreat of the moon from the earth, the inference that this retreat is occurring was already drawn by George Darwin in 1897 (in the same paper as that in which he suggested the earth's rate of rotation to be slowing down). Darwin pointed out that, as the earth's rate of rotation is reduced, it suffers a loss of the energy known as angular momentum. Energy is indestructible, however, and must therefore be transferred elsewhere. The transfer takes place by way of the tidal torque applied to the earth by the moon. Just as the moon is exerting a backward tug upon the tidal protuberance of the earth, so also the rotating tidal protuberance exerts a forward tug upon the moon. The moon responds like a slingshot, flying ever faster in its orbit and, therefore, in accordance with Newton's and Kepler's laws, is moving outward into successively larger orbits appropriate for its greater velocity.

Darwin predicted that this process will continue until the day and the month both equal 47 of our present days. The effects of the solar torque will then become predominant. Later still, if the solar system still exists, the transfers of angular momentum will operate in such a manner as to return the moon to the near vicinity of the earth. Jeffreys believes that the final fate of the moon will be to suffer disruption. Such will occur when the moon is located at a distance of about 2.5 earth radii from the earth. This is the distance, known as Roche's limit, where the tidal forces are so strong as to disrupt a satellite. The moon would then become a ring of fragments around the earth, like the rings of Saturn.

10

WINDS, CURRENTS, AND WATER MASSES

The oceanic gyres / Relationship of surface circulation to the oceanic winds / Drift bottles / Vector components of the biosphere / Effect on the dispersion of marine organisms / George Hadley / Tidal currents / Monsoon winds / Early Chinese voyaging / Hippalos crosses the Indian Ocean / Greek voyages to the North Pacific / Discovery of the planetary winds of the North Atlantic / Trades and antitrades / Causes of the monsoon cycle / Parameters of the planetary airflow / The global pattern emerges / Hadley's synthesis / Ocean currents driven by the winds / Cause of the gyral rotation / Temperature of an ocean current / Some regional currents / Sir Hans Sloane discovers the North Atlantic drift / Seeds dispersed by ocean currents / Labrador and East Greenland currents / Ocean currents during the past 100 million years / Deep-sea circulation / *Discovery* investigations of the water masses / The bottom water / Intermediate water / North Atlantic deep water / Antarctic bottom water / Antarctic surface water / Antarctic intermediate water / The Antarctic convergence / Atlantic central waters / Movement of water masses / Energy exchanges

One of the best-known features of the oceans is the constant circulation of the water, forming the **ocean currents.** The first oceanic current to be reported seems to have been the great westward drift across the Atlantic from Africa to the Caribbean, known today as the north equatorial current of the Atlantic. The observer was Christopher Columbus, and he deduced the existence of the current because his astronomical observations disclosed that his positions on the outward voyage proved to lie further to the west than was to be expected from the dead reckoning given by his log. The northern part of the North Atlantic is traversed by a current in the reverse direction, flowing from the Americas toward Europe, the detection of which is often incorrectly attributed to Benjamin Franklin. Research in the Harvard archives shows that the discovery was made by Sir Hans Sloane, who published his observations in a letter addressed to the Royal Society in London in 1696.

Viewed in its simplest form, the ocean current of the North Atlantic may be thought of as having a slow clockwise rotation, a complete circuit requiring approximately one year. Each of the other great oceans rotates similarly, clockwise in the case of the Northern Hemisphere oceans and counterclockwise in the Southern Hemisphere oceans. These facts are evidently interrelated, and a common explanation is therefore to be sought. Such an explanation must also take account of the fact that oceanic winds tend to present a rather constant relationship to the ocean currents: The air masses move mainly from east to west in the tropics and from west to east in the northern and southern temperate belts of the planet. The paths of ocean currents may be determined by releasing floating bottles that contain a card to be returned by the finder (Fig. 10.1).

Viewed in the context of the biosphere, the sea winds and ocean currents are the vector components of the biosphere and constitute a special category of planetary influences on the biota characterized by **directional** properties. Whereas the scalar components govern the diurnal and seasonal activities of feeding, sleeping, growth, and sexual activity, the vector components influence the **dispersion** of organisms. There are two dominant vector fields, that of the winds and that of the ocean currents. The two fields are causally related but are by no means identical.

The first scientific interpretation of the causes of winds and ocean currents was developed in 1735 by an English astronomer, George Hadley, who invoked two other discoveries by his contemporaries to explain the observed facts. The two principles were (1) the conservation of angular momentum, propounded by Isaac Newton, and (2) Boyle's law of gases. Hadley deduced, we believe correctly, the motions of the atmosphere produced by the operation of these laws. Before examining Hadley's findings, however, we may pause to glance over the history of man's ventures on the high seas: Apart from their intrinsic interest, these experiences throw light on the probable effects of the global winds and currents upon the dispersion of many kinds of living organisms that, unlike man, do not keep written historical records of their travels.

In ancient times mariners were familiar with currents; these were not ocean currents, however, but local manifestations, such as tidal streams through narrow straits that alternated in direction as the tide came in or went out. In special cases, as the Straits of Messina between Italy and Sicily, the reversing current occasionally produces a dangerous whirlpool, an example being the much-feared Charybdis, said to suck ships into its vortex. However, ancient mariners did encounter planetary currents when once they summoned the courage and skill to cross an open sea.

Regional airflow patterns known as monsoon winds aided these bold men. Probably the first Polynesian voyages began during the first millenium B.C., at first restricted to the Indonesian region and later spanning half the world. During the Han Dynasty (206 B.C.–A.D. 221) China assumed in the east a commanding role paralleling the contemporary rise of the Roman Republic in the west. By the year 140 B.C., when the Wu emperors obtained the dragon throne, Chinese navigation

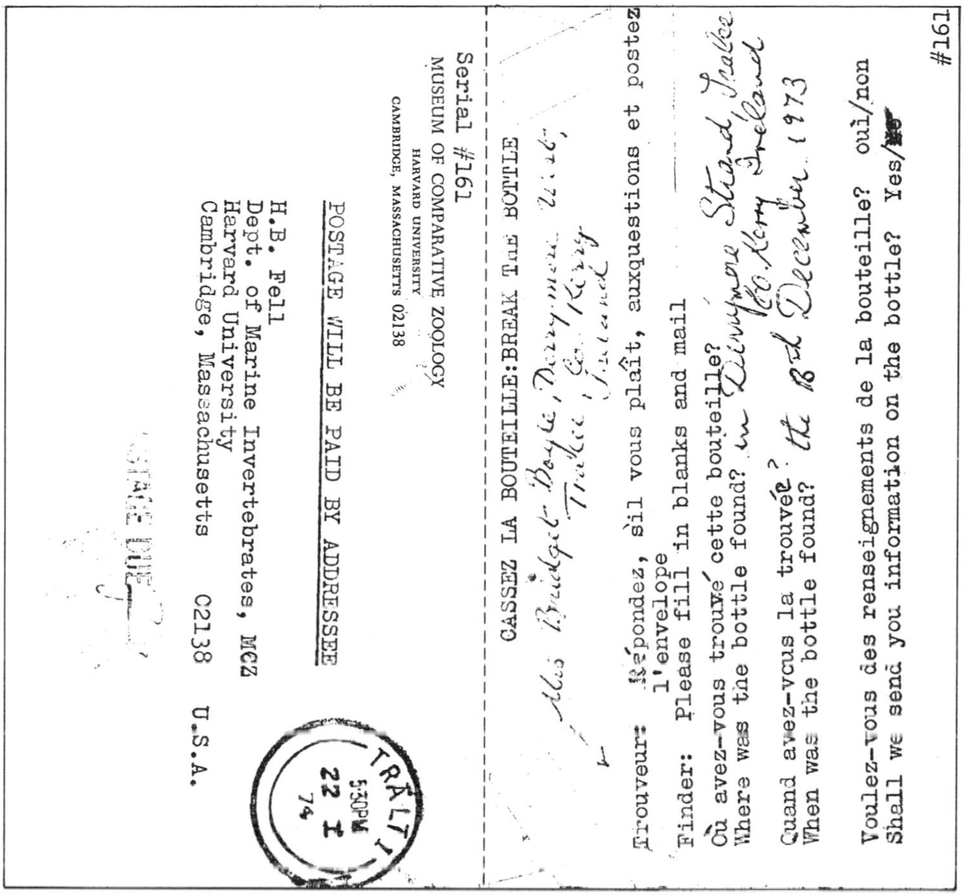

CASSEZ LA BOUTEILLE:BREAK THE BOTTLE
Trouveur: Répondez, sìl vous plaît, auxquestions et postez l'envelope
Finder: Please fill in blanks and mail
Où avez-vous trouvé cette bouteille?
Where was the bottle found?
Quand avez-vous la trouvée?
When was the bottle found? the 18th December 1973
Voulez-vous des renseignements de la bouteille? oui/non
Shall we send you information on the bottle? Yes/No
#161
Serial #161
MUSEUM OF COMPARATIVE ZOOLOGY
HARVARD UNIVERSITY
CAMBRIDGE, MASSACHUSETTS 02138
POSTAGE WILL BE PAID BY ADDRESSEE
H.B. Fell
Dept. of Marine Invertebrates, MCZ
Harvard University
Cambridge, Massachusetts C2138 U.S.A.
TRALI
5.30PM
22 I
74

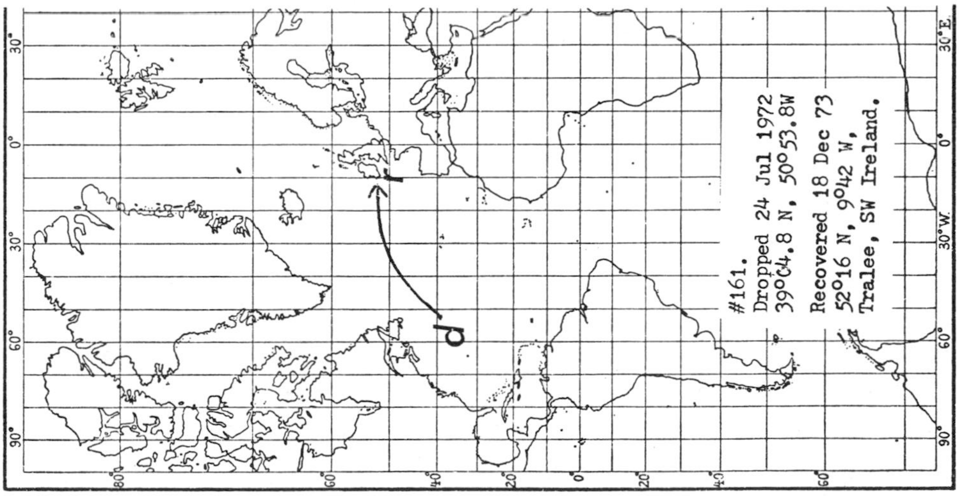

Fig. 10.1 Drift card from a bottle released in the northwest Atlantic and subsequently recovered on the Irish coast.

had reached the point at which the first thalassic voyage could be mounted. The emperor himself led the expedition, accompanied by most of his court officials. Leaving the capital, Ch'ang An, in northern China, the ships sailed 3000 miles southward to the Straits of Malacca, then they rounded the Malayan Peninsula setting their prows northwest to cross the 1200-mile open stretch of the great Bay of Bengal. Landing at "Huang-chih" (which modern Chinese scholars believe to have been Conjeveram, near Madras), the emperor marketed his cargo of gold and silk and returned to China with pearls, crystal, and precious gems. We are not told by the historian (Pan Ku, A.D. 2–92) how the voyage was accomplished, but it seems probable that somehow the Chinese utilized the seasonal winds called the monsoon. It was, in all events, the greatest voyage in the history of man. The Han court subsequently was disrupted by civil revolution, and China withdrew from the international scene for nearly 300 years.

Hippalos Crosses the Indian Ocean

Around A.D. 30 a certain sea captain named Hippalos was engaged in the spice trade between Egypt and northwestern India. His name suggests to scholars that he was an Alexandrian Greek. Like his predecessors for centuries since the time of Ezekiel, he followed the coastal route around Arabia Felix and then sailed northeast round the Persian Gulf, following the coasts of Pakistan, to reach the delta of the Indus River. In the course of his protracted voyages (the round trip lasted three years), Hippalos became aware of the general trend of the Asiatic coasts and realized that, theoretically, a direct route to India could be achieved by sailing northeast from Aden. One advantage of such a precarious venture would be to avoid paying the imposts levied by the Arab chieftains along the west Asian coasts. The practicality of such a voyage would depend upon whether favorable (i.e., following) winds might be encountered.

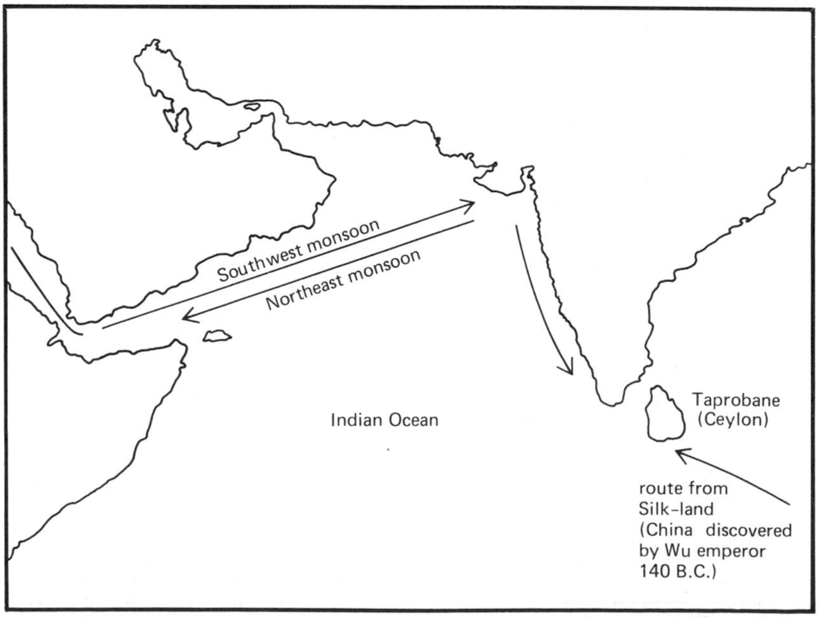

Fig. 10.2 The monsoon route to India pioneered by Hippalos in A.D. 30.

To this question Hippalos applied a novel answer, reinforcing his arguments by his long experience of the seasons of the Indian Ocean. He pointed out that in the summer months the wind blows with remarkable regularity from the southwest, whereas in the winter the wind reverses its direction to blow with equal regularity from the northeast. He had, in fact, detected the main features of the monsoon cycle, although its causes were to await explanation. Putting his theories and the courage of his crew to the test, Hippalos sailed one summer on the southwest monsoon, arriving in India in the unheard of time of about three months; the return voyage was performed on the northeast monsoon, and the ship returned to Egypt within the course of a year for the entire outward and homeward voyage (Fig. 10.2). Almost overnight the trading routes were revolutionized, with up to 100 ships setting out for India in a single season. By the time of Claudius and Nero, some Greek and Roman captains were sailing directly to Taprobane (Ceylon), and Roman currency flooded the East, as the innumerable coins found in India and Ceylon bear mute testimony. In Ceylon the Greek traders encountered a mysterious new fabric, silk, attributed by the Sinhalese hosts to a distant land to the East, whence traders brought it to India.

But before the Western visitors had time to contemplate their next step, the Chinese acted first. According to the contemporary historian Fan Yeh, the Chinese emperor An Ti in the year A.D. 120 had been entertained by a troupe of Roman musicians and circus performers, thoughtfully directed to him by the king of Burma. This visit, reinforced by favorable reports of the Roman world sent back from Antioch by an earlier overland embassy, must have prompted the Chinese to open the way for Western captains to receive friendly guidance across the Bay of Bengal, through the Malaccan Straits, and so on to the Gulf of Tonking. An Egyptian Greek named Alexander is reputed to have performed the first passage, landing at Haiphong. Contact with the Imperial officers was established at Cattigara, now called Hanoi. By A.D. 166, in the reign of the Han emperor Huan Ti, an embassy from Rome arrived bearing gifts from the Roman emperor An Tun, by which Western scholars recognize the name of Antoninus (Marcus Aurelius Antoninus was emperor in A.D. 166). No trace of these transactions remains in the scattered records of the Roman world, and we are wholly dependent upon the Chinese historians.

Consequences of Greek Voyages to the Pacific

The eventual collapse of civilized contacts between East and West as the two empires disintegrated all but obliterated the cultural results of the voyages. One lasting result, however, was the compilation of the *Geographies* of Strabo and the *Almagest* of Claudius Ptolemy, a Greek astronomer of Alexandria and a contemporary of Marcus Aurelius. Latitudes and estimated longitudes were tabulated for numerous distant Eastern cities and coasts, and these provided the view of the world that the Moors and Hebrews brought to Spain and that Columbus used as the basis of his projected voyage westward to the Indies.

The Discovery of the Planetary Winds of the Atlantic

Like Hippalos before him, Columbus was inspired by his previous experience of oceanic winds. His voyages on the west African coasts had taught him to expect with virtual certainty a steady wind from the east when sailing south from Spain and approaching the Canary Islands, at about 30°N latitude. Knowing that occasional reversals occur, he reckoned on being able to perform a westward passage on the east wind and to use fortuitous reversals of the wind direction to sneak back to Europe again. In fact, he found no such thing. The east wind turned up, as expected, when he reached the Canaries, and it carried him right across the Atlantic in 70 days. But on attempting to return by the same route, Columbus found no west wind. Forced to tack, he gradually beat his way painfully northeastward until he neared the vicinity of Bermuda and crossed the 30th parallel. Almost at once he picked up the longed-for west wind and was swept safely back to Spain. As soon became plain from the voyages that followed, Columbus had established the main outlines of the planetary wind system—a great flow from east to west over the equatorial belt between 30°N and 30°S latitude and two great belts of west-to-east winds north of 30°N and south of 30°S. The navigations that followed during the next two centuries disclosed that a similar pattern of winds exists in the Atlantic, the Indian, and the Pacific oceans. The tracks of the voyages show how effectively the sea captains learned to use these winds, planning their navigations to extract the best possible advantages of the airflow of the planet. Soon the equatorial east wind was called the **trade wind,** and the west winds to the north and the south were called the **antitrades.** By the eighteenth century scientific explanations were forthcoming as to the causes of the circulation.

Causes of the Monsoon Cycle

In summer the dry desert and plateau region of Asia becomes heated, the overlying air mass expands and rises and then flows outward at a higher altitude. The low-pressure region remaining below is therefore subject to the inflow of lower-level air, mainly from the cooler air mass over the Indian Ocean. This northward-moving air near the air–sea interface begins its journey from a part of the earth's surface near the equator where both the air and the earth's surface initially have a west-to-east velocity of 1000 miles per hour imparted by the earth's rotation. But as the air flows northward, it enters regions where the west-to-east motion of the earth's surface progressively decreases (the surface velocity of the earth falls off as the cosine of latitude, becoming zero at the poles). Since the initial angular momentum of the air cannot be destroyed, the equatorial air, as it moves northward, acquires a relatively increased west-to-east component of motion with respect to the surface. Thus, instead of the incoming cool air arriving from the south, it appears to be flowing from the southwest. This air, laden with oceanic moisture, reaches India as the southwest monsoon.

The converse is true of the winter months. The central Asian highlands become cold, high-pressure regions, whence air flows outward near the surface of the earth. The portion flowing southward toward the Indian Ocean enters regions of increasing angular velocity, and so it has insufficient velocity to keep pace. Lagging behind the rotating surface of the earth, its direction appears to change from being a north wind to become a wind from the northeast. This is the dry northeast monsoon.

A monsoon is a regional wind that exhibits properties of a planetary wind; that is, its directional qualities, or vector components, are related to the earth's rotation and are determined by the principle of conservation of angular momentum.

Some Parameters of the Planetary Airflow

As already noted (page 74) the parallels of 30°N and 30°S serve to divide the oceanic air into three main zones: namely, an equatorial belt of east-to-west winds and a northern and southern zone each of west-to-east winds. Following are relevant clues to the nature of this system of airflow.

The system conserves angular momentum, because:

1. At any instant as much air is flowing from west to east as is simultaneously flowing from east to west. This proposition can be illustrated by considering the properties of a sphere. The area of the surface of a sphere is $4\pi r^2$. The area of that part of the surface between 90° and 30° is πr^2 for each hemisphere. The area between 30° and 0° is also πr^2 for each hemisphere. If we assume the atmosphere to have a uniform depth, then it follows that equal volumes of air move in the two opposed directions. Thus the total airflow neither adds nor subtracts from the total angular momentum of the rotating planet.
2. The angular momentum of the moving air is a function not only of its velocity but also of its radial distance from the earth's axis. Thus, in order to balance, the velocity of the air of the equatorial belt (which lies farthest from the earth's axis) would have to be less than the velocity in the reverse direction of the air in the two west-wind belts. This is observed to be the case: The trade winds are gentle and steady, whereas the winds of the antitrades are strong and often of near hurricane strength (as in the southern latitudes called the Roaring Forties).

Vertical airflow lies between zones of horizontal airflow, because:

1. The doldrums, or so-called windless region, form a narrow belt along the equator where the barometric pressure is low. A barometric low is commonly produced when air is moisture laden because air containing water vapor weighs less than dry air. But over the ocean all air is moist (unless it has just arrived there from some dry region such as a desert). Thus moisture alone cannot be the cause of the barometric low pressure of the doldrums. The only other possible cause is that the low pressure is produced by rising air. This is highly plausible because equatorial air receives most solar energy and therefore becomes hotter than other air, decreases in density, and floats upward.
2. The analogous regions of so-called windless air, the so-called horse latitudes, register high barometric pressure. In the same latitudes (30°N and 30°S) over land, dry desert air is present. So it seems that descending dry air reaches the surface of the earth from a higher level of the atmosphere.

THE GLOBAL PATTERN EMERGES—HADLEY'S SYNTHESIS

The facts and inferences noted above were synthesized into a simple cohesive theory by George Hadley in 1735. Hadley's theory states that:

1. Heated air in the equatorial region over the ocean expands, rises, and flows outward to the north and south at a higher level in the atmosphere. After being cooled in the upper levels, and thereby losing its moisture content, it descends as dry air in latitude 30° north and south of the equator.

2. Surface air flows toward the equator from the south and from the north to replace the air that has risen. Conservation of angular momentum necessarily means that such air acquires an east-to-west component of motion because it is entering a region of higher velocity of the earth's surface in the west-to-east direction. This effect produces the easterly trade winds.
3. The cold air descending at the horse latitudes spreads to the north in the Northern Hemisphere and to the south in the Southern Hemisphere. Since this air is entering regions of lower surface velocity of west-to-east rotation, the west-to-east component already present in the northward and southward spreading air exceeds that of regions it is entering. The conservation of its angular momentum causes it to assume the character of a west wind, since it is already rotating faster than the earth beneath it.

Hadley's theory not only makes sense of the atmospheric motions, but it also explains all the major features of the global circulation of the oceans.

The Ocean Currents Are Driven by the Winds

All major ocean currents flow in the directions we would expect if they are wind driven.

In particular cases this may be proved easily. For example, in the northern Indian Ocean, where the regional monsoon wind reverses its direction every six months, the underlying waters of the ocean correspondingly reverse their direction of flow to keep in phase with the wind. Thus the Somali current, which flows along the northeast coast of Africa, reverses its direction shortly after each reversal of the monsoon. During the southwest monsoon, the Somali current flows from the southwest. When the northeast monsoon begins, the Somali current gradually slackens its flow, comes to a halt, and then slowly begins to pick up speed in the southwest direction, that is, begins to flow from the northeast. The frictional effect of moving air over the surface of the sea is called the **air–sea couple.** It is exactly analogous to the effect produced when one blows across a dish of water, setting the water in motion.

The Oceanic Gyres

The oceanic circulation (Fig. 10.3) can be simplified into a global pattern of five great rotating whirlpools, called **gyres** or gyrals. Two of the gyres lie in the Northern Hemisphere, and rotate in the clockwise direction. They occupy the North Atlantic and North Pacific oceans, respectively. The remaining three gyres lie in the Southern Hemisphere and rotate in the counterclockwise direction. They occupy the South Atlantic Ocean, South Indian Ocean, and South Pacific Ocean, respectively. The North Indian Ocean is of very limited extent at the present stage in earth history; and, as already noted, it carries a monsoon-driven current. During most of geological history the North Indian Ocean has been much larger than it is now, covering much of Asia. At such epochs it would carry a rotating gyre like the other oceans.

Cause of the Gyral Rotation

Each ocean basin is bounded by eastern and western continental shorelines. Each has an overlying equatorial air mass that moves from east to west. Nearer the

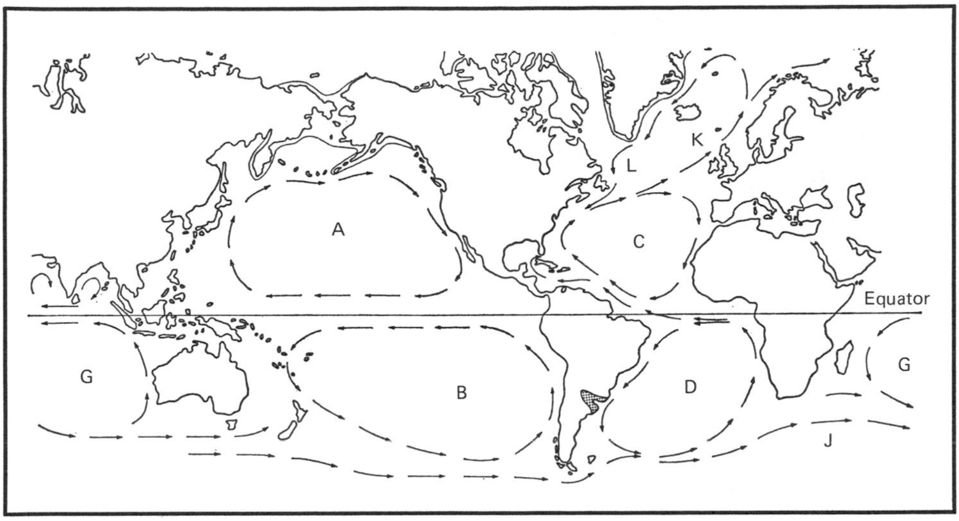

Fig. 10.3 Main oceanic gyres at the present epoch. Compare with Figs. 10.4 and 10.5. A, North Pacific gyre; B, South Pacific gyre; C, North Atlantic gyre; D, South Atlantic gyre; G, South Indian gyre; K, North Atlantic drift; L, Labrador current. Note that the North Indian gyre, E in Figs. 10.4 and 10.5, is extinct; also note in Fig. 10.4 the former sea Tethys, labeled F.

poles, each has an overlying air mass that moves from west to east. Under such conditions it is obvious that northern oceans must inevitably develop clockwise rotating gyres and southern oceans must develop counterclockwise rotating gyres. The separate parts of a gyre—the eastern side, the equatorial part, etc.—are commonly designated by particular names, for example, the Gulf Stream or the Kuroshio. Table 10.1 lists the more important of these currents. Near its center a gyre rotates slowly, so the currents have little velocity; nearer the edge of a gyre the velocities are usually highest. The speed commonly falls within the range of a quarter-knot (about 400 m per hour) to 1 knot (1.85 km per hour).

Temperature of an Ocean Current

The water mass of a gyre gradually becomes warm as it passes through the equatorial part of the gyre. Water holds its temperature for a considerable period of time, so the western sides of oceans (which are supplied with water coming from the equatorial region) are also warm. The water loses its heat whilst passing through the part of a gyre lying nearer the pole; thus the eastern sides of oceans are bound to be cooler because they are fed by waters arriving from the cold part of a gyre.

Coral Reefs and Ocean Currents

An important effect produced by the temperature distribution is that coral reefs tend to grow on the eastern margins of continents, for these are warmed by the western water mass of the gyres. The western margins of continents have only impoverished reefs of very limited extent, for the waters are too cool to promote coral growth.

Table 10.1 shows the more important ocean currents distinguished by mariners and oceanographers.

Table 10.1 Major Ocean Currents

GYRE	DIRECTION OF ROTATION	NAMES OF PARTICULAR CURRENTS CONSTITUTING THE GYRE			
		Southern Limb	*Western Limb*	*Northern Limb*	*Eastern Limb*
North Atlantic	Clockwise	North Equatorial current	Gulf Stream	North Atlantic drift	(Diffuse)
South Atlantic	Counterclockwise	West-wind drift	Brazil current	South Equatorial current	Benguela current
North Pacific	Clockwise	North Equatorial current	Kuroshio	North Pacific current	Californian current
South Pacific	Counterclockwise	West-wind drift	(Diffuse)	South Equatorial current	Peru, or Humboldt, current
South Indian	Counterclockwise	West-wind drift	Algulhas current	South Equatorial current	West Australian current

NOTE: The inferred evolution of the system is illustrated in Figs. 10.4 and 10.5.

SOME REGIONAL CURRENTS

North Atlantic Drift

The northern continuation of the Gulf Stream is the North Atlantic equivalent of the west-wind drift of the Southern Hemisphere. The North Atlantic drift bifurcates off western Europe, with one part flowing northeast into the Arctic Ocean and the other part continuing southward along the North African coast, to rejoin the Equatorial current.

On page 70, reference was made to a letter from Sir Hans Sloane that was published by the Royal Society of London (*Transactions*, September–October, 1696). Here are the parts of it that relate to the North Atlantic drift:

> An Account of four sorts of strange beans, frequently cast on Shoar on the Orkney Islands, with some conjectures about the way of their being brought thither from Jamaica, where three sorts of them grow.
>
> I had several times heard of strange beans thrown up by the Sea on the islands on the North-west parts of Scotland, especially those of them who are exposed to the Waves of the great Ocean: they are thrown up pretty frequently in great Numbers, and are no otherwise regarded than as they serve to make Snuff boxes. . . . Lately Dr. George Garden of Aberdeen did me the favor to send me four sorts of them, very fresh and little injured by the sea. Three of these beans grow in Jamaica where I gathered them, and I mentioned them in my Catalogue of the Plants of the Island, *Catalogus Plantarum quae in Insula Jamaica sponte proveniunt:* Hans Sloane, M.D., London, Dan Browne at the sign of the Swan, 1696.

After discussing each of the beans and identifying them as to their species and the localities where they grow, Sloane goes on to say:

> How these several Beans should come to the Scotch Isles, and one of them to Ireland, seems very hard to determine. It is very easy to conceive, that growing in Jamaica in the Woods, they may either fall from the Trees into the Rivers, or be in any other way conveyed by them into the Sea. It is likewise easie to believe, that being got to Sea, and floating in it in the neighborhood of the Island, they may be carried from thence by Wind and Current, which meeting with a stop in the main continent of Am[erica] is forced through the Gulph of *Florida*, or Canal of *Bahama*, going thence constantly E[ast] and into the *N American Sea;* for the . . . Sargasso Weed grows on the Rocks about Jamaica, and is carried by the Winds and Current (which for the most part go impetuously the same way) towards the Coast of Florida, and thence unto the Northern Am[erican] Ocean, whereas I mention p. 4 of my Catalogue it lyes very thick on the surface of the Sea. But how they should come to rest of their Voyage I cannot tell, unless it be thought reasonable that as Ships when they go South expect a trade Easterly Wind, so when they come North, they expect, and generally find, a Westerly Wind, for at least two parts of three of the whole Year, so that the Beans being brought North by the Current, from the Gulf of Florida, are put into these Westerly Winds way, and may be supposed by this means at last to arrive in Scotland.

It would seem highly probable that Sloane's paper was known to George Hadley (also a member of the Royal Society) when he proposed his theory of winds and ocean currents in 1735.

Labrador Current and East Greenland

These tongues of cold water expelled from the Arctic, presumably to compensate for the entry of North Atlantic drift water into the Arctic. The cold currents follow the eastern seaboard of the Canada and U.S. Atlantic states, excluding them from the moderating influence of the Gulf Stream, which therefore lies offshore.

Equatorial Countercurrents

These narrow eastward-flowing seasonal currents, lying on or near the equator in all oceans, divide the westward-flowing equatorial waters into a northern and a southern section. It has been supposed that the countercurrents are analogous to rivers, comprising water that is flowing downhill from the western sides of oceans, under the doldrums, where no air is flowing horizontally. The elevated western sea level of each equatorial oceanic region is attributed to the piling up of wind-driven water carried westward by the trade winds. Such water can flow backward (i.e., in an easterly direction) along the downslope produced by the lack of trade winds along the equator. This explanation has been disputed.

Ocean Currents During the Past 100 Million Years

For much of geological past, the Indian Ocean has covered extensive areas of Asia and Europe by a shallow epicontinental sea known as Tethys. During the Cretaceous period a large part of North America was also covered by a shallow epicontinental sea known as the Sundance Sea. During the greater part of the past 100 million years a broad seaway has separated North America from South America. About 30 million years ago Tethys was bisected by the elevation of the middle east region. The western part of Tethys then became the Mediterranean Sea, and other residual fragments are represented by the Caspian and Black seas. The eastern half of Tethys vanished to become dry land. Thus the Indian Ocean lost its northern gyre. With the closure of Tethys, the northern gyre of the former northern Indian Ocean was disrupted. One of the residuals is considered (Fell, 1967) to be the present North Atlantic drift complex. The Gulf Stream system supposedly developed after the closure of the Mediterranean from the Indian Ocean. These inferred changes in oceanic distribution, and in the ocean currents, are illustrated in Figs. 10.4 and 10.5.

DEEP-SEA CIRCULATION

With the initiation of deep-sea research programs in 1872 by the H.M.S. *Challenger* (see Chapter 28), it was discovered that there is a marked fall in temperature of the seawater as the depth increases. Later work, particularly that carried out on board the *Discovery* and her successor *Discovery II*, led to a recognition of **water masses.** Water masses are extensive layers of the deeper water of the sea in

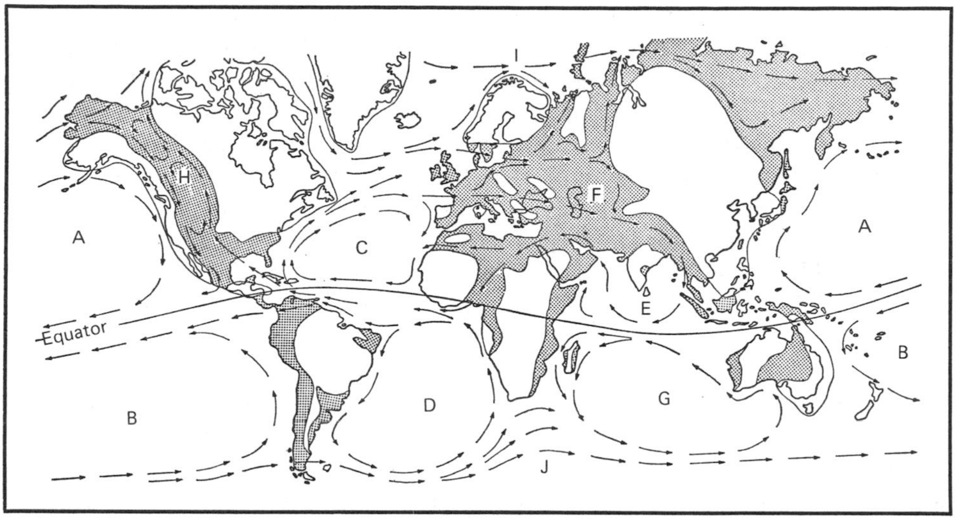

Fig. 10.4 Inferred oceanic circulation in Cretaceous times. See Fig. 10.4 for gyre names.

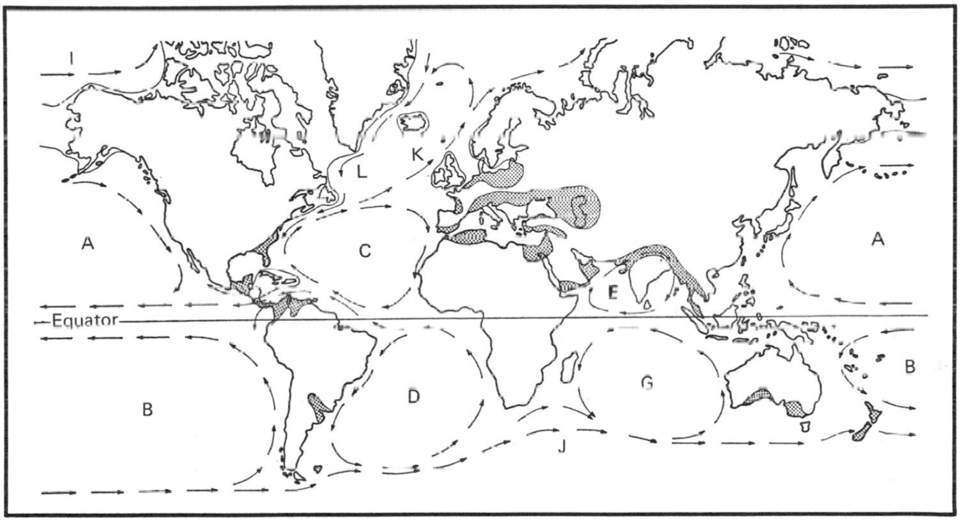

Fig. 10.5 Inferred oceanic circulation in mid-Cenozoic times, after the disruption of Tethys. See Fig. 10.4 for gyre names.

which there is, in any given water mass, a relatively constant ratio of dissolved salts and a correspondingly constant temperature. These conditions could not exist if the gyral motions of the surface waters comprised the entire ocean, so it is apparent that the pattern of deep-sea circulation must differ considerably from that of the surface waters.

The Bottom Water

The relatively warm surface water does not extend to any great depth. In the tropics where the surface water is warmest, with a mean annual temperature of about 26°C, the temperature falls off rather rapidly with increasing depth, and it is only about 10°C at a depth of 250 fathoms. In other parts of the world the warm surface water is even more restricted in the vertical plane, and lower temperatures occur at shallower depths as one approaches the polar seas. Just over the ocean floor itself the temperature of the water is only about 3° to 4°C, and thus very near to freezing. The low temperature of bottom water is one of the significant characteristics of the deep-sea environment. Since we know that heat is radiated from the interior of the earth and that heat must therefore be flowing outward through the seafloor into the overlying water, we can only account for the extreme coldness of seafloor water by inference that some system of ocean currents must be responsible for bringing down cold water from elsewhere. The universal distribution of cold bottom water can only be explained on the assumption that it comes from the polar regions. This supposition has been confirmed by studying the content of oxygen, salt and other chemical substances present in solution in bottom water. The proportions present show that bottom water is indeed derived from the Arctic and Antarctic regions, where the offshore seawater has a closely comparable chemistry. Evidently, therefore, bottom water spreads across the floor of the deep sea from the polar regions, flowing southward from the Arctic and northward from the Antarctic. However, the chemical substances dissolved in polar waters occur in detectably different proportions in the Arctic and in the Antarctic, and we can therefore distinguish each of the two water masses wherever we encounter them. Early studies soon disclosed that there are certain differences separating the distribution of Arctic and Antarctic water masses over the deep seafloor.

Intermediate Water

Between the water resting on the seafloor and the surface waters lie other intermediate water masses. These several intermediate layers can be characterized by their temperatures, salinities, oxygen content, and other features, and they are spread over very large areas of the ocean. These water masses seem to be relatively stable, with little vertical mixing, except in certain more or less well-defined areas. The interrelationships of the various superimposed water masses has been most extensively studied in the Atlantic Ocean, but it is already evident that similar features occur in the Pacific and Indian oceans. Therefore we can probably take the Atlantic Ocean as a legitimate model upon which we may base inferences as to the corresponding mechanisms in the other great oceans.

North Atlantic Deep Water

In the Arctic Atlantic most of the cold surface seawater is derived from an admixture of ice melt and effluent river water from the northern land areas. Since these sources yield fresh, or almost fresh, water the resultant salinity of the superficial layer of the sea is relatively low. These waters float upon the more saline—and hence denser—intermediate water mass and drift south until in the vicinity of the 50°N parallel they encounter the water mass of the North Atlantic drift. The latter is, of course, derived from the Gulf Stream and is therefore much warmer than the Arctic surface water. Even though the North Atlantic drift is relatively saline, it is less dense than the Arctic surface water, on account of its high temperature. Thus, in the region of the so-called (and rather ill-defined) Arctic Convergence (see Chapter 27) the southward-flowing and northward-flowing surface waters meet, and the lighter, warmer North Atlantic drift water overrides the colder, denser Arctic water. So the water mass of surface origin in the Arctic is now forced to continue its southward flow below the North Atlantic drift water, thereby becoming what is called the North Atlantic deep water. It does not immediately overlie the bottom of the ocean, however, for it is separated from the seafloor by yet another water mass, one of greater salinity (and hence of greater density) and is known to have been derived from the coastal margin of the Antarctic continent, as we shall see shortly. The North Atlantic deep water ultimately returns to the surface, upwelling in the Antarctic Ocean after having traversed the entire Atlantic Ocean from north to south. So we have the remarkable fact that water that was precipitated from the atmosphere as rain or snow upon the lands bordering the Arctic Ocean and that flowed down to the sea as rivers or glaciers must travel some 7000 miles as a submarine flow beneath the surface of the Atlantic, only emerging as it approaches the continent of Antarctica.

Antarctic Bottom Water

The ocean surrounding Antarctica is the coldest in the world. In winter it is subject to such low temperatures that very extensive surface ice forms on the sea. The process of freezing water excludes salts, therefore ice contains very little salt, so the residual, unfrozen seawater becomes increasingly saline as more and more ice forms. Increasing salinity implies increasing density, and so the residual unfrozen water must sink and, being also very cold (—1.9°C), it is the densest of all known seawater. It is this water that descends to the ocean floor and, flowing northward, comes to lie between the southward-flowing Arctic water of the North Atlantic deep water and the seabed. The northward-flowing sheet of dense bottom water is called the **Antarctic bottom water.** It eventually crosses the equator to mingle with the deep water masses of the North Atlantic.

Antarctic Surface Water

Farther out to sea from the Antarctic coast another water mass forms, one having a low density. It owes its origin to the constant precipitation into the sea of snow and rain and, in the summer months, to the melting of sea ice that has drifted north. As it is of lower density than other Antarctic water it floats, in spite of the fact that it is also very cold and might otherwise have been expected to sink. This **Antarctic surface water** flows northward and, therefore, has to be replaced. Its replacement is in fact that same southward-moving sheet of North Atlantic deep water, whose vicissitudes we have already traced in a previous paragraph, and which originated in the Arctic region. So there is an upwelling of deep water some hundreds of miles off the coast of Antarctica. During its passage southward at great depths the North Atlantic deep water acquired dissolved mineral salts from showers of plankton falling through the water column. This enables it to support a rich population of zooplankton, which feed on the phytoplankton. The food chain erected upon the plankton bloom culminates in the great mammals of the Antarctic Ocean, the whales and seals, and those great birds, the emperor penguins. Meantime, the Antarctic surface water, of lower salinity and lower density, which has been spreading northward at the surface, encounters in latitudes around 50° to 55°S another surface water mass, part of the west wind drift, originating from lower latitudes, and it is therefore warmer than the Antarctic surface water, and less dense. Consequently, the west wind drift surface water overrides the northward-flowing Antarctic surface water which, accordingly, is forced to sink. It continues its journey northward below the surface as the **Antarctic intermediate water.** The region where the two water masses meet and where the Antarctic surface water sinks to become the Antarctic intermediate water is known as the **Antarctic convergence.** Like the corresponding water derived from the Arctic regions, the Antarctic surface water, having become the Antarctic intermediate water, now flows toward the equator at a depth of about 1000 fathoms and eventually mingles with the cooler water of low salinity that underlies the warm, saline **Atlantic central waters** of the Sargasso Sea region.

In the course of these interchanges of water masses between the North and South Atlantic, some 500,000 million tons of seawater flow across the equator in either direction every 24 hours. In the space of a year the interchange amounts to some 70,000 cubic miles of seawater, equivalent to an aqueous asteroid some 50 miles in diameter. The energy required to effect the transfer arises partly from solar radiation falling upon the earth, producing the temperature–density differences we have noted, and partly from the earth's kinetic energy of rotation and its convected heat energy received through the seafloor. These immense operations apply only to the Atlantic Ocean—far greater exchanges must occur in the Indo-Pacific Ocean.

11

PHYSICAL AND CHEMICAL FACTORS OF SEAWATER

Variation of temperature with depth and latitude / Bipolarity of polar species / Stenothermal and eurythermal organisms / Temperature indicator species / Biological thermometry / Salinity / Effect of Congo and Amazon efflux in Atlantic / Euryhaline and stenohaline organisms / Effect of Antarctic summer ice melt / Water masses and TS characteristics / Origin of sea salt / Evaporites / Geoelectric fields / Depth and pressure / Fishes with a gas bladder / Eurybathic and stenobathic fishes / Solubility of calcium carbonate at high pressures / Brittlestars on deep-sea floor / Color and depth / Vertical migration and color / Illumination / Turbidity / Iridescent cornea of bottom-living shelf fishes / Positive and negative phototropisms / Oxygen / Anoxic condition / Pollutants / Oil spills / Organic mercury / Other toxic pollutants / Thermal pollution / Aerial mapping procedures

The determination of the chemical and physical properties of seawater lie properly within the domain of physical oceanography, but they are important to the marine biologist, since they determine to some extent the habits, and to a much greater extent the distribution, of marine organisms. The term parameter may be used to include any variable, such as temperature, pressure, or salinity, or illumination. Any one of these measurable quantities helps to determine the overall conditions of life in the sea in some way. Following are illustrations of how some of these parameters operate.

TEMPERATURE

The **temperature** of the sea at any place varies according to the **season**, the **latitude**, and the **depth** at which the measurement is made. The warmest seawater is found at the surface near the equator, where the average temperature of the water over the whole year is about 26.6°C (see Table 6.1). The coldest seas are those of the polar regions, with a minimum possible value of —1.9°C (when the most saline seawater freezes). The total range of marine water temperatures is thus about 30°C, a modest figure as compared with the fluctuations observed in terrestrial environments. So in marine animals we do not find any pronounced adaptations for protection against temperature changes or extremes, such as are observed in polar terrestrial organisms or desert organisms. Many marine animals are tolerant of only a minor range of temperatures. Coral reefs develop only in or near the tropics, where the water temperatures do not drop below about 18°C, and some marine plants, such as the large brown kelps, grow only in cool latitudes and are killed if accidentally swept by currents into tropical seas. Thus the South African kelp *Ecklonia* is carried by the Benguela current northwest to the island of Saint Helena, where it is washed ashore dead. On the other hand, eurythermal starfish that cling to the kelp survive the journey and have colonized Saint Helena.

Cold water is denser than warmer water over most of the temperature ranges encountered, therefore, in any latitude the temperature decreases with increasing depth. This has one interesting effect on distribution: It is possible for polar species to exist in deep water in quite low latitudes. Some species occur in surface waters of both the Arctic and Antarctic seas, and have a continuous distribution, occurring in progressively deeper water nearer the tropics, and crossing the equator in very deep water. Obviously temperature is an important parameter for such organisms because their distribution shows that they are obliged to conform to appropriate isotherms without regard to depth. This means that animals living in the deeper parts of the North American shelf also inhabit the more northern and colder seas around the North Pole. Therefore, the quickest route to the Arctic Sea for a New Englander or Californian is simply to dive. For the Bostonian, the dive can be well within the capacity of a scuba diver of 60 or 80 feet certification; at that level animals can be found that also live immediately beneath the Polar ice. In 1953, I well remember the interest aroused when, on a visit to Narvik Fjord, specimens of such Arctic sea snails as *Buccinum undatum* and others of the same northern family first appeared in our collections. I took specimens back to New Zealand so that my students there could see what Arctic animals look like. Years later, on a tropically hot day of the Massachusetts summer, when shallow-water dredging was in progress on the New England shelf, I was astonished to see these old friends of the Norwegian Arctic coming up from below—a vivid lesson on how near at hand are Arctic denizens off our local coasts.

An example of this phenomenon is seen in the crinoids or feather stars. It is known from the work of A. H. Clark (1915, 1937) that boreal species of the genus *Heliometra* follow the appropriate isotherms as their distribution is plotted along the west American coasts, becoming progressively more bathyal toward the equatorial belts. Thus *Heliometra glacialis* presents the remarkable feature of enduring depths and pressures 135 times as great at the lower extremity of its range as at the upper limit. Thus it is evident that an extensive and detailed examination of the thermal parameters of existing species in their natural environments can yield information that elucidates problems of biogeography.

Organisms that tolerate a wide range of temperatures evidently derive their adaptability from some physiological character distinguishing them from related intolerant species. Thus P. A. Cook and P. A. Gabbott (1972) have found evidence suggesting that there may be a relation between biochemical characters of the tissues of the barnacle *Balanus balanoides* and the ambient temperatures of the environment, possibly enhancing the tolerance of cold by this species. Note that the natural variation of any environmental factor is termed **ambient,** a word derived from the French **ambience** (environment).

Long-term climatic changes are known to occur, as evidenced by the ice ages, and there is reason to believe that the temperature of the seas is changing at the present epoch. The mean temperature of the sea as measured over a span of years has been observed to change, as for example by A. J. Southward and E. I. Butler at Plymouth, England (1972). See Fig. 11.1.

Biological Thermometry

Biota that are restricted to a limited range of temperatures are said to be **stenothermal,** those that tolerate wide temperature ranges **eurythermal** (Fig. 11.2). Stenothermal marine organisms can be grouped according to their preference for

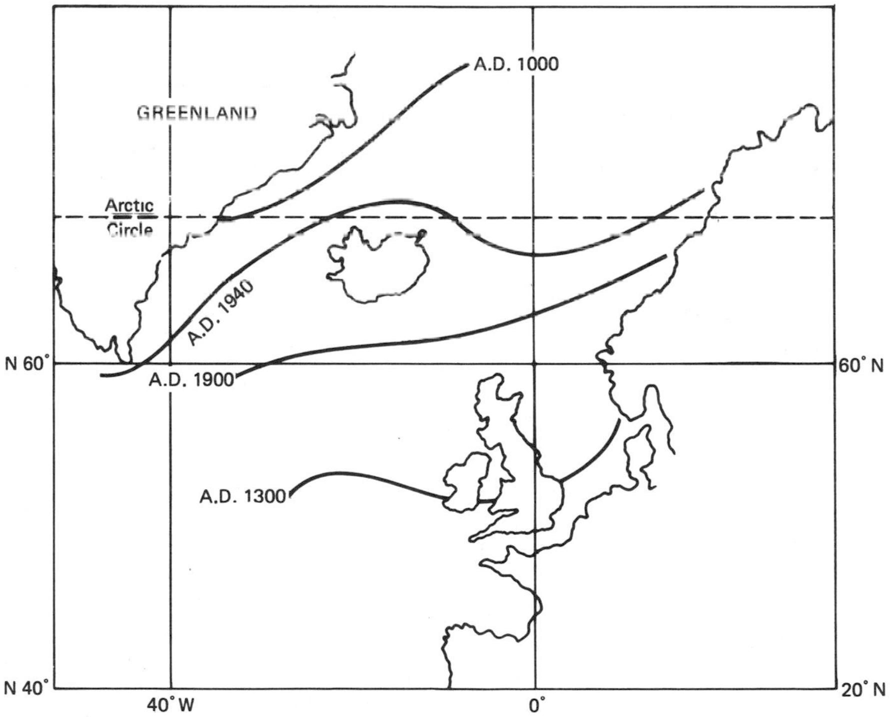

Fig. 11.1 Changing water temperatures during the past thousand years are believed to be responsible for the regional displacement of the herring fishery in European waters. Present higher temperatures have led to the reestablishment of breeding herring stock off Iceland and Greenland. This appears to have been the case in Viking times. In the Middle Ages the fishery lay in the North Sea.

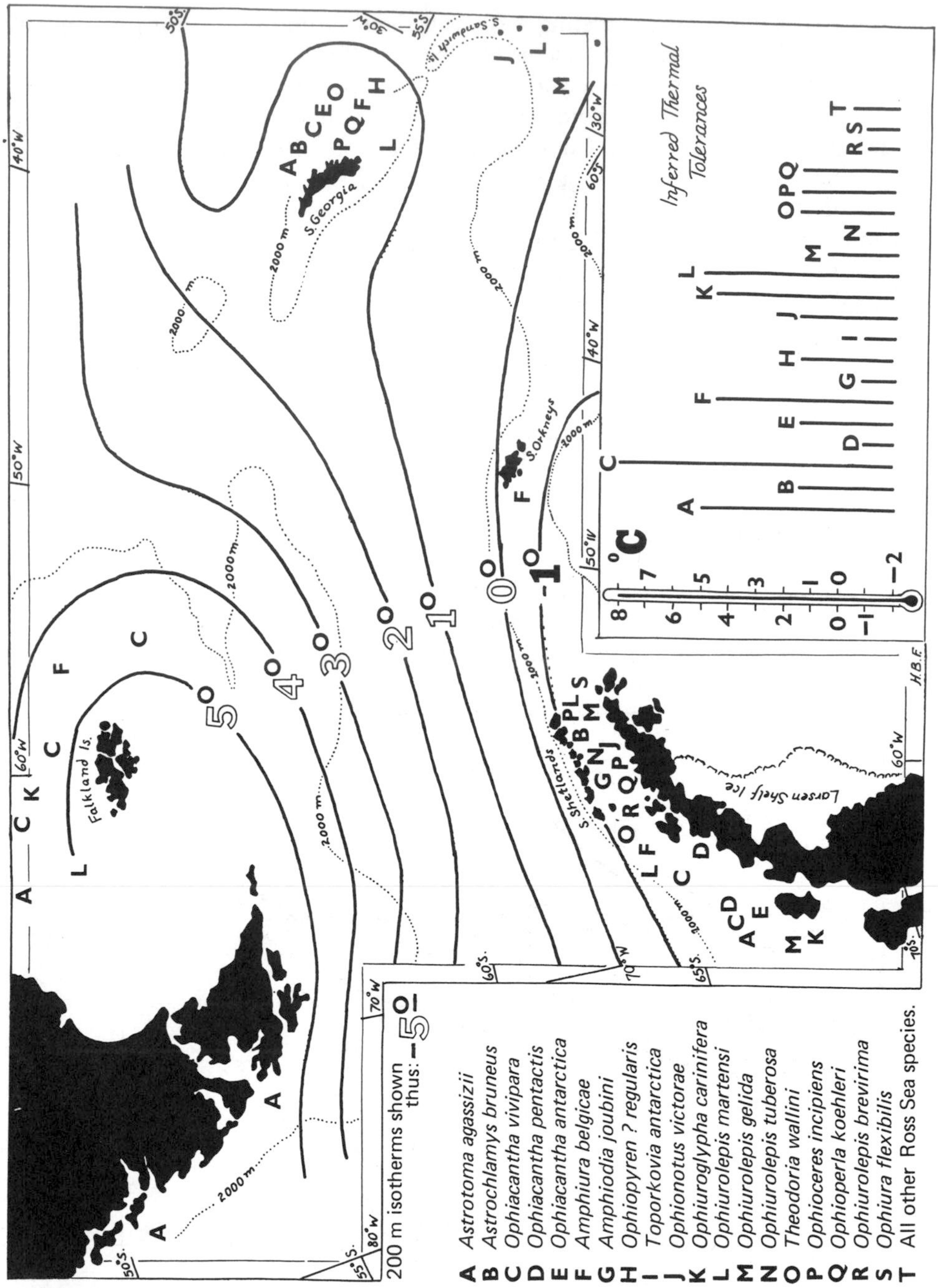

Fig. 11.2 Marked differences in the latitudinal distribution of brittle stars in the Southern Hemisphere apparently reflect differences in temperature response. Eurythermal forms are able to disperse along the west drift and also enter Antarctica, whereas stenothermal Antarctic forms do not transgress the Antarctic Convergence.

cold or warm water, and in some groups the various species in a genus may differ markedly in this respect. Such stenothermal species provide a kind of biological thermometer by which we can measure sea temperatures in former epochs if we know the temperature tolerance of living species that also occur as fossils. Species that are useful in this regard are termed **temperature indicators.** For example, when the climates of the earth became colder, polar stenothermal species dispersed toward or into the tropical zone. On the other hand, warm-water species moved nearer to the poles during warm phases. Pelagic species are preserved in layered sequence in the sediment on the seafloor, provided they have durable skeletons. So, by studying cores of sediment collected on the seafloor we can often detect evidence of temperature change in the overlying waters.

Various species of pelagic forams serve as valuable indicators of surface sea temperatures in seafloor cores. The commonest of the equatorial species is *Globigerinoides sacculifera*; other warm water indicators are *Globigerinoides rubra, Globorotalia menardii,* and *G. truncatulinoides.* Of the cold water forms the commonest are *Globigerina bulloides* and *G. inflata.* In sediments from the deep seafloor the forams are generally lacking because their remains tend to dissolve under high pressure in water. Here the sediments comprise red clays and radiolarian ooze.

SALINITY

The salinity of the sea is its best-known characteristic and also one of the most interesting and important environmental parameters. Average seawater contains about 33 parts per thousand (33‰) of dissolved salts, mostly sodium chloride. Enclosed seas, such as the Baltic Sea, where the influx of rivers is relatively high have reduced salinities (Baltic Sea 8‰, Aral Sea 10‰, Caspian Sea 12‰, Black Sea 18‰). Closed basins that have become isolated are subject to excessive evaporation and may have high salinities, as for example the Dead Sea, the Caspian Sea, the Great Salt Lake. But the oceans at large do not vary so widely—the lowest salinities are usually around 27‰ where equatorial rains are heavy and rivers such as the Congo and Amazon discharge, and highest salinities occur under the opposite circumstances, where evaporation is high and rainfall low, as in the Red Sea, with salinities reaching about 40‰.

These varying salinities are important parameters restricting the life of marine animals. For example, nearly all echinoderms are intolerant of lowered salinities and avoid such regions. Because of their lack of tolerance for low salinities, coastal echinoderms of the north Atlantic are prevented from dispersing south of the Congo and Amazon, for these rivers discharge so much fresh water that the upper-level salinity of the tropical Atlantic is reduced. Similarly the echinoderms of southern South America do not disperse northward to the Caribbean.

Animals tolerant of varying salinity are termed **euryhaline,** and those that are not are **stenohaline.** Euryhaline fishes include salmon and eels, both groups spending a part of their lifespan in fresh and saltwater. Some fishes, for example, sharks, freely range from the sea into rivers and even enter freshwater lakes, such as communicate with rivers. Some of the worst shark attacks in Australia have occurred in rivers or even on river banks, hundreds of miles inland. The ability to tolerate a wide range of salinities depends on the water-balance physiology of each fish, mediated by the kidneys. Puffer fishes, flounders, and eels are other examples of euryhaline fish. The migrations of salmon from the sea to the rivers and of young salmon from rivers to the sea are other well-known cases of such tolerance. Most marine fishes do not tolerate such changes in salinity. The fishes of the Baltic, Caspian, and other seas of low salinity are essentially freshwater species together with a few tolerant marine species.

Among invertebrates, some species of crabs, for example, *Carcinus mediterraneus*, can be acclimated to low salinities (C. Lucu, 1973). Like other echinoderms, brittlestars are almost invariably intolerant of low salinity; no freshwater echinoderms are known, but a very few species tolerate brackish conditions. In the Baltic Sea *Ophiura albida* encounters salinities as low as 10‰, but it cannot exist east of the Belts (in central Denmark) where the water is almost fresh (Brattstrom, 1941). Thomas (1961) found that the amphiurid brittlestar *Ophiophragmus filograneus* of Florida occupies waters of lower salinity than is usual for echinoderms. This species can tolerate salinities as low as 7.7‰, believed to be a unique low for any echinoderms. The low salinities in the occupied area resulted from freshwater runoff from the Everglades and the slight influence of the tides in the estuarine conditions of the region. These periods of low salinity last for weeks. Apparently *O. filograneus* is able to tolerate these conditions for extended periods, but salinities continued to decrease after 1959, and by 1961 the water had become almost fresh; no echinoderms survived in the sample areas by that date.

Euryhaline animals occur on the Antarctic shelf, where seasonal freezing and melting of ice must involve temporary alterations in the salinity of waters on the upper part of the shelf, and near the margin of the floating ice shelves. Crinoids are certainly present in these waters, and presumably they are adapted to withstand the changes long enough to enable their temporary migration elsewhere.

The oceans can be shown to be stratified into various layers, or water masses, having varying salinities and also varying temperatures, such that any particular water mass can be defined as having a stated *temperature–salinity* (TS) range (see Chapter 10). These defined water masses are believed to originate from steady-flow processes in the sea, by which seawater of, for example, Antarctic origin is conveyed to particular parts of the Atlantic and Pacific. As a result particular species of fishes, which favor particular TS values, are found in specific water masses. The relationship is so pronounced that an ichthyologist studying the contents of a trawled sample of fishes from a given depth can often predict correctly what the TS values of the seawater must be. Facts such as these show that the physical parameters of the oceans are often important in determining the characters of local marine faunas. Plants are even more susceptible to limitations set by salinity, so much so that very few plants can grow in both fresh and saltwater. It is a commonplace observation that seaweed does not occur in lakes, and lake algae do not grow in the sea. A. F. Mohsen et al. have shown a relationship between salinity, growth rate, fat and sugar content, and metabolism in the chlorophyte alga *Ulva* (1972). A. J. Jordan (1972) has shown that decreased salinity is associated with increased branching and vesiculation in *Fucus vesiculosus*.

The origin of sea salt is to be found in the minute quantities of dissolved salts carried down to the seas by rivers. The amount of dissolved salt delivered each year by the world's rivers can be estimated. When the total sea salt is compared with the annual delivery rate, it appears that it would take 100 million years to produce the observed salinity. When this figure was first calculated it was taken to be a measure of the age of the oceans (and hence of the earth). It is now realized that 100 million years is the average time that a salt molecule spends in solution before its constituent atoms are extracted by natural processes, converted into other minerals, and deposited, given the observed conditions. Evidently, therefore, the oceans acquired their present salinity soon after the ocean began to form 4.5 billion years ago, and they must have retained the same salinity, more or less, ever since.

It is often stated that the amount of salt in the blood of land animals is a perpetuated salinity of the oceans at the time when land vertebrates left the sea some 300 million years ago; such ideas are clearly incorrect because the salinity of seawater then and now is much too high to justify this.

When an arm of the ocean is isolated as an inland sea, as may happen as a result of seismic events or the slow buckling of the earth's surface, the isolated sea will either remain a body of water if the stream influx and the precipitation by rainfall equals or exceeds the rate of evaporation, or else it will slowly evaporate if the reverse is true. The beds of salt left behind when an ancient sea has dried up are called **evaporites.** The salts are deposited in layered sequence, according to their solubility, with sodium chloride always most abundant; gypsum (calcium sulfate) is another common evaporite. Ancient evaporites on the floor of the Red Sea are now being heated by the magma below; they are subject to re-solution by the seawater wherever it can penetrate to the beds. Thus more than 30 "hot pools" of highly saline, dense water are now known to exist at the bottom of the Red Sea (Backer and Schoell, 1972).

GEOELECTRIC FIELDS

Seawater contains dissolved salts, ionized particles with electric charges. When these charged particles (ions) are moved by an ocean current through the geomagnetic field, there is an interaction with the field which pulls the positive ions in one direction and the negative ions in the opposite direction. This differential movement of the ions creates geoelectric fields in the sea. The greater the speed of the water current the stronger the generated electric field. In absolute units these fields are extremely weak, less than one millionth of a volt across one centimeter The direction of the electric fields is perpendicular to the direction of the ocean current. Recent experiments (Rommel and McCleave, 1973) have shown that American eels (*Anguilla rostrata*) and Atlantic salmon (*Salmo salar*) can detect electric fields as weak as those generated by ocean currents. Eels and salmon migrate long distances (from 1000 to 9000 km) in order to spawn. During the open-ocean portion of the migrations it is believed the fish are able to detect the direction of these electric fields and use them in navigation. However, it is not known which anatomical structures in the named species detect the electric fields.

DEPTH AND PRESSURE

Pressure does not exert any influence on marine organisms that do not have gaseous components in their bodies; the fluids that permeate the tissues suffer no more compression than that of seawater itself. Consequently, when deep-water species are brought to the surface, provided they have not suffered injury from other causes, they remain active in an aquarium tank at sea level that is maintained at the same temperature as the bottom water.

Most marine invertebrates fall in this category. However, a few, such as the pearly nautilus, have a gas gland that secretes air into the chambers of the buoyant shell. If the animal is moribund, too much gas is secreted, and it then rises to the surface and dies. In the case of some deep-water fishes, many of which have a gas gland in the abdomen (used for adjusting the specific gravity of the body to permit rapid movement), and air-breathing marine animals, all experience serious physiological difficulty if transferred out of their normal depth. Such animals are, of course, distributed in such a way as to conform to their depth requirements or **bathymetric parameters.** A deep-water **stenobathic** species, when brought too rapidly to the surface in a net, suffers disruption of the internal organs, which are forced out of the mouth by the expansion of the gas bladder in the celom. Sharks, which have no gas bladder, are **eurybathic,** passing readily from one depth to another.

Contrary to common belief, most crinoids are shallow-water forms, and the majority are also stenobathic. This is well illustrated by Zenkevitch et al. (1955), who have examined the bathymetric distribution of various marine groups, and by Zenkevitch, Beliaev, Birstein, and Filatova (1959) who analyzed the bathymetric ranges of 615 crinoid species. Of these, only 19 species penetrate to a depth of 3 km, 12 species to 4 km, 4 species to 5 km, 2 species to 8 km, 1 species to 9 km, and none to a depth of 10 km.

This marked bias on the side of shallow-water distribution may be contrasted with similar figures presented by the same authors for the Holothuroidea. Of 1100 holothurian species, 143 species reach a depth of 3 km, 100 reach 4 km, 41 reach 5 km, 19 reach 6 km, 15 reach 7 km, 9 reach 8 km, 3 reach 9 km, 1 reaches a depth of 10.19 km. Thus in the lower abyssal and hadal zones the holothurians (reckoned by species) outnumber crinoids by factors of between 3 and 20; and, when total biomass is considered, Zenkevitch et al. (1955) have shown that in the Kurile-Kamchatka Trench the holothurians comprise more than 90 percent of the total biomass at depths between 9 and 10 km. Using the same sampling procedure, these authors found that crinoids did not constitute a significant part of the abyssal and trench fauna.

Polar species tend to follow their isotherms, becoming progressively more bathyal with increasing distance from the poles. Some species achieve a bipolar distribution in this way, occurring on the Arctic and Antarctic shelves, although restricted to deep water in the intervening oceans. An example is the brittlestar *Toporkovia antarctica* (Fell, 1961; Beliaev and Ivanov, 1961).

Brittlestars as a group exhibit a very wide range of depth tolerance, even though most of the species are stenobathic. Six families range below a depth of 2 miles; and the genera *Ophiura, Amphiophiura,* and *Ophiacantha* range below 4 miles. The shallow-water forms hide among and under shells and other debris or bury their disk in the soft substrate. Deep-water forms lie in or on the bottom, adhere to corals, or adhere to the spines of sea urchins. Of the eurybathic brittlestars, such common littoral species as *Amphipholis squamata* range down to 480 meters in the north Atlantic (Mortensen, 1927) and to 300 fathoms (550 m) in submarine canyons off New Zealand (Fell, 1958), provided other factors such as food availability, oxygen availability, and water movement remain substantially unaltered.

Another factor favoring brittlestars as colonists of deep-sea habitats is their constant secretion of massive calcite skeletal plates. Seawater at high pressures tends to dissolve aragonite (the form of calcium carbonate in mollusk shells) so that few mollusks occur in the deepest zone. Those that do have paper-thin shells. On the other hand, brittlestars are relatively immune, having calcite plates interpenetrated by living tissue that replaces the lime lost in solution. This has not been confirmed experimentally however.

Color and Depth

Deep-water planktonic organisms, such as crustaceans, tend to be red in color, even though they are found at depths where no light penetrates. On the other hand, the most common benthic animals on deep-water seafloors, the ophiuroids, are usually white or pale yellowish. This correlation of color with depth is interpreted as follows. Benthic organisms, being always in darkness in the abyss, suffer no disadvantage by being light-colored because they cannot be seen, and therefore pigmentation is not favored by natural selection. But since the swimming forms are red, a color that becomes invisible in blue-light (the wavelength that penetrates deepest in the sea), it is very probable that such red abyssal plankton rise into the dimly lit **dysphotic** zone from time to time in search of available food and that natural selection has therefore favored the red forms. This inference also helps us to understand the method by which a flow of energy to the lowest levels is maintained, for, after finding food in upper levels, the satiated wanderers return to their normal deep zone.

ILLUMINATION

The amount of suspended insoluble matter in the oceans depends on the distance from the effluence of rivers, which deliver suspended particles carried in their swift currents. It also depends on the degree of disturbance of the sea itself, disturbed waters sweep up silt from the bottom. The relative **turbidity** of the sea affects the penetration of light and, hence, sets a limit to the depth (ca. 200 m, the **euphotic** zone) at which light-demanding plants can grow. In addition some animals, such as crinoids, require clear water, for suspended silt seems to clog their respiratory systems. A measure of turbidity can be made by determining the visibility of standardized sets of spaced lines, or similar devices, suspended in the water and observed from measured distances through the water. An example is the Secchi disk (Fig. 11.3). Photography through turbid water also requires special techniques (Fig. 11.4).

Fishes that frequent the seabed in illuminated depths have an iridescent layer on the cornea. J. N. Lythgoe (1971), who has studied this feature, believes that its function is to reflect extraneous light from above away from the retina, thus permitting a more sensitive reaction by the retina to the weaker light from the seabed itself where food organisms are to be found.

Many shallow-water invertebrates react to strong light by moving away from it (negative phototaxis). Thus reef and rock-pool brittlestars shelter under algae, shells, or stones, within crevices, or in the cavities of sponges. Experiments show that when brittlestars are placed in a tank in strong illumination they congregate in clusters on the side away from the light. Such clustering is seen in shallow-water areas if there is little available cover. In conditions of darkness they emerge and become active. In conditions of permanent darkness, as on the floor of the deep sea, photographs show that ophiuroids are scattered over the surface of the substrate with their disk exposed and their arms fully extended.

Fell (1961) reported that strong light had a narcotic effect on young stages of *Ophiomyxa*; all movement ceased, and death followed protracted exposure. In indirect daylight specimens were active on the floor of an aquarium tank, although most took shelter under stones.

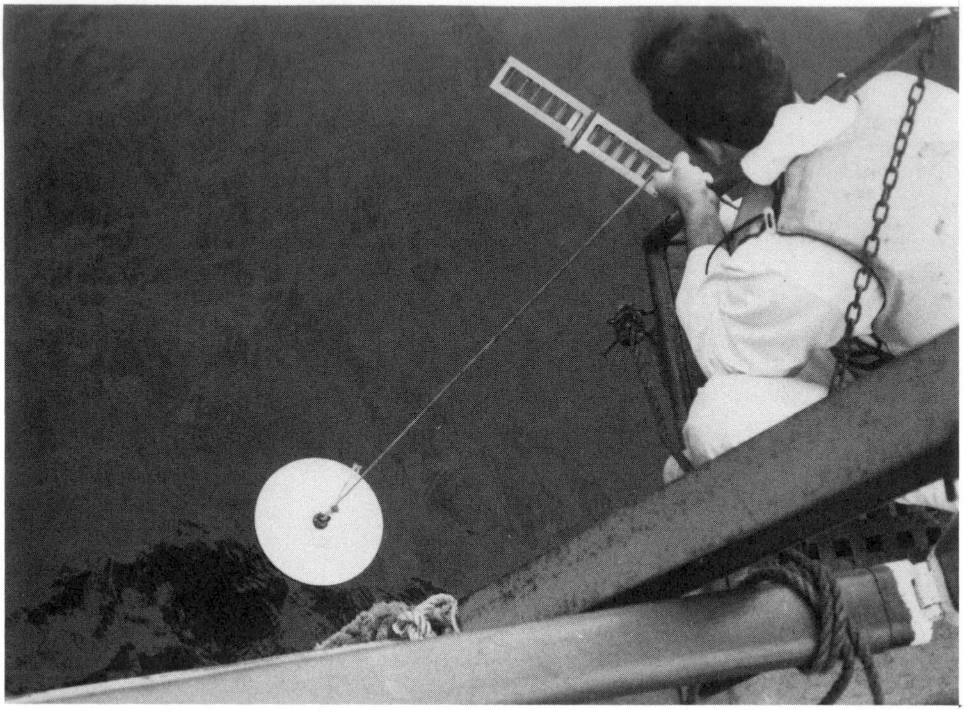

JAMES F. CLARK

Fig. 11.3 Turbidity measurement by means of lowering a Secchi disk until it vanishes from sight, the length of line being an inverse function of the turbidity. The color of the seawater is estimated by comparing the apparent color of the white disk, at a depth of 1 m, with the Forel scale.

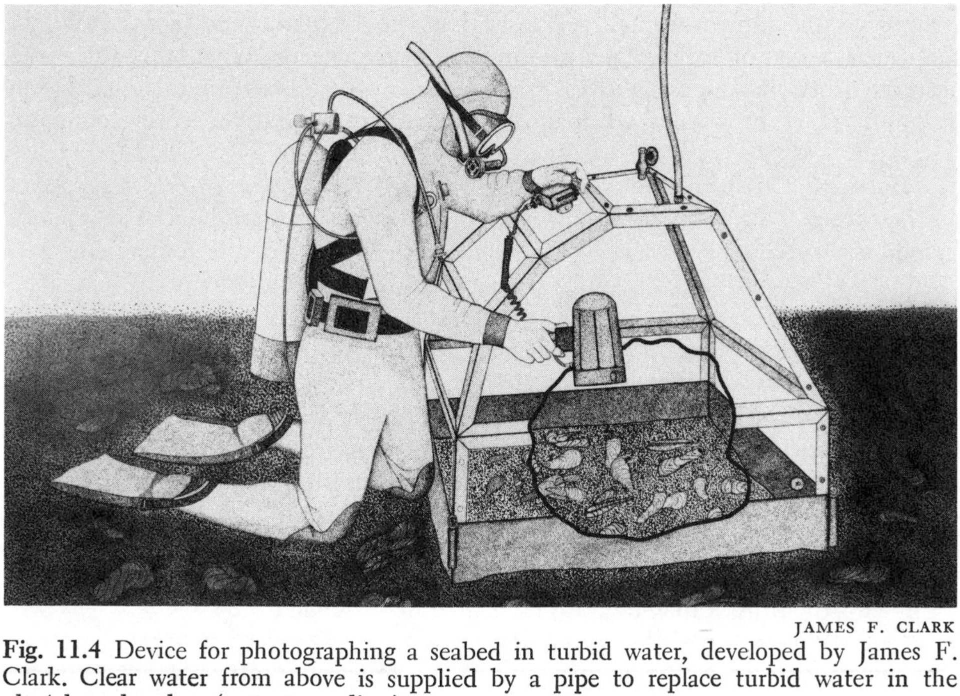

JAMES F. CLARK

Fig. 11.4 Device for photographing a seabed in turbid water, developed by James F. Clark. Clear water from above is supplied by a pipe to replace turbid water in the plexiglass chamber (patent pending).

Although many marine organisms withdraw from strong light, many are attracted by weak light during darkness. Predatory fishes, such as plaice, are attracted at night by phosphorescent prey organisms.

OXYGEN

All organisms require a source of energy, and most obtain it by oxidizing substances such as sugars or fats in their body tissues. Thus oxygen in solution is an important environmental parameter in seawater. In some enclosed seas, such as the Mediterranean and parts of the Caribbean, the deeper levels are deficient in oxygen, resulting in stagnant lifeless zones. The pollution of the sea by man has exacerbated this condition, and it seems likely that seas such as the Mediterranean will become almost totally lifeless unless sewage pollution is brought under control.

Some coastal inlets, partly cut off from the open sea by the development of a spit, may periodically develop anoxic features if the basin is deep enough to be unaffected by the normal tide or river currents. In such cases the anoxic conditions may persist until exceptional circumstances, such as flood tides or seasonal changes in the level of the thermocline, bring about renewal of the water in the deeper zone (J. J. Anderson and A. H. Devol, 1973).

It is believed that anoxic conditions have developed in enclosed seas such as the deep basins of the Caribbean as a result of changes in the circulation of the oceans following climatic change.

POLLUTION

As noted above, the entry of toxic pollutants into the sea produces or exacerbates effects analogous to that of anoxia. Regions become temporarily uninhabitable to all organisms save a few anaerobic forms, such as bacteria. In addition complex side effects, such as oil-induced pneumonia following injury to the plumage, may affect seabirds that have been in oily waters. According to a U.S. House Committee report issued in September 1972, oil spills on the U.S. coast were occurring at a rate of 8000 per annum. Of sea birds affected in one estuary (grebes, scoters, loons), 80 percent were killed despite treatment to remove the oil from the plumage. Some of the dispersant agents employed in the removal of spilled oil cause damage in turn by their own toxicity (E. J. Perkins et al., 1973). Although mineral oil is slowly lost from a polluted water mass by evaporation of the lighter constituents and bacterial degrading of the heavy fractions, the oil is ingested or otherwise absorbed into the body tissues of organisms, surviving or arriving at the edge of the lethal zone. The pollutant may also persist through its absorption into the soft seafloor sediments. Thus studies by Farrington and Quin (1973) demonstrated such persistence of hydrocarbons in the bottom sediments and in the tissues of clams.

Organic mercury has become one of the most widely dispersed toxins. In addition to its well-known occurrence in the tissues of dominant marine predator fishes, such as tuna and swordfish, it occurs in other predators, for example in fish-eating seabirds (Dale et al., 1973) and in harbor seals (*Phoca*) off Maine (D. E. Gaskin et al., 1973). Mercury is also taken up by filter-feeding invertebrates, such as oysters in Australian waters (M. Hussain and E. L. Bleiler, 1973). It has been suggested that mercury be artificially removed from fish tissues before they are marketed or from fish protein concentrates (Regier, 1972).

Other Toxic Chemical Pollutants in the Sea

Lead (Shiber and Ramsay, 1972), cadmium, cobalt, nickel, and zinc are also found in the coastal waters of the northern Atlantic (Windom and Smith, 1972). Eisler and Gardner (1973) report acute toxicity of cadmium, copper, and zinc salts in estuarine fish populations, and cobalt-60 is reported by Kimura and Ichikawa (1972) to accumulate in the tissues of gobies.

Pollutant-induced anoxia of enclosed marine waters results from eutrophication; that is, the overgrowth of plant life following contamination by nitrogenous wastes, with a resultant lack of nocturnal oxygen caused partly by decay of the vegetable material and partly because plants require oxygen at night.

Thermal Pollution

A previously unknown and potentially serious form of pollution has recently been noticed as a consequence of the establishment of atomic power stations. The operation of these stations requires the diversion of cold water from an adjacent river or stream to pass through the reactor as a coolant. In so doing the temperature of the water is raised to such an extent that an artificial tropical environment is created wherever it is discharged. If the heated waters reach the coast, as is often the case, the natural cool-water populations are destroyed or driven away. But that is not all. Passing ships that have recently been in the tropics commonly carry semidormant tropical biota, such as wood-boring shipworms or barnacles. Normally these importations die in the first winter, but if a heated discharge occurs in the area of settlement, the tropical animals survive and introduce new problems. For example, Ruth Turner has found that the destruction of large numbers of wood structures (boats and wharves) on the New Jersey coast is a direct consequence of the establishment of a tropical faunule following the discharge of heated water from a power plant at Oyster Creek (Turner, 1973).

Aerial Mapping Procedures

Infrared sensitized film may be used by aircraft in plotting the extent and distribution of thermal effluents. Dyes released into the effluent as it leaves the source are another means of studying its dispersion.

Interestingly, the development of techniques for following thermally polluted water may lead to advances in fishery procedures. Certain fishes school in warmer water masses, where the plankton may be richer, and the same aerial mapping techniques could facilitate the detection of such schools.

12

THEORY OF POLAR WANDERING

Coral reefs as indicators of the position of the tropics / Fluctuating width of the coral-reef zone through time / Effect of ice ages / Changing inclination of coral-reef belt through time with respect to the existing equator / Polar wandering theory / Hooke's hypothesis / Leonardo da Vinci / Variation of latitude / The Chandler wobble / Inferred Ordovician poles / Strakhov and climate-related rocks / Biological astronomy

Corals are organisms related to sea anemones but differing in secreting a lime skeleton. In the tropics, where corals are abundant in shallow waters on the continental shelf, a special kind of hard substrate develops from the dead coral skeletons. These, accumulating over millions of years, build up massive coral reefs upon which a rich benthos lives. At the present time reefs extend to about 30° north and south of the equator, where the surface water of the sea is never colder than about 18°C. Thus, the reefs form a girdle around the tropics (Fig. 12.1). Fossil reefs are found in some regions where the coast corals are not living. This is taken as evidence of changing climates, which was discussed briefly in Chapter 11. Some further aspects of this topic will now be considered.

Fluctuating Width of the Reef Zone

During the glaciations of the past two million years the overall temperature of the oceans fell, resulting in contraction of the width of the coral-reef belt. This means that the more northern and more southern reefs must have died at that time. They are now living again, which is evidently a result of subsequent recolonization of the dead reefs by living larvae from more equatorially placed reefs. The larval

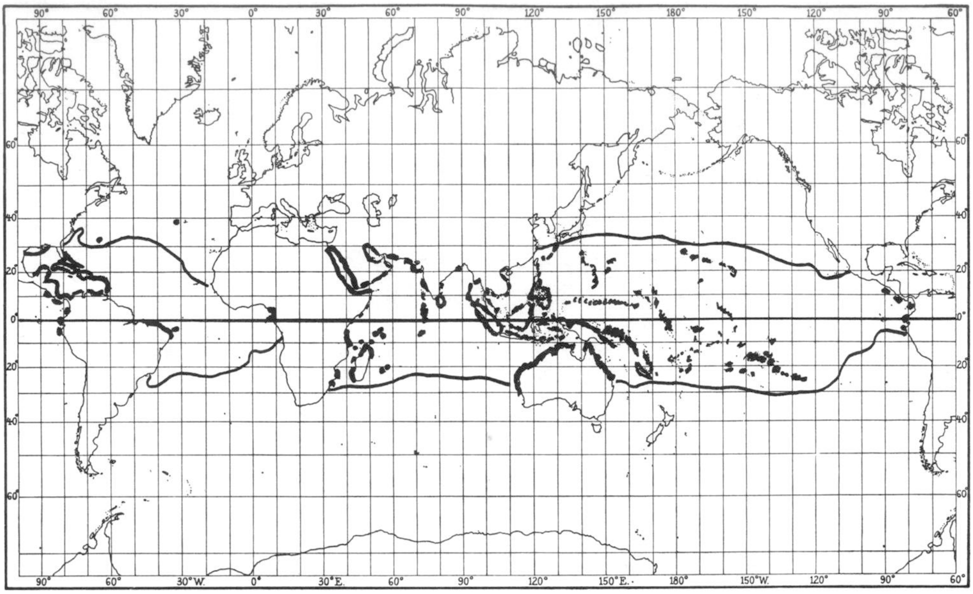

Fig. 12.1 Present distribution of reef corals forming an east–west belt that is nearly symmetrical with respect to the equator. (Fell, 1968)

stages float and can therefore be carried far and wide by ocean currents. There are still some dead submerged reefs along the northern and southern margins of the existing belts; apparently these are old reefs that are not yet reactivated, presumably because the oceans are not yet as warm as they have been in the past. During the Miocene period, about 20 to 25 million years ago, the whole earth was warmer than it is today, and living reefs extended as far south as New Zealand, in 42°S latitude. These data, and other similar distribution data on terrestrial plants and animals, tell us of periodic warming and cooling of the earth, when tropical animals sometimes spread far beyond the limits of the present tropics.

Polar Wandering Theory

If we carry the investigation further back in geological time, a different kind of variation becomes discernible. Instead of simple north and south fluctuations in distribution, we find that the east–west parallelism varies too. In the Cretaceous period, for example, some 100 million years ago, the coral-reef belt of North America and of Europe extended northward to about 50° north latitude, and cold-blooded animals, such as crocodiles, also ranged much further north than today. Yet in the southern part of the Western Hemisphere the coral-reef belt seems to have extended only to about 20° present south latitude. Therefore the belt of coral reefs seems to have been about as wide as it is now, only displaced toward the north. On the opposite side of the world, the reverse seems to have been the case, the displacement being toward the south.

Many investigators believe that facts such as these imply that the earth's axis of rotation then lay in a different position from its present one, the North Pole lying on the north coast of Siberia and the South Pole nearer the Palmer Peninsula, south of the Magellan Straits. If facts like these are interpretated in a similar manner for still earlier geological periods (Fig. 12.2) we find, for example, that in

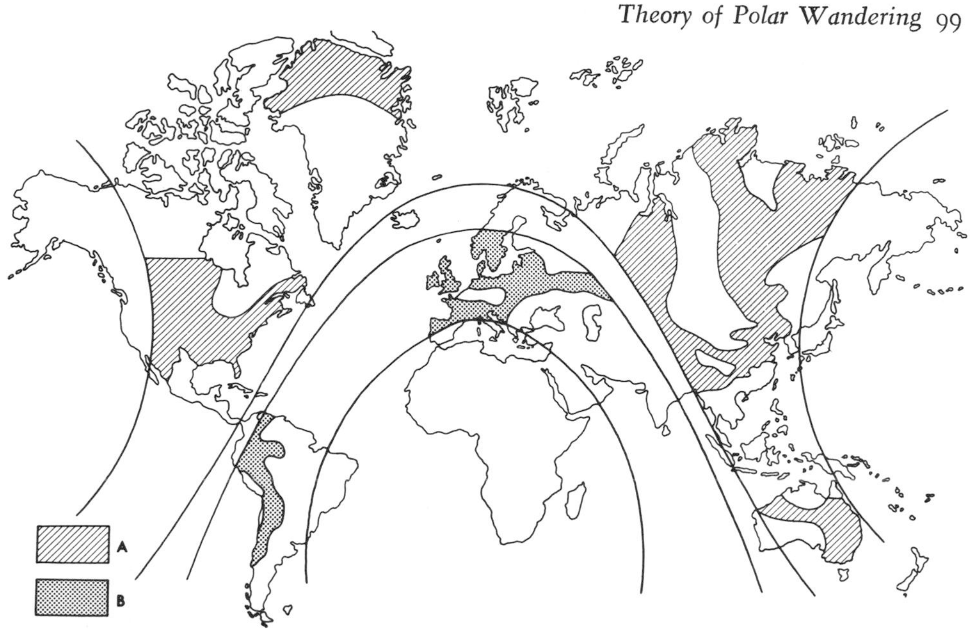

Fig. 12.2 Apparent distribution of Ordovician marine fossils suggesting a former tropical belt comprising a northern (A) and a southern (B) assemblage. (Fell, 1968)

Ordovician times, some 500 million years ago, the corals seem to form a belt tilted so steeply with respect to the present belt as to girdle the earth by swinging along the east coast of North America, passing through Greenland and northwestern Europe, then south through India to reach Australia. This suggests to me a North Pole near Hawaii and a South Pole off West Africa! (See Fig. 12.3.) Geologists, such as Strakhov in Russia, find that desert sandstones and other climate-related

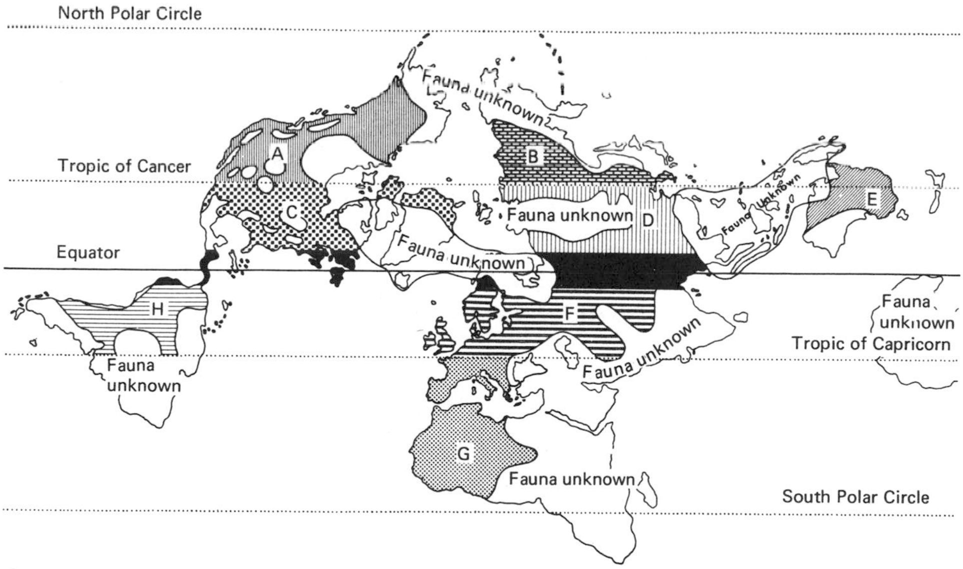

Fig. 12.3 Interpretation of Fig. 12.2 by conversion of geographical coordinates to suit an inferred North Pole near Hawaii and South Pole off West Africa (Fell, 1968). A, B, supposed northern faunas; C, D, E, F, H, supposed tropical faunas; G, supposed southern fauna.

rocks also seem to girdle the earth in positions that suggest poles in the positions indicated by the corals. Unfortunately we lack data from much of the Pacific (where no large land surfaces occur to yield fossil reefs of such great age). Interestingly, the idea that the earth's poles may have changed their positions in the course of time is by no means new, and a truly scientific exposition of the matter was given by Robert Hooke nearly three centuries ago.

Hooke's Hypothesis

After the Great Fire of London in 1666, Sir Christopher Wren and Robert Hooke were commissioned by Charles II to design and build the major architectural monuments of London. One of Hooke's responsibilities was to inspect the stones sent from the Portland quarries for the construction of Saint Paul's Cathedral. He records that one day his attention was drawn to the impression on one of the blocks of stone of what looked like a giant sea snail. He traveled to Portland, and when the quarrymen showed him their exposed rock face Hooke soon realized that the rock was indeed the hardened remains of a former seabed, with the remains of sea animals imbedded in it as fossils dating from a time when the rock was soft sediment.

Further inquiries among the quarrymen brought to light bones that Hooke recognized as similar to those of turtles and crocodiles in the collection of tropical specimens of the Royal Society of London. In a paper that he read to the Royal Society, Hooke theorized that England had ". . . at a certain time for Ages past lain within the Torrid Zone . . . I think it not improbable that there may be some such motion of the Earth's Axis as may alter the Latitude of Places. . . ."

At a later meeting of the Society in 1696, Hooke set up a heavy sphere of lignum vitae suspended by a cord. Spinning it, all present could see the vertical axis of rotation. Hooke then attached an iron weight to one side of the sphere and set it spinning again. He was able to demonstrate that the change in the center of gravity of the globe now caused its axis of rotation to change its position in the direction of the weight; the rotating ball wobbled around its new axis instead of spinning smoothly as before.

Leonardo da Vinci

Now, Leonardo da Vinci (who did not publish his speculations) had entered in his notebooks a century before Hooke's experiment an observation to the effect that rivers constantly sweep sediment from land to sea, and that therefore the position of the center of gravity of the earth must change with time. A similar idea must have occurred to Hooke, but, unfortunately, Sir Isaac Newton resolutely declined to believe that the axis of the earth might move. So Hooke's ideas were forgotten until 1830, when the British geologist Lyell drew attention to them.

Variation of Latitude

In 1885 Seth Chandler of Harvard Observatory noticed that some measurements of the altitudes of stars were discrepant with other measurements he had made some months earlier. On examining the matter he eventually discovered that all stars vary slightly in their apparent positions, the variation being such as to imply that the earth slips about or wobbles upon its axis of rotation. In fact, the latitude of Cambridge was found to be varying, and hence the apparent altitude of stars varies. Observations repeated in other parts of the world eventually showed that when the apparent altitude of a star is at a maximum on one side of the world, it is at a minimum on the other side, and so Chandler's interpretation was proved.

The Chandler Wobble

Chandler went on to discover that the variation is complex and is made up of the summation of two variations: One variation is annual, and apparently reflects changes in the distribution of air and water masses as the Northern and Southern Hemispheres exchange water and air as a consequence of ice melt in their respective summers; the other variation had a period of 14 months, and Kelvin showed that this is the natural vibration period of a body the size of the earth if it has the elasticity of steel (most of the earth is composed of a steel alloy of iron and nickel). The lowest common multiple of 12 and 14 is 84, which means that every 84 months, or 7 years, the two variations reinforce each other, with intervening periods when they counteract each other. This effect, known as the **Chandler wobble**, causes the earth's poles to move irregularly over an area about the size of a football field.

However, taking into account Leonardo's observation and also recent measurements of the effect of major earthquakes upon the earth's rotation, it becomes apparent that after a long lapse of time the slowly moving center of gravity of the earth is bound to cause the poles to wander outside the area they presently traverse. After a very long span of time such wandering can become highly significant.

What seems to be happening is that the earth's axis itself remains fixed at an angle of 23½° to the plane of the earth's revolution about the sun, but the **whole body of the earth** is slowly slipping around its fixed axis. For an observer on the earth this must mean that the poles appear to move across the surface of the planet over a long span of time.

If the inferences as to the position of the poles are correct for the Ordovician period, then apparently during the 500 million years since the Ordovician, the North Pole has moved from Hawaii northwest (by modern reckoning) across the Pacific to pass through Kamchatka about 300 million years ago, and thence on to the north coast of Siberia by the Cretaceous period, and after that across the Arctic Ocean to assume its present position. The future motion of the North Pole will take it into Greenland and then southward toward the northeastern part of the United States, although that would not occur till some hundreds of millions of years after now.

Biological Astronomy

The discussion above shows that marine organisms can throw light on astronomical problems, a topic that is also discussed in Chapter 29. There seems that the fossil coral-reef belts of our planet may ultimately prove to be reliable datum zones for other geophysical and astronomical measurements. This subject, however, is still in initial stages of study.

13

PLANKTON

Phytoplankton / Zooplankton / Remains of plankton in deep-sea sediments / Biocoenoses / Thanatocoenoses / The plankton net / Decreasing concentration of plankton with increasing depth / The *Valdivia* expedition / Application of scuba diving to plankton study / Plankton collection from stomachs of deep-sea fishes / The scattering layers / Food chains / The trophic pyramid / Marine pollution / Diatoms / Spring diatom increase (SDI) / Coccoliths / Dinoflagellates / Red tides / Toxicity of some dinoflagellates / Sargasso weed / Epiplanktonic dispersion / Zooplankton classification / Foraminifera / Handedness of forams as climatic indicator / Radiolaria / The Swedish *Albatross* expedition / Invasion of tropical Atlantic by cool-water plankton during glacial stages / Siphonophorida / Scyphozoa / Ctenophora / Polychaeta / Copepoda / Mysidacea / Hyperiidea / *Phronima* / Euphausiacea / Larval stages / Ianthinidae / Heteropoda / Pteropoda / Chaetognatha / Salpida / Pyrosomatida / Fish larvae / Energy transfers in the sea

Plankton comprises all floating organisms. Marine plankton occurs at all depths in all oceans. It consists of **phytoplankton** (floating plants) and **zooplankton** (floating animals). The phytoplankton is mainly concentrated near the surface of the sea, within the first hundred fathoms, where the sunlight is strongest; it includes various kinds of unicellular minute plants, especially those called diatoms, together with some larger seaweeds (algae) equipped with flotation organs. The animals of the plankton, on the other hand, occupy all depths and, instead of depending upon flotation organs alone, are commonly provided with muscular swimming organs. The planktonic animals are very varied in structure and relationships, and most of them are unfamiliar to the average reader.

When a planktonic organism dies its body begins to sink. On the way down it may be eaten by a deep-water animal, but the skeleton or shell is likely to reach the seabed eventually. Thus samples of the seafloor, including the floor of the deepest oceans, yield sediments that invariably contain a high proportion of remains of planktonic organisms that formerly occupied the water mass of the column above the seabed. The shells of surface-living organisms such as forams or diatoms may, therefore, constitute a large part of the sediment on the bottom, where other kinds of forams also live and die. Seafloor sediments, therefore, exhibit a mixture of species of diverse origin that do not mix when they are alive. In studying seafloor sediments we have to distinguish **biocoenoses**, true communities of organisms associated when alive, from **thanatocoenoses**, fortuitously associated remains of organisms that did not necessarily live in association.

Plankton is usually collected by towing a fine-meshed net through the sea (Fig. 13.1). Zooplankton commonly occupies the surface waters during the night, sinking to greater depths during the daylight hours; the richest hauls made in darkness are made if the net is operated at the surface. During the day rich hauls

JAMES F. CLARK

Fig. 13.1 One-meter plankton net being lowered from hydro platform.

can generally be made if the net is weighted to operate at a depth of 35 to 50 m (20 to 30 fathoms) or some greater depth. The plankton net is usually towed at speeds of about 1 knot (at speeds of over 1.5 knots the water tends to spill out instead of passing through the mesh).

Plankton becomes progressively scarcer with increasing depth, so much larger plankton nets must be used in order to filter larger volumes of water. A large plankton net, such as may be employed by an ocean-going research vessel, may measure up to about 5 m in diameter. The techniques of sampling deep-water plankton were developed in the closing years of the nineteenth century by the German *Valdivia* expedition, which brought to light many kinds of planktonic animals previously unknown. One of the outstanding discoveries which followed the development of these techniques was that made by a Danish expedition of 1908–1909, led by Johannes Schmidt; namely, that common freshwater eels descend into the depth of the ocean in their breeding season, for the youngest larval stages of eels were found only in such locations. To make a deep-water plankton sample requires about 2 hours' towing, plus several additional hours for the lowering and raising of the net.

Scuba Diving for Plankton Study

A new method of studying living plankton was developed in 1973 by William Hanmer, working in the waters of Bahamas. Noting that scuba divers frequently experience disorientation and nausea if deprived of familiar landmarks, such as the visible seabed beneath them (a disability that has prevented the use of this diving technique in the open ocean), a compensatory reference landmark was found in a system of floating pipes. The pipes are let down into the sea to provide divers with a means of visual orientation when the floor of the sea is too deep to be seen. A safety man is placed in the middle of the system, to keep watch for sharks and to keep untangled a set of safety lines by which each of the free divers is kept in physical connection with the pipe mesh. The system is applicable to clear ocean environments where the visibility distance under water may be so great as 150 feet. Using this system, Hanmer and his associates have initiated on-site study of the living activities of planktonic organisms, such as the tunicates, observing their manner of trapping and capturing food and other features of their life history that previously was not understood.

Certain animals habitually feed on plankton, so examination of their stomach contents often yields interesting results, although we cannot of course tell at what depth the sample originated. Oceanic birds, such as petrels, feed at night on deep-water plankton that rises to the surface. Small pelagic fish, such as herring and some larger species, such as mackerel, often feed on plankton near or at the surface. Deeper down, fishes such as *Alepisaurus* feed on deep-water plankton, and their stomach contents may be exceedingly varied and informative. Figure 13.2 illustrates a food chain based on plankton.

The Scattering Layers

When echo-sounding by means of some equipment was introduced for bathymetric work (see Chapter 12), the observations were sometimes found to include reflections from mysterious layers in midwater. Later work has established the

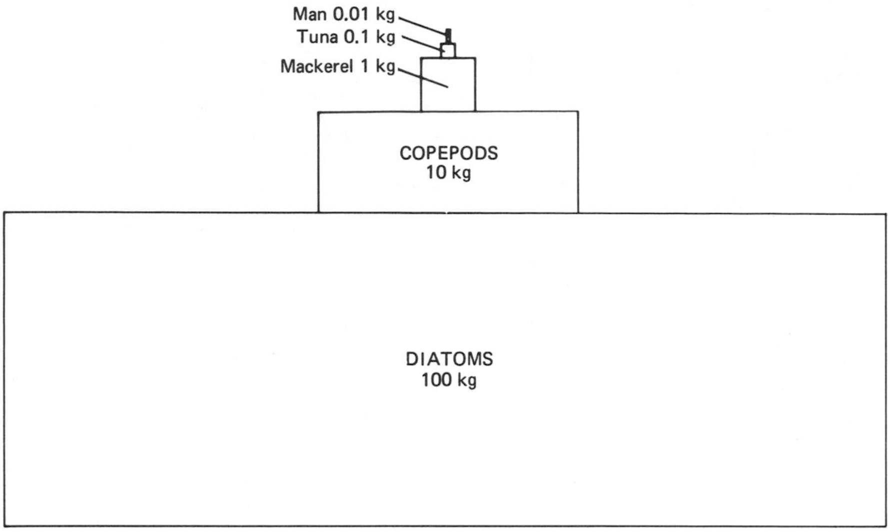

Fig. 13.2 Trophic pyramid for a dominant predator such as tuna. Because it takes about 100 kg of diatoms to produce 0.1 kg of tuna flesh and the same amount of diatoms produces 1 kg of mackerel, a more productive harvest would be that concentrating on mackerel. Mercury and other toxins not excreted by the kidneys become concentrated in the higher tiers of the pyramid, so tuna flesh can be dangerous for man to consume.

cause of the scattering of the sound waves to be shoals of planktonic organisms in some instances and schools of fish in others. The level at which the reflections are produced varies with the time of day, thus showing that the organisms concerned rise and sink in the course of 24 hours. Several layers are sometimes present, the shallow and deep layers produced by different species of organisms.

All marine food chains are based ultimately on primary producers (autotrophs), nearly all of which form a part of the phytoplankton. Even in the deepest parts of the ocean the bottom-dwelling organisms depend indirectly upon the plankton, for their only source of energy lies in the input of materials derived from the sunlit layers far above. The small particles that sink to the seafloor are utilized by bottom-dwelling or bottom-visiting animals. However, some food chains are located exclusively in the upper layers of the open ocean, and these chains involve plankton elements in all but the final one or two stages (Fig. 13.3).

Food Chains

Diatoms and other minute floating plants are eaten by somewhat larger floating organisms called copepods (remotely related to shrimps). These in turn are eaten by mackerel, and that fish species is commonly eaten by tuna, which in turn may be fished and eaten by man. Any such sequence is termed a food chain. Measurements show that it commonly takes about 10 times the weight of mackerel to produce the equivalent weight of tuna flesh, and this ratio holds for other parts of the food chain. So all the combined weight of a food chain makes a pyramid, in which the broad base is made up of the large amount of initial phytoplankton, and the apex is the relatively small weight of resultant tuna flesh (or human flesh). Pollutant substances such as DDT or mercury are retained in the tissues of organisms. Thus each level in the food chain, or **trophic pyramid,** has an in-

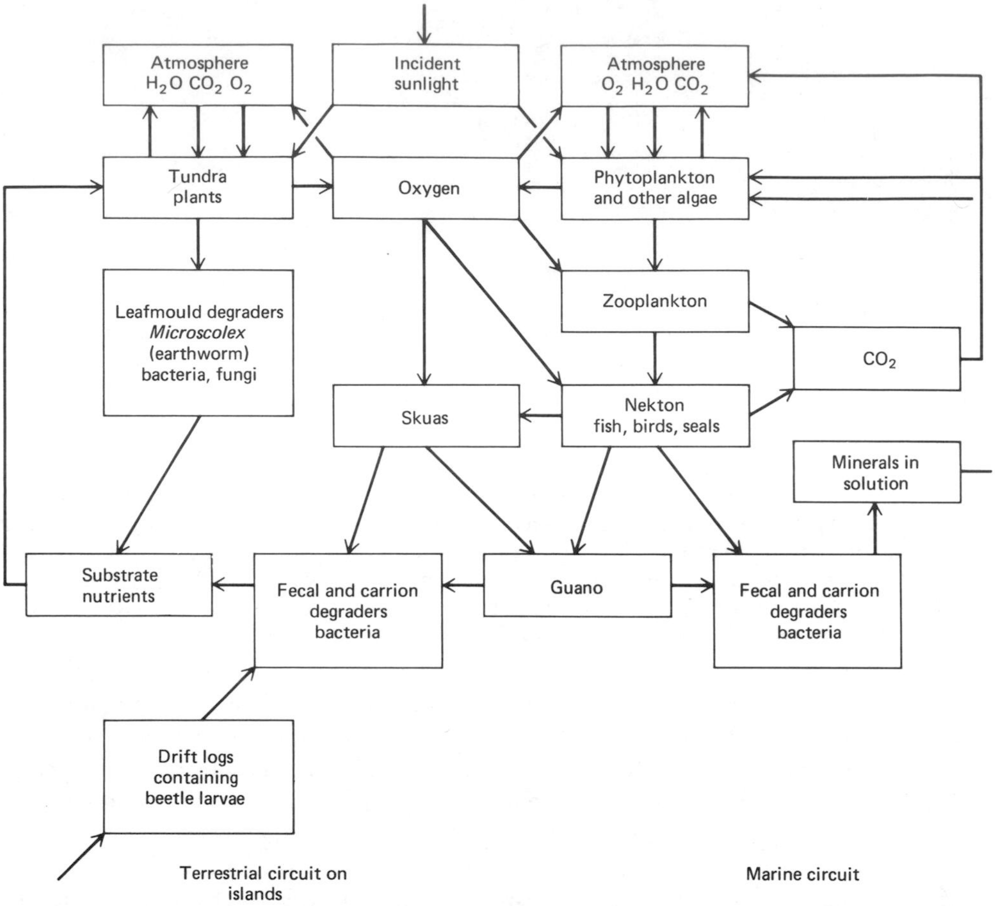

Fig. 13.3 Example of a food chain in the Antarctic Ocean. Plankton forms the basis of the marine cycle shown to the right of the diagram. Tundra plants on oceanic islands (such as Macquarie Island) are the main base of the terrestrial cycle, shown on the left. Grazing animals are lacking on the land. Guano, derived ultimately from the plankton, contributes to land fertility.

creased concentration of toxic pollutants, the increase factor being about 10 times for each stage. This is why tuna may be dangerous for human consumption in the present polluted state of the oceans. Table 13.1 sets out the classification of phytoplankton.

PHYTOPLANKTON

Diatoms are single-celled plants that utilize a small part of the incident solar radiation energy to produce carbohydrates and oils. The synthetic process is mediated by chlorophyll, a pigment contained in small bodies in the diatom called plastids. The plastids also contain a yellow pigment, xanthophyll, and carotenes. Characteristic is the two-valved siliceous capsule that a diatom secretes about its cell body. Many diatoms form chains. Marine diatoms are the most abundant life on earth, exceeding in mass all other living organisms in the sea and on land.

Table 13.1 Key to the Phyla of Marine Phytoplankton

DIAGNOSTIC CHARACTERS	PHYLUM
Minute floating one-celled plants showing detail only under a microscope	
Nonmotile cells, lacking flagella, each with a 2-valved siliceous capsule, showing detail under ×50	Chrysophyta
Motile cells, with 1 or 2 flagella, not aggregating in chains, usually with cellulose epithecal plates, showing detail under ×500	Pyrrophyta
Macroscopic seaweeds showing detail to the naked eye	
Floating brown seaweed buoyed by air-filled bladders or attached to floating logs	Phaeophyta
Green seaweed epiphytic on floating brown seaweed or attached to floating logs	Chlorophyta
Red or purple seaweed epiphytic on floating brown seaweed or attached to floating logs	Rhodophyta

Spring Diatom Increase (SDI)

Diatoms of temperate seas bloom in spring, when concentration may reach 0.5 gm per liter of sea water, or 0.5 percent, imparting a pale golden-brown tint to the surface. Diatoms are classified as a phylum of algae named Chrysophyta, defined by the characters noted in Table 13.1. They form part of the plant subkingdom Thallophyta, defined as plants without roots, leaves, or vascular tissue. The term algae includes those phyla of Thallophyta that posses the pigment chlorophyll.

Coccoliths are a second group of the phylum Chrysophyta found in large numbers at the surface of the sea (Fig. 13.4). They are symmetrical single-celled algae with lime platelets in the cell wall; the platelets are usually cemented together so they remain intact after the plant dies and sinks to the seafloor. The skeleton is remarkably symmetrical, although the whole organism seldom exceeds a few thousandths of a millimeter in diameter. They are found in association with *Globigerina,* and on the seafloor about one third of the so-called globigerina ooze is actually made up of the remains of coccoliths.

Dinoflagellates are single-celled plants also, but they differ from diatoms in several respects. (See Fig. 13.5.) They are motile, swimming by means of one or two contractile threads called flagella. They lack the siliceous capsule and have instead one or more cellulose plates on the outer surface of their cell body; some kinds have a naked cell body. Dinoflagellates are much smaller than diatoms and can only be seen distinctly under the higher powers of a microscope. They have some brown pigments accompanying the chlorophyll and occasionally a red pigment. Enclosed seas where dinoflagellates bloom periodically include the Red Sea and the Vermilion Sea in California, both taking their names from the red dinoflagellate blooms. Dinoflagellates are probably much more ancient than diatoms. Diatoms did not become abundant in the earth's oceans until the Cretaceous period, about 100 million years ago. Probably, therefore, earlier food chains may have depended upon dinoflagellates as primary autotrophs.

The name **red tide** is often applied to patches of sea in which the concentration of dinoflagellates is temporarily increased following a bloom. Dinoflagellates are sometimes very toxic to fishes, and mass mortality of fishes is a cyclic and little-understood phenomenon connected with dinoflagellate ecology. Dinoflagellates constitute the phylum Pyrrophyta in the subkingdom Thallophyta.

Epiplanktonic Dispersion

Some of the massive marine algae, or **seaweeds**, are phytoplankton. In the tropical Atlantic, a bottom-giving brown seaweed named *Sargassum* (Fig. 13.6) breaks loose from the seabed and floats with the Gulf Stream. This Sargasso weed was discovered by Columbus in the mid-Atlantic in 1492. In the Southern Ocean, and to a lesser extent in the northern Pacific, large bottom-growing brown kelps break loose in storms and float away from the coast by means of their air bladders, borne on the current. They remain alive for long periods and drift thousands of miles on the open sea. They often carry a whole miniaturized ecosystem of on-board passengers, underswum by a cloud of kelp fishes. In all oceans objects such as logs drift to sea and float for great distances with the ocean currents. Such logs may carry growing kelp and other seaweeds. The seaweeds in turn may carry a variety of epiphytes, both plant and animal. Benthic animals, such as sea urchins and starfishes, have been taken in situations pointing very strongly to the probability that they have crossed ocean gaps as passengers on board such rafts (Fig. 13.7).

ZOOPLANKTON

Zooplankton is much more varied than phytoplankton, for almost every phylum of animals is represented in the marine plankton. Table 13.2 gives the classification of the main groups of zooplankton.

Foraminifera

Forams are marine protozoa characterized by their external calcareous, chambered shell. The single-celled animal that secretes the shell resembles an amoeba with more numerous and more slender pseudopodia (called *filopodia*). The filopodia are extended through small holes (*foramina*) in the walls of the chambers, and they serve to capture diatoms as food. Most foraminifera live on the bottom of the sea

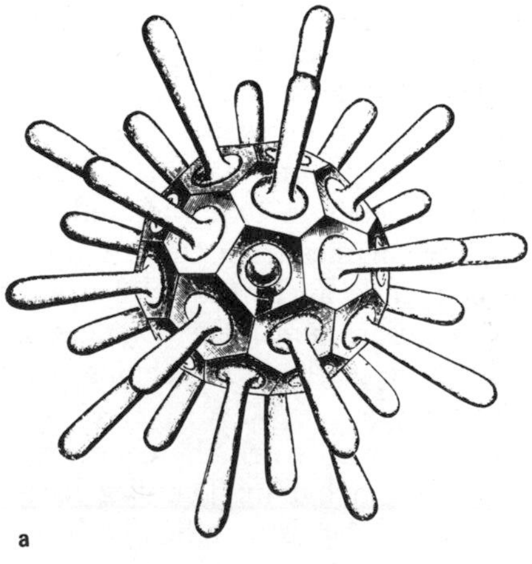

a

b

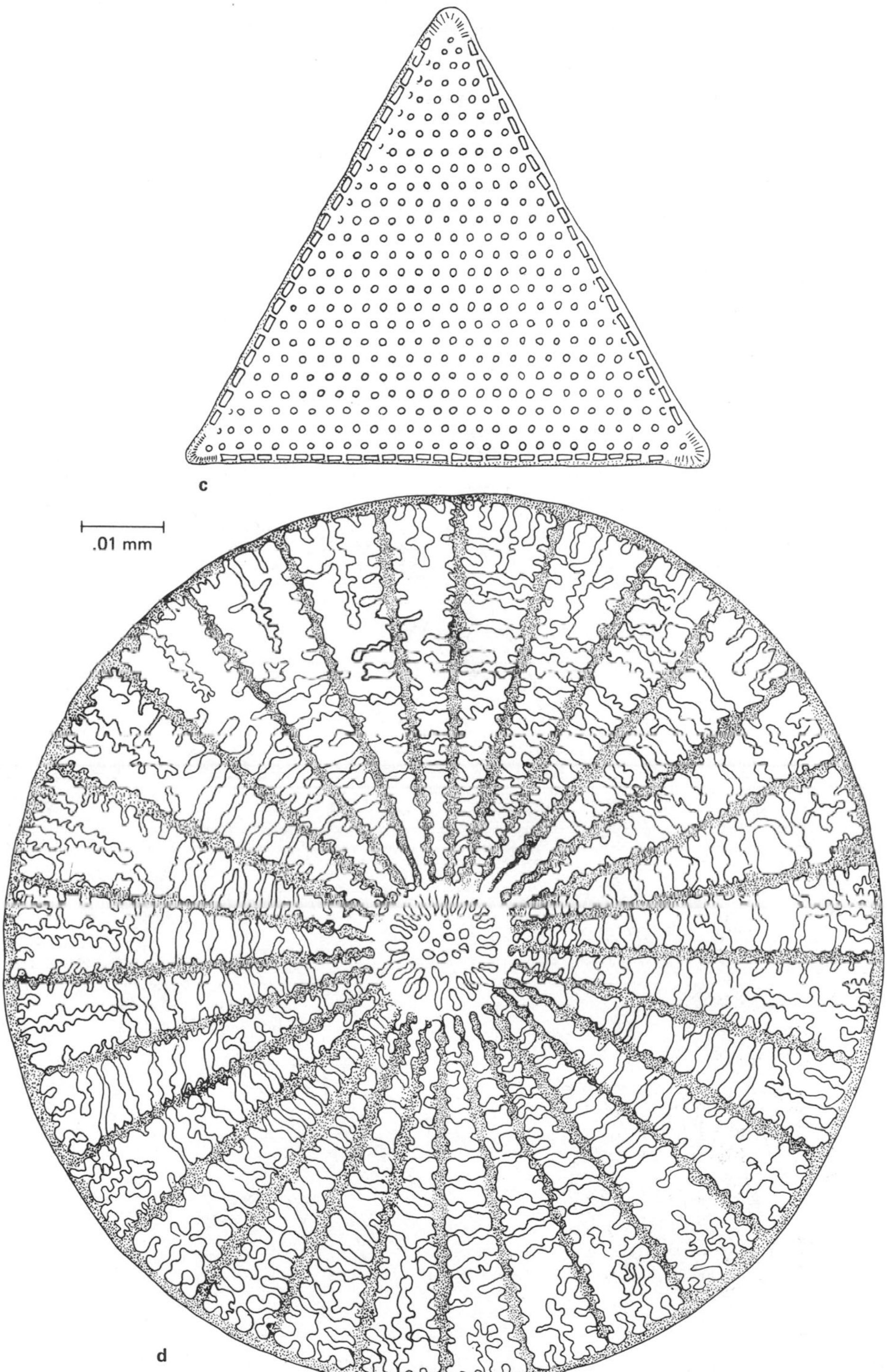

Fig. 13.4 [Opposite and above] Examples of Chrysophyta in the plankton. (a) *Rhabdosphaera* and (b) *Discosphaera*; both coccoliths, ×500. (c) *Triceratium*, ×100, and (d) *Arachnodiscus*, ×50; both diatoms.

and thus belong to the benthos; however, about 2 percent of forams are planktonic. The planktonic forams belong to two families:

1. Globigerinidae, in which the capsule looks like a collection of small spheres of various sizes stuck together. *Globigerina* is the best known genus and also the common planktonic genus of forams.
2. Globorotaliidae, with the type genus *Globorotalia*, in which the chambers of the shell are arranged in an expanding flat spiral.

When surface-dwelling forams die they sink to the ocean floor, where their lime skeletons accumulate as the so-called globigerina ooze of the deep sea.

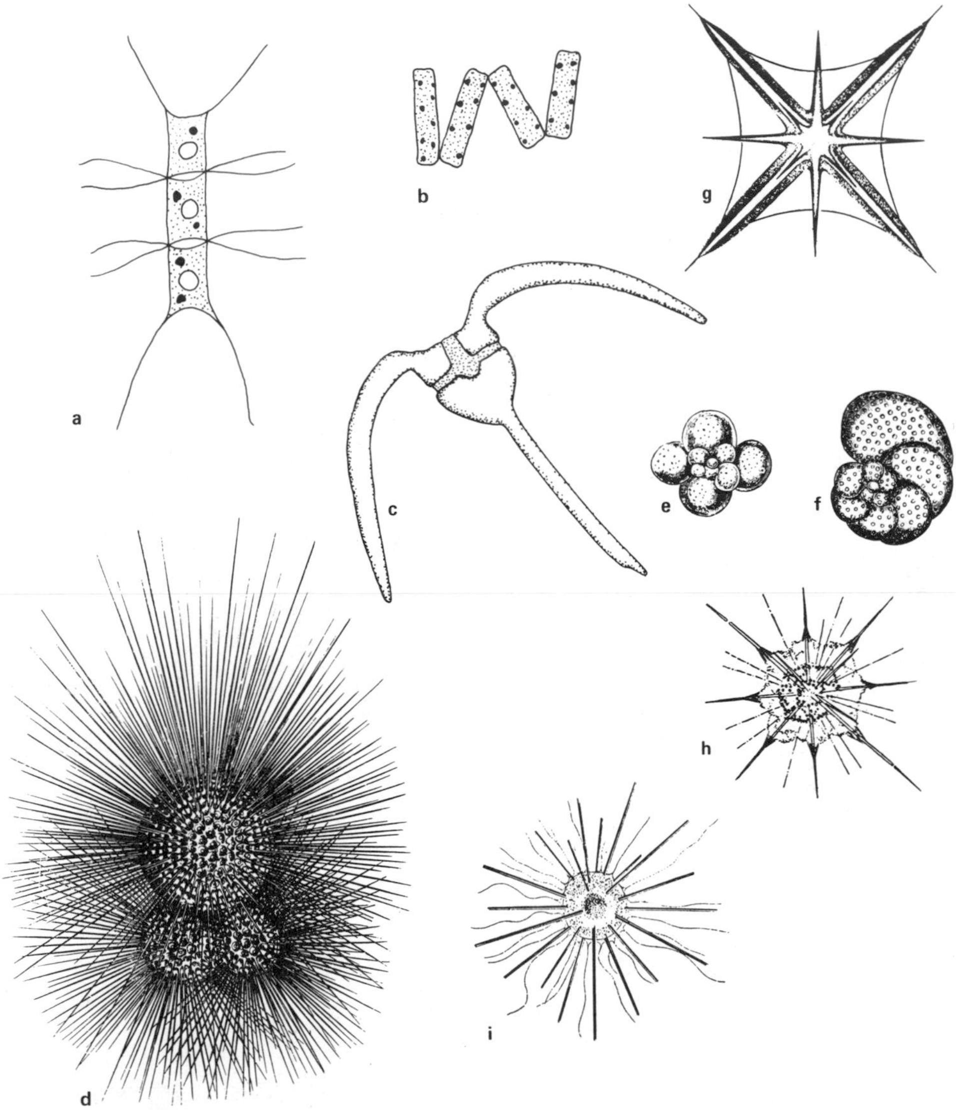

Fig. 13.5 Plankton. (a) *Chaetoceras,* ×200, and (b) *Thalassiothrix,* ×400; both diatoms of the open sea. (c) *Ceratium,* ×250, a dinoflagellate. (d) *Globigerina* with filopodia extended, ×10. (e) *Globigerina* and (f) *Globorotalia,* shells; both forams of the open sea. (g) *Acanthostaurus,* (h) *Acanthometron,* (i) *Acanthonidium*; all radiolarians of the open sea, ×200.

Table 13.2 Key to the Phyla of Zooplankton

DIAGNOSTIC CHARACTERS	PHYLUM
Microscopic animals, usually transparent, lacking chlorophyll	
Body one-celled, with an external capsule of lime or silica	Protozoa
Body multicellular, with no capsule (larval forms, see p. 119)	
Animals at least 3 mm long, often much larger, easily visible to naked eye	
Body circular, saucer-shaped or cup-shaped or conical, with a ring of tentacles around the margin, or with a hollow float from which hang long stinging tentacles	Coelenterata
Body globular or bell-shaped with 8 meridianlike bands	Ctenophora
Body elongate, slender, wormlike	
Body clearly segmented, with fleshy fins on each segment	Annelida
Body unsegmented, with paired horizontal finlike flanges	Chaetopoda
Body sluglike, often with a lime shell, sometimes with a pair of fleshy winglike fins, often brightly colored	Mollusca
Body shrimplike, often red, with paired jointed limbs	Arthropoda
Body transparent, barrel-shaped and open widely at either end, or torpedo-shaped (sometimes adhering in chains); or resembling a flannel ice cream cone; or tadpolelike	Chordata

Shape of Forams as Climatic Indicator. It is a remarkable fact that there is a link between environmental temperature and the direction in which the shell of members of the Globorotaliidae is found to coil. Analogous differences also occur between species of the Globigerinidae.

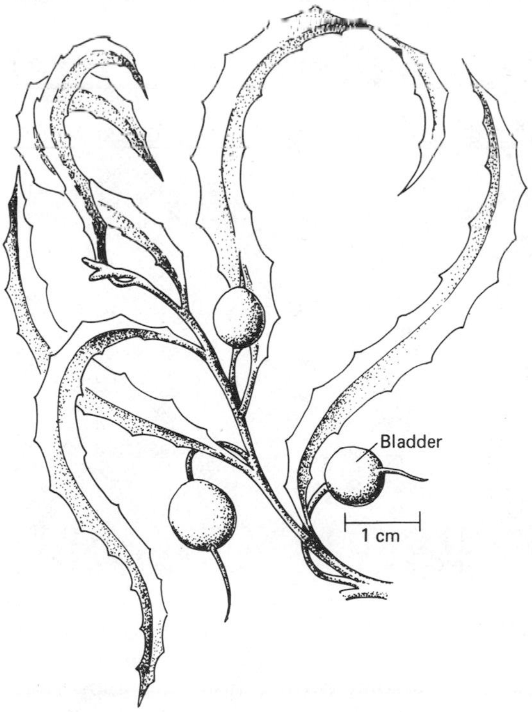

Fig. 13.6 Piece of Sargasso weed, *Sargassum*, natural size.

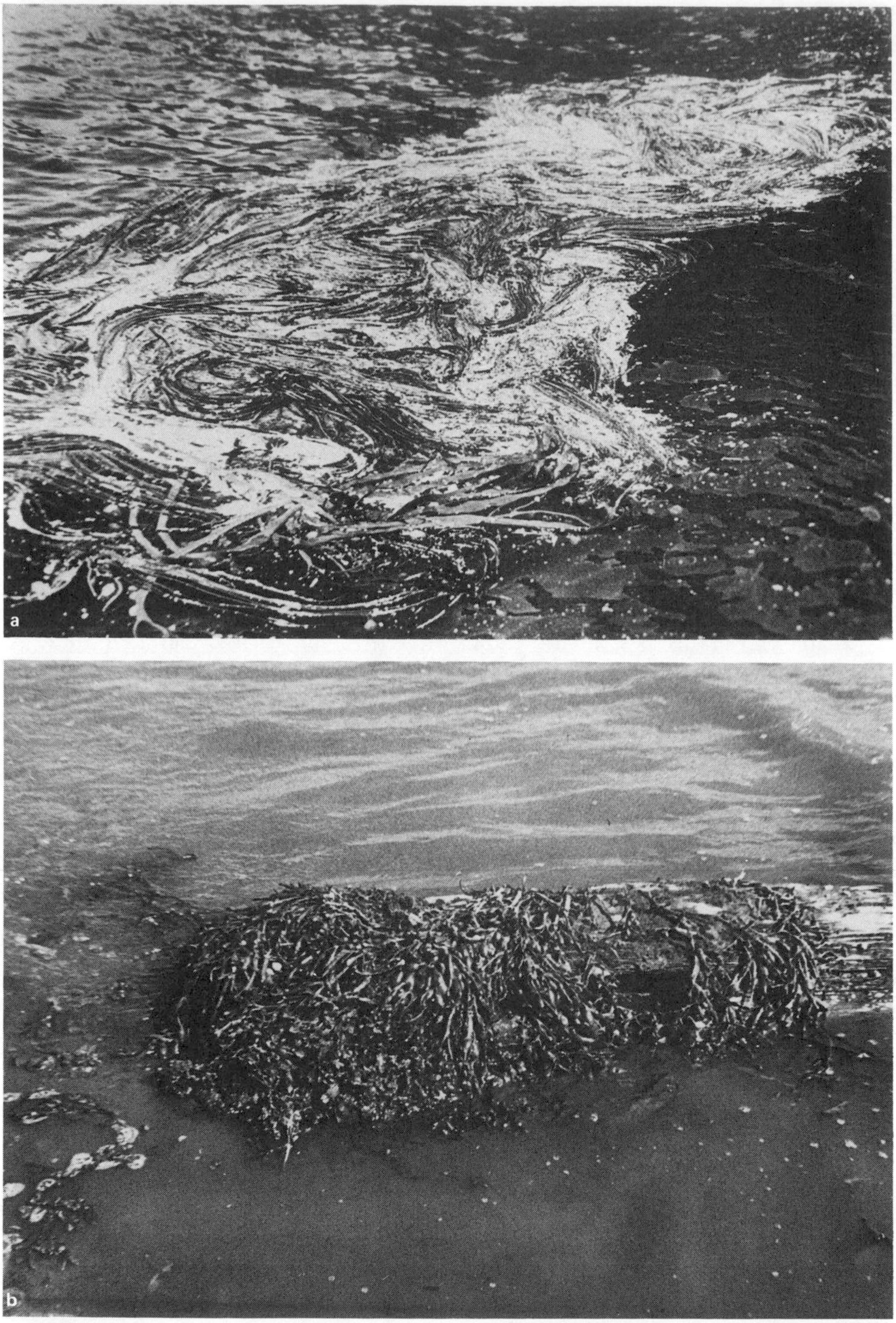

(a) ALAN BAKER (b) PETER GARFALL

Fig. 13.7 (a) Natural raft of giant kelp *Durvillea* floating in the west-wind drift of the ocean, 20 miles northwest of New Zealand. Rafts of similar type, when examined, have been found to carry a considerable onboard population derived from the upper continental shelf at the place where the algae were originally anchored to the seabed. (b) Stranded drift log on the Maine coast carrying a rich flora and fauna.

If you hold a sea snail so that the pointed spire of the spiral shell is directed away from you and the aperture of the shell is directed upward, then the aperture is almost invariably found to lie on the right side of the shell. Such specimens are said to be right-handed, or **dextral;** and the coiling direction, when viewed from the spire end, is clockwise. Very few specimens can be found where the aperture lies on the left side, and the coiling direction runs counterclockwise; such rare specimens are said to be left-handed, or **sinistral.** A few species of sea snail are known where the above rules are reversed. An example is the west Atlantic whelk, *Busycon contrarium,* in which nearly all individuals are left-handed. A sinistral individual has all the organs of the body arranged like a mirror image of his dextral equivalent; chance reversals are believed to arise when the first cell division of an embryo occurs at right angles to the usual plane of division, although it is not known why this happens. On the analogy that a left sole becomes a right sole if turned over through three-dimensional space, some puckish mathematician has suggested that a three-dimensional dextral animal gives rise to a sinistral one if, through some accident, it is "turned over" through the four-dimensional time–space continuum.

In the case of Globorotaliidae, where the minute shell simulates that of a sea snail, a coiling rule is apparent too, but the details differ. For example, in the genus *Globorotalia,* several species are known where both right-handed and left-handed forms are very common. It would seem to be a matter of indifference whether the shell coils clockwise or counterclockwise. If, however, attention is paid to the latitude at which specimens are collected, a new fact becomes apparent. Thus, in the species *Globorotalia pachyderma,* Ericson and Bandy (1962) discovered that the direction of coiling is correlated with sea temperature. Populations in the far northern parts of the Northern Hemisphere seas have a marked preponderance of sinistral individuals, and the same is true of populations of the far southern part of the Southern Hemisphere. On the other hand, populations inhabiting latitudes nearer the tropics prove to be predominantly dextral, with only a minority of sinistral individuals. Similar distinctions have been observed in the cases of several other species of *Globorotalia.* These finds are interesting and a little mystifying, but they assume a considerably enhanced importance if time is brought into the picture. This was done by Bandy, who, in 1960, was able to show that fossil *Globorotalia pachyderma* in Californian sedimentary deposits spanning the past 10 million years exhibited periodically reversing directions of coiling as one passes from lower time horizons to upper (i.e., later) ones; he concluded that the evidence implied a sequence of climatic changes.

Similar investigations concluded in New Zealand in 1971 by D. Graham Jenkins have yielded results consistent with those reported by Bandy. Jenkins found evidence of coiling reversals of *Globorotalia pachyderma* in southern seas during the past 10 million years. There seems to be evidence for nine successive reversals, and the reversals can apparently be correlated in corresponding sedimentary columns at several different sites. The findings, taken in conjunction with other independent evidence from fossils found in the same beds, point to climatic change of much the same kind as was inferred by Bandy. The last of the coiling reversals converted the majority of New Zealand members of the species into the dextral camp, and it is consistent with this observation that living specimens in New Zealand seas at this time are predominantly dextral. Thus it would appear that the last climatic change was a swing toward warmer temperatures. Some puzzles remain, however. It is not yet proven (though generally assumed) that the Southern Hemisphere changes took place simultaneously with those of the Northern Hemisphere. There appears to be one anomaly in the New Zealand results: The population at the close of the warm Miocene period was sinistral, which ought to indicate colder conditions. Jenkins does not discuss this point. Perhaps the samples from the earlier part of the range may have been from sediments deposited in deeper, and therefore colder seas, not necessarily reflecting temperatures at the surface.

If man were to follow the curious rules of *Globorotalia*, most Canadians would be left-handed, whereas most people of the United States and Mexico would be right-handed, except during ice ages, when all Americans would become predominantly left-handed.

Radiolaria

Like forams, radiolarians extrude long slender filopodia. With these they capture copepods, diatoms, and other minute organisms. The capsule is siliceous, with highly developed three-dimensional symmetry; the filopodia are extruded through the foramina in the siliceous mesh. In parts of the ocean where radiolarians are abundant, the underlying seafloor may be covered by a sediment called radiolarian ooze, made up of a mixture of empty capsules, meteoritic dust, and other sedimentary matter. Such sediments accumulate very slowly so that drill cores in the seafloor can recover successive layers of sediment spanning 150 million years of continuous deposition. Some few radiolarians produce strontium sulfate skeletons.

The Swedish *Albatross* Expedition

During the years following the end of World War II a Swedish deep-sea expedition on board *Albatross* succeeded in obtaining the first long cores of deep-sea sediments, using a new explosive corer developed by Kullenberg. Some of the cores were believed to represent the accumulation of sediment over periods as long as a million years. The sediment, largely made up of forams and radiolarians, was studied by Wolfgang Schott (1952) and Fred B. Phleger (1948). Both investigators reported a vertical succession of alternating thanatocoenoses corresponding to alternations of warm and cold surface waters in the Atlantic. These evidences of climatic change, corresponding to the glacial and interglacial stades of the Pleistocene, showed periodic invasion of the warm Atlantic zone by subarctic species during cold stades.

Siphonophorida

Siphonophores are common in warm seas. An example is the well-known Portuguese Man o'War, *Physalia physalis*, so called because the early European voyagers first encountered them as they crossed the edge of the continental shelf off Portugal. The common species is blue, bleaching to white in preservation. Each specimen is actually a colony, the float representing one member, and the dangling portion several hundred other members of two or more types. One type (dactylozoid) mainly comprises a stinging tentacle for the capture of fishes and other prey; another type (gastrozoid) consists mainly of a mouth and digestive system, serving as the feeding elements for the whole colony. *Physalia* captures prey with the aid of a paralytic toxin secreted by cells in the tentacles, and some species are deadly to man. The buoyancy of the gas-filled float permits *Physalia* to live at the interface between the sea and the air. *Physalia* is subject to wind dispersion. *Velella* and *Diphyes* are other genera of siphonophores. Siphonophores are classified with sea anemones, corals, and jellyfishes to form the phylum Coelenterata.

Scyphozoa (Jellyfishes)

Nausithoe is a flattened, saucer-shaped jellyfish with stinging tentacles around its margin. Scyphozoans swim by slow rhythmical muscular contractions; the swimming movements are not strong enough to permit swimming against a current, so scyphozoans are treated as part of the floating plankton. Scyphozoans feed on other planktonic animals, which are engulfed through a mouth opening on the lower side of the umbrella. Jellyfishes form the main part of the diet of ocean sunfish and of some other fishes. *Periphylla* is another genus scyphozoans related to *Nausithoe.* It is distinguished by its high conical body and by the purple pigment lining the umbrella. Some species occur in deep water, and others range into the polar seas.

Ctenophora (Comb-Jellies)

Ctenophores are related to coelenterates, but because they have certain structural differences, they are usually treated as forming a distinct phylum. All are marine, and all are planktonic. Distinctive are the eight bands of ciliated swimming combs arranged like meridians across the body. Some species are nearly spherical, and in these the comb bands do indeed occupy meridians. *Beroe* is a genus in which the body has a flattened form with a bell-shaped outline and its mouth being at the broad end. Ctenophores are carnivorous, feeding on planktonic animals and also on fishes. Ctenophores are themselves eaten by fishes. They are sometimes brightly colored. Some species live in the surface waters of the oceans, others range down to 1000 fathoms or more.

Polychaeta (Bristle Worms)

Nearly all polychaetes live on the seafloor, but a few, such as *Alciopa,* occur in the plankton. Many are carnivorous, seizing other animals by means of jaws at the anterior end of their body. They have segmented bodies, as do earthworms and leeches, to which they are related and with which they are collectively classified in the phylum Annelida. (See Fig. 13.8.)

Copepoda

Copepods are small crustaceans related to shrimps and crabs. (See Figs. 13.9 and 13.10.) Copepods are the most common planktonic animals, constituting about 95 percent of the zooplankton. They occur in all seas in a wide range of depths. The most prevalent genus is *Calanus.* Species from deeper water are usually much larger than those that live near the surface. Surface-water copepods usually measure about 3 mm long, whereas in deeper water they are likely to measure about 1 cm long. Copepods feed on diatoms in the upper waters, and in lower waters they feed on smaller copepods (larval stages) and other young zooplankton. Copepods are themselves eaten by many other marine animals, including mackerel. The class

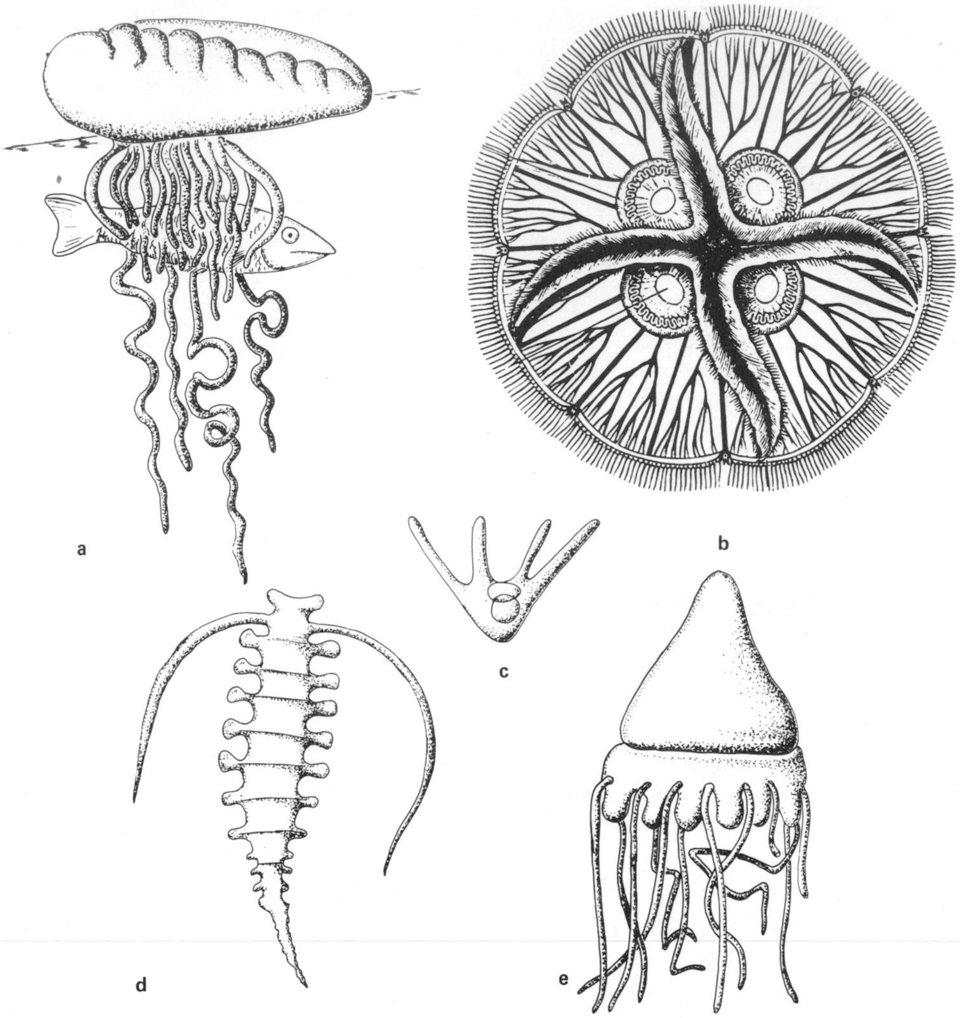

Fig. 13.8 (a) *Physalia* with captured fish, ×0.25, Siphonophorida. (b) *Aurelia,* ×0.25, Scyphozoa. (c) Larval stage (ophiopluteus) of a brittlestar, ×50. Echinodermata. (d) *Tomopteris,* ×10, pelagic bristle worm, Annelida. (e) *Periphylla,* ×2, Scyphozoa.

Crustacea, together with other animals with analogous structure (such as insects and spiders) comprise the phylum Arthropoda. All members of this phylum have jointed appendages.

Mysidacea (Opossum Shrimps)

Opossum shrimps are easily recognized by two characters: In the female there is a conspicuous transparent pouch beneath the thorax where the young are carried, and the thorax is completely covered by a cloaklike layer of soft tissue called the carapace. Most specimens are females, but males can be recognized by their evident similarity to females in the same sample. The soft carapace functions as the respiratory organ and is permeated by blood vessels; it is only loosely attached to the thorax, and can be lifted up to disclose the thoracic segments beneath. There are no gills. (Gills are filamentous structures in most crustaceans, carried at the

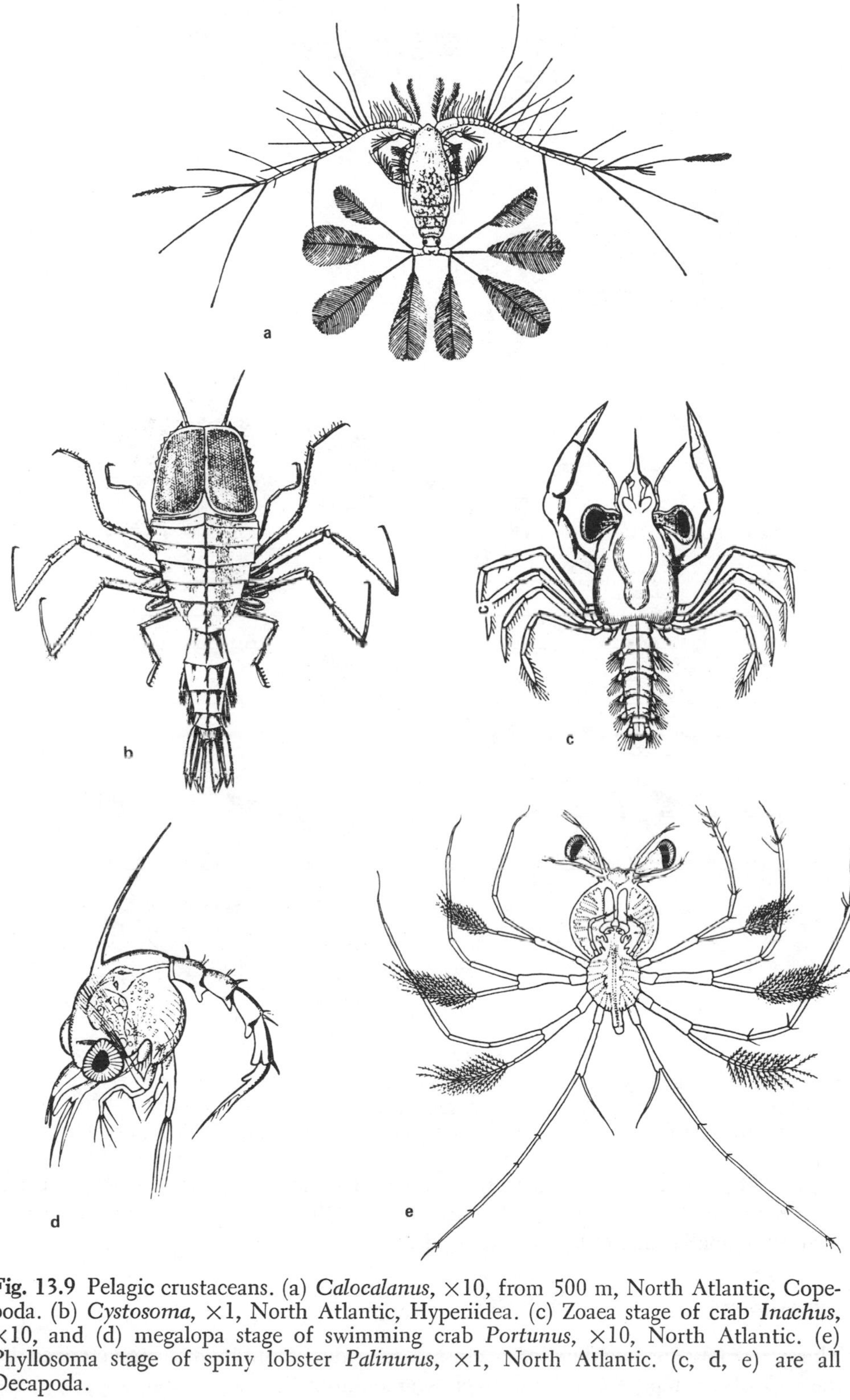

Fig. 13.9 Pelagic crustaceans. (a) *Calocalanus*, ×10, from 500 m, North Atlantic, Copepoda. (b) *Cystosoma*, ×1, North Atlantic, Hyperiidea. (c) Zoaea stage of crab *Inachus*, ×10, and (d) megalopa stage of swimming crab *Portunus*, ×10, North Atlantic. (e) Phyllosoma stage of spiny lobster *Palinurus*, ×1, North Atlantic. (c, d, e) are all Decapoda.

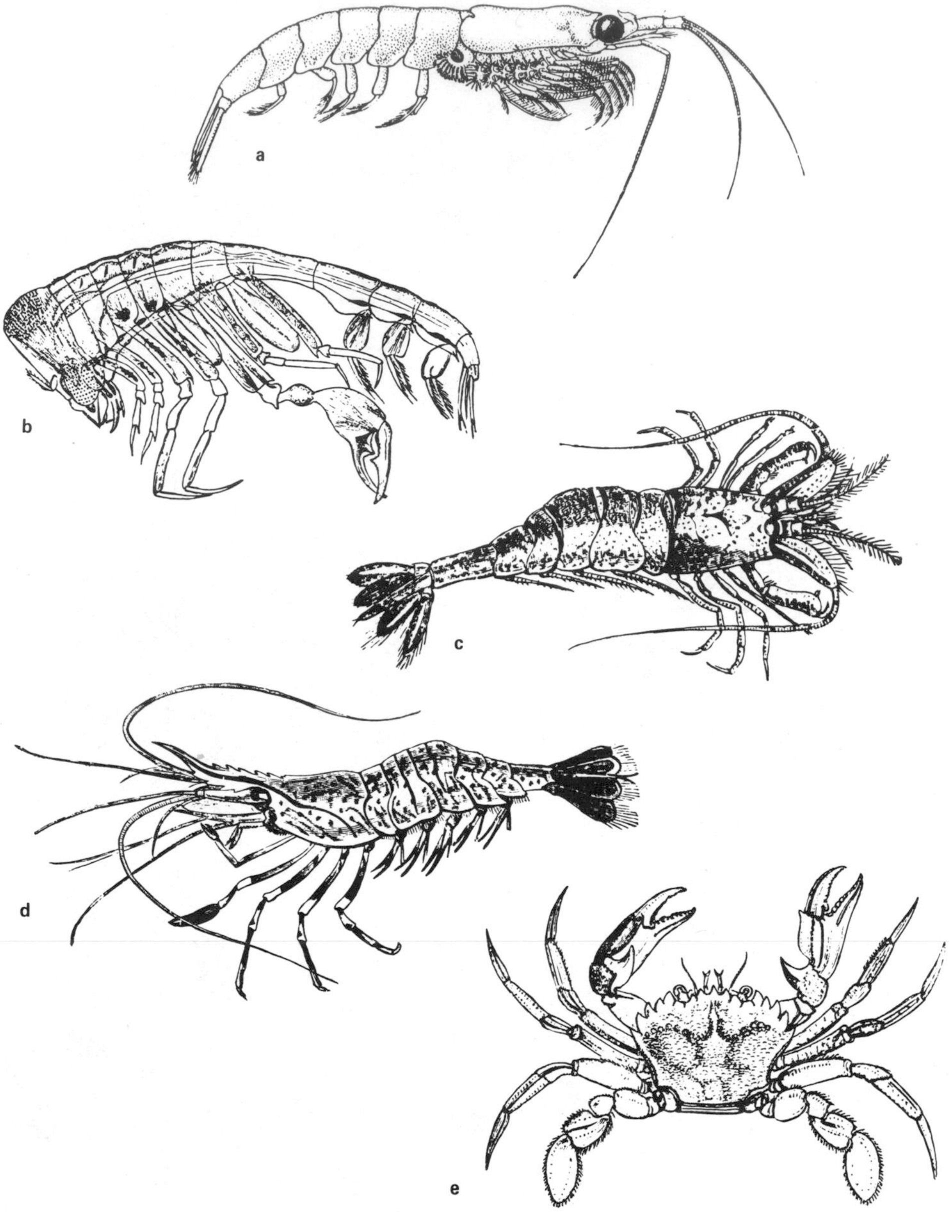

Fig. 13.10 Pelagic crustaceans. (a) *Meganyctiphanes*, ×1, North Atlantic, Euphausiacea. (b) *Phronima*, ×2, North Atlantic, Hyperiidea. (c) *Crangon*, ×0.5, neritic. (d) *Palaemon*, ×0.5, neritic. (e) *Portunus*, ×0.5, swimming crab, mainly shallow water. (c–e) Decapoda.

bases of the thoracic limbs where they pass under the carapace.) Mysids are eaten by fishes. Some mysids are benthic and live on the bottom. The food of mysid comprises smaller plankton, such as copepods.

Hyperiidea

These animals are members of a crustacean order Amphipoda, characterized by the lack of a carapace (so the thoracic segmentation is visible externally) and the compression of the whole body (i.e., flattened in the vertical plane). The best-known members of the Amphipoda are the flealike sandhoppers of sandy beaches and intertidal regions. Most amphipods are bottom dwellers, but the suborder Hyperiidea forms an important part of the plankton of the open ocean. These animals have very large, unstalked eyes. They live most of their lives in the faintly illuminated zone below 100 fathoms. Most hyperiids parasitize jellyfishes of the zone they inhabit. Hyperiids are eaten by deep-water fishes such as *Alepisaurus*.

Phronima is a genus of Hyperiidea of interest on account of its peculiar ecology. Spending the greater part of its life in the deeper regions, *Phronima* swims to the surface of the ocean when sexual maturity is reached; there it captures a doliolid or sometimes a salp. The captor kills and eats the captive, then it creeps inside the floating, transparent, barrel-shaped test of the dead doliolid and takes up residence there. In the surface waters the mating and shedding of gametes occurs. The young hatch out at the surface, later sinking down to the depths normal for the species. Thus *Phronima* has a **vertical migration cycle.**

Euphausiacea (Lantern Shrimps)

Euphausians are strictly planktonic organisms, unlike mysids (which they somewhat resemble). Their characteristic features are feathery gills that are placed at the bases of the thoracic limbs and that are fully exposed to view because their carapace does not reach all the way down sides of the thorax. This characteristic, by the way, distinguishes the euphausians from the true shrimps (*Decapoda natantia*), where the gills are similarly developed but are completely covered by the sides of the carapace. Phosphorescent organs, which give the euphausians their distinctive name of lantern shrimps, are visible in preserved specimens as red spots—two pairs on the thorax and four pairs on the abdomen. Living lantern shrimps emit brilliant flashes of light visible at night when they rise to the surface. There are only a few species of euphausians represented in the oceans. Each species prefers a particular type of water mass defined by temperature and salinity. Several species girdle the earth in the Southern Ocean, and each band lies at a particular latitudinal range. They constitute a major food item for the large filter-feeding pelagic vertebrates, such as baleen whales. Euphausians feed on copepods.

Larval Forms

Larval stages of both planktonic and benthic animals are found in the plankton, usually at or near the surface. Most probably this occurs because larval stages are small and, therefore, require a plentiful supply of small particles of food. Small particles are rare in deep water and at the bottom, however, because small particles fall more slowly than large ones (Stokes' law) and are more liable to be eaten up by small organisms. Therefore, organisms should increase in size within limits as depth increases—the limits being set by the total availability of food. We can surmise that larval stages of deep-water animals must rise to the surface in order to survive the risk of starvation at lower levels. When large enough, they can return to the region where their parents lived. The life cycle of euphausians includes a *Furcilia* larva, which resembles a copepod but differs by having two long flotation rods, one in front of the thorax and one behind. Similar larvae occur in the life cycle of lobsters and crabs, where the larva is called a *Zoaea*. The *Furcilia* larva grows into a *Cyrtopida,* characterized by the remarkably long eyestalks. At stages it changes into another form by shedding its skin. Numerous **ecdyses,** or sheddings, occur in one animal's lifetime. The youngest stages feed on diatoms, gradually changing to copepods as they grow older and larger. The bizarre appearance of larval stages such as these puzzled the early oceanographers, who supposed them to be adult animals and who gave them their special names originally as generic names.

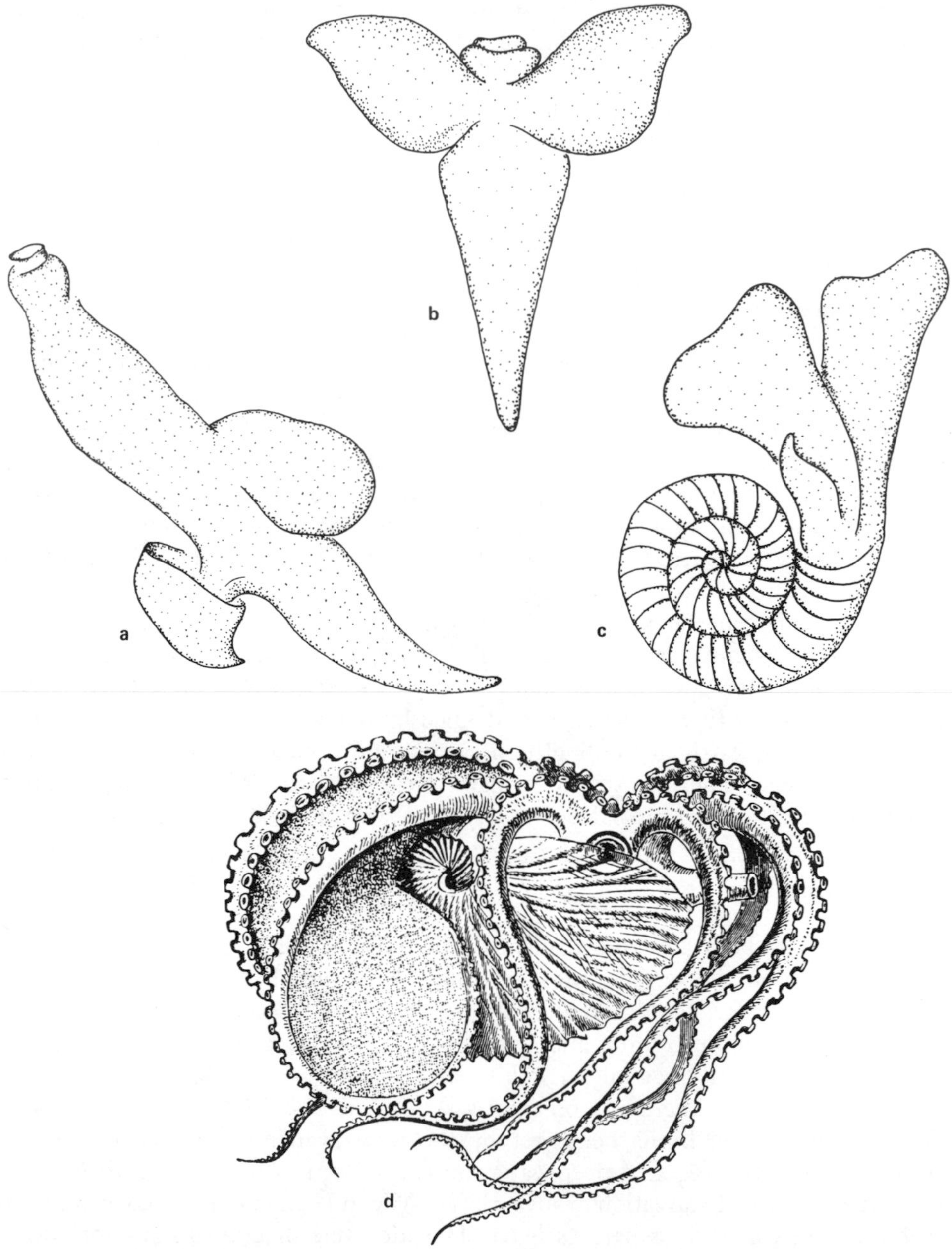

Fig. 13.11 Pelagic mollusks. (a) *Carinaria,* ×0.5, tropical Atlantic, Heteropoda. (b) *Clione* and (c) *Limacina,* ×4, tropical Atlantic, Pteropoda. (d) *Argonauta* (female), ×0.5, tropical oceans, Cephalopoda.

Ianthinidae (Violet Snails)

Violet snails belong to the molluscan genus *Ianthina,* mainly restricted to tropical and subtropical waters. Their delicate shells are swept by currents on to distant shores, including Cape Cod. They inhabit the sea–air interface, held there by rafts of bubbles and slime that they secrete. *Ianthina* lives in swarms, and it is a predator, attacking similar swarms of *Physalia.* Each snail creeps aboard the bladder of *Physalia* colony, abandoning its own raft. It remains on the captured vessel until it has killed and eaten all the zooids in the colony. The snail then spins a new raft and takes off again in search of another *Physalia.* Siphonophores have existed for the past 500 million years, as the fossil record discloses, while *Ianthina* ranges back for at least 100 million years. Thus peaceful navigation preceded piracy. The violet snails, together with the heteropods and pteropods, are classified in the class Gastropoda of the phylum Mollusca.

Heteropoda

The large sea snails constituting this order have a glassy transparency and a pair of very conspicuously pigmented eyes. The small shell, shaped like a Trojan cap, is carried on a similarly shaped visceral hump. Heteropods are mainly found in tropical waters, but apparently they are carried northward by the Gulf Stream, for they have been collected near the Azores Islands, in 40° north latitude. Heteropods swim rapidly with the aid of muscular contractions, using finlike lateral flanges of the body. They swim while inverted and capture and eat small fishes and jellyfishes. They are seldom caught in plankton nets, presumably because they can swim out of the net.

Pteropoda (Sea Butterflies)

Pteropods are really sea snails, and they are classified, therefore, among the gastropods of the phylum Mollusca (Fig. 13.11). All pteropods do not have shells, but, where the shell is lacking, the body still has a general similarity to that of the shelled forms. Thus their characteristic pair of winglike fins make them fairly easy to recognize. Usually their body is brightly colored. *Pteropoda* live in swarms in the ocean, somewhat resembling flying butterflies as they pass ships. Some species have the silvery mirrorlike surface that fishes of the lower zones adopt (in conditions of even low-level illumination, a mirror becomes invisible). Shell-bearing pteropods inhabit warmer regions of the ocean, where their shells may lie on the seafloor in such profusion as to give the sediment the name pteropod ooze.

Chaetognatha (Arrow Worms)

The arrow worms have no known close relatives, except only some Cambrian fossils that are about 600 million years old. They are therefore placed in a phylum by themselves. They are highly predatory animals, ranging the oceans from the polar regions to the equator. They have a pair of jaws at the anterior end of their body, and a pair of tail flukes at the opposite extremity. There are paired horizontal finfolds along the sides of the body, reminiscent of stabilizers, which they may be. Only the tail fins are motile. They swim feebly, but are voracious and are able to swallow a vertebrate fish whole, even though it may be several times their own bulk. Examples are the deep-water and polar genus *Eukrohnia*, and the warm, shallow-water genus *Sagitta*. Arrow worms are themselves eaten by fishes. Some arrow worms are associated with the sonic echoes produced by schools of organisms at particular depths. Arrow worms rise to the surface at dusk and at dawn and occupy deeper zones at noon and midnight.

Salpida

Salps are elongated, torpedo-shaped planktonic organisms of a glassy transparency, which are usually colorless. Some specimens are asexual; these are called nurse stages. The nurse produces babies by the convenient device of growing a cord of undifferentiated tissue out of her body, the cord (called a genital stolon) subsequently divides into a chain of babies, all joined side by side like mass-produced Siamese centuplets. As her babies grow larger, they form a train in which the patient nurse is the locomotive, and her progeny trail behind as she swims by the usual muscular contractions of planktonic creatures. Eventually the young salps become sexually mature, and they then tend to break up into smaller chains of 10 or 20 individuals, each one of which is sometimes male and sometimes female. They shed their ova and sperm into the seawater, and the resultant abandoned progeny grow up to be the sexless nurses. Thus there is an alternation of generations, with nurse stages and chain stages appearing in turn. The internal structure of a salp shows that it is related to other animals with archaic features of vertebrate ancestors; thus salps are classified in the phylum Chordata (Fig. 13.12). *Doliolida* are related to salps but have even more extraordinary reproductive aberrations. Here the nurse doliolid grows a tail along which the outer two rows become food gatherers and community stomachs, the members of the middle row do nothing but reproduce sexually, while the nurse degenerates into a tractor for the whole of her varied progeny. The offspring in the middle row of zooids (gonozoids) become tadpoles, which, after swimming around for a while, redesign themselves into proper doliolids and assume the sexless state of the nurse.

Pyrosomatida

Members of this group are related to the salps and doliolids, but differ in having a permanently colonial habit and no alternation of generations. Some also have phosphorescent properties. There is no vernacular name in general use, but in New Zealand the fishermen call them Mermaids' sox. They look like cones made of gray flannel, more suited to be ice-cream cone covers, although occasional giant specimens might fit a very small mermaid. The individuals in the colony are arranged side by side so that their mouths all lie on the outside of the cone. The most common species is *Pyrosoma atlanticum atlanticum*, which despite the insistence of the name, is actually widespread throughout the Pacific. It is considered good eating by a few fish, including *Alepisaurus*, but most fish ignore it.

Fish Larvae

Plankton hauls and the stomach-contents of plankton-eating fishes commonly yield some tadpolelike young stages of various vertebrate fishes. Adult fishes can swim

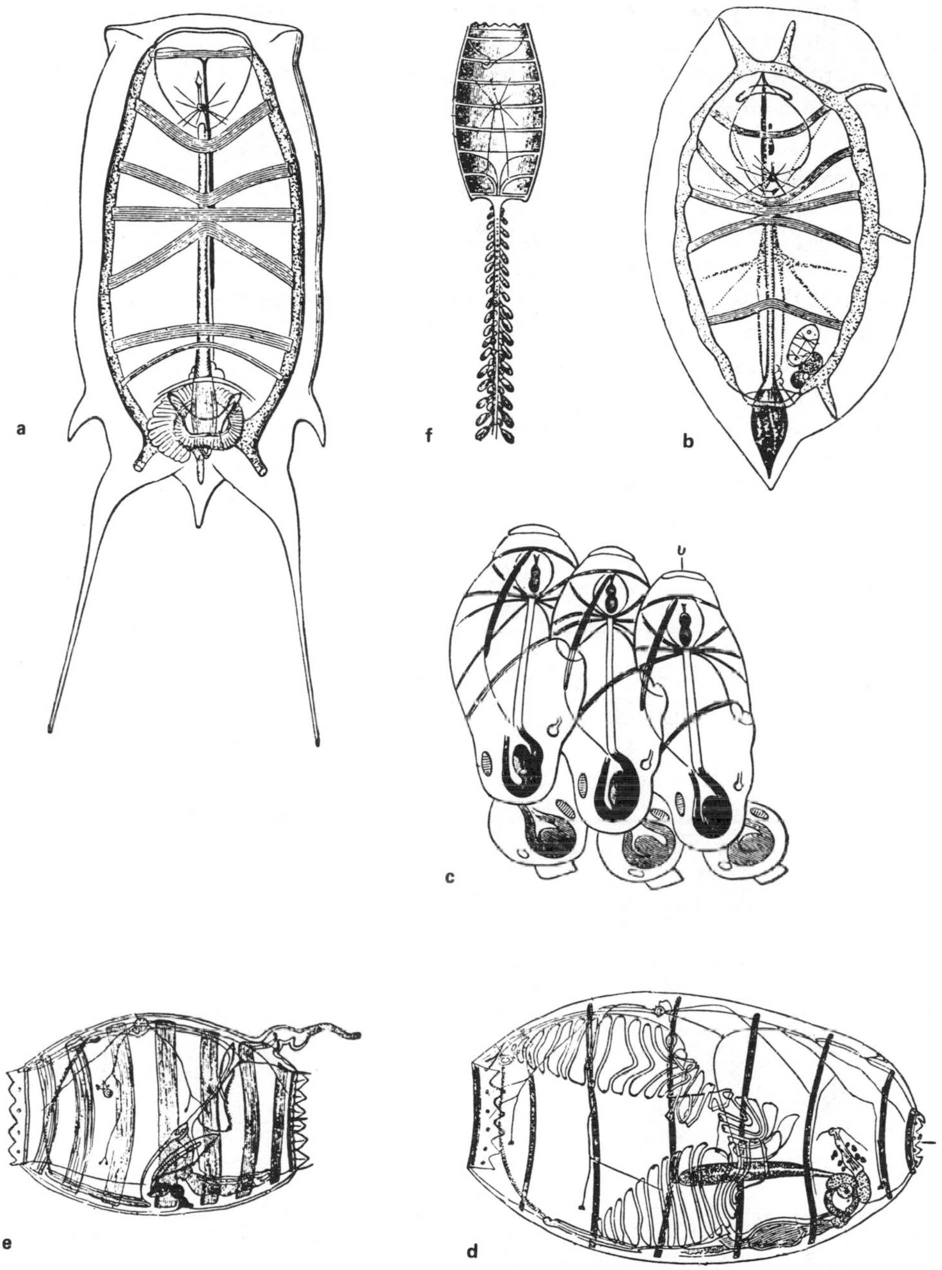

Fig. 13.12 Pelagic protochordates. (a) *Salpa democratica* and (b) *Salpa mucronata*, ×0.5; both worldwide, Salpida. (c) Part of a young chain of *S. democratica*, ×1. (d) *Doliolum*, ×0.5, worldwide, sexual stage. (e) Nurse stage. (f) Nurse stage trailing colony.

against the current and thus they are classified as nekton because they are not liable to passive dispersion. Their young stages, however, lack the mobility of the adults and drift with currents. They probably constitute a substantial part of the diet of other larger plankton because the infant mortality of fishes seems to be astronomical. For example, a female mackerel produces 500,000 eggs at one spawning, yet the mackerel population is not known to be increasing. So almost all of the young probably die at a very early stage. Because mackerel are plankton eaters and their progeny are eaten by the planktonic organisms that adult mackerel eat, the recycling of materials could lead to high concentrations of the nonexcretable toxic pollutants in these fishes.

Energy Transfer in the Sea

Because the phytoplankton alone captures radiant energy from sunlight and can exist only in the illuminated upper waters of the ocean, it follows that the deeper regions, in darkness, acquire energy needed for vital activities of deep-water animals by stealing it from above—that is to say, by feeding upon organisms whose food chains lead back to the surface of the sea. But surface organisms do not wander far from the surface, and the animals of the deep sea do not visit the surface. The manner in which this energy transfer is effected is discussed in Chapter 28.

14

FISHES

Nekton / External characters of diagnostic value / Locomotion / Swim bladder / Amphibious fishes / Diurnal periodicity / Nocturnal fishes / Protective structures / External armor / Surgeonfishes / Venomous fishes / Scorpion fishes / Stonefish / Adaptive coloration / Cryptic coloration / Demersal fishes / Sematic coloration / Disguise coloration / Territorial behavior / Commensal relationships / Predator–prey relationships / Population explosions / Systematic review / Cyclostomata / Sharks and rays / Distribution of sharks / Man as prey to sharks / Sturgeons / Herrings / Flying fishes / Eels / Sea horses and pipefish / Flatfishes / Anglerfishes / Perchlike fishes

Nekton comprises all those organisms able to disperse in any direction in water, and fishes are the most abundant members. Dispersion is independent of ocean currents and independent of the substrate. Thus swimming organisms such as mysids or scyphozoans are not part of the nekton, for their swimming powers are insufficient to counteract ocean currents, and so they drift with the water masses and are counted with the plankton. Fishes, on the other hand, may drift with ocean currents; but, if carried into unfavorable environments, they can facultatively swim against the currents. Similarly, fishes can perform annual migrations northward and southward, although the ocean currents they traverse continue to flow in the same direction. Of the other marine vertebrates, the marine mammals behave similarly to the fishes, and obviously fall in the category of nekton. Although some marine birds such as penguins lack wings, they also form part of the nekton (Table 14.1). Other marine birds can both swim and fly; as a matter of convenience, they are considered here to be part of the marine nekton. A few marine invertebrates, such as squid and some pelagic octopuses, have such highly developed swimming powers as to place them in the category of nekton. No plants can be classified as nekton (except in science fiction). Table 14.2 gives the common categories of nektonic distribution.

Table 14.1 The Classes of Marine Vertebrates Forming the Nekton

DIAGNOSTIC CHARACTERS	CLASS	EXAMPLES
Vertebrates breathing by means of gills in water, fishlike animals		
Body eel-shaped, with no paired limbs, and a circular suctorial mouth, lacking jaws	Cyclostomata	Lampreys and hagfishes
Body typically fishlike, with paired pectoral and pelvic fins, and jaws; skeleton of cartilage only	Chondrichthyes	Sharks and rays
As above, but skeleton includes bone elements	Osteichthyes	Bone fishes
Vertebrates breathing by means of lungs in air, four-footed animals		
Marine reptiles	Reptilia	Turtles, crocodiles, and sea snakes
Marine birds	Aves	Shore and oceanic birds
Marine mammals	Mammalia	Furred animals, whales, dolphins, and so on

External Anatomy of Fishes

To identify a fish you need to know certain anatomical terms used in standard descriptions. These external features are known as **characters of diagnostic value** (see Figs. 14.1 and 14.2). Fishes have the body divided into a **head, trunk,** and **tail;** the anterior half of the trunk is termed **thoracic,** the posterior part **abdominal.** There are two pairs of **paired fins** normally present, the anterior or **pectoral fins,** and the posterior or **pelvic fins.** The pelvic fins may be set far back on the body in an abdominal position; or they may lie just below and behind the pectorals, in which case they are termed thoracic in position; or they may be attached just in

Table 14.2 Vertical Stratification of the Nekton

DESCRIPTIVE TERM	DEPTH AT WHICH NEKTON OCCURS (IN METERS)	DEFINITION
Littoral	0–5	Frequenting inshore water
Neritic	0–200	Frequenting waters overlying the continental shelf
Epipelagic	0–200	Frequenting the upper waters of the open ocean
Demersal	Any depth	Nektonic, but habitually resting on the seafloor, as, for example, flounder
Mesopelagic	200–1000	Frequenting pelagic waters at depths corresponding to the archibenthos
Bathypelagic	1000–2000	Frequenting pelagic waters at depths corresponding to the bathyal benthos
Abyssopelagic	2000–4000	Frequenting pelagic waters at depths corresponding to the abyssal benthos
Hadopelagic	4000 and deeper	Inhabiting the waters of the deepest trenches

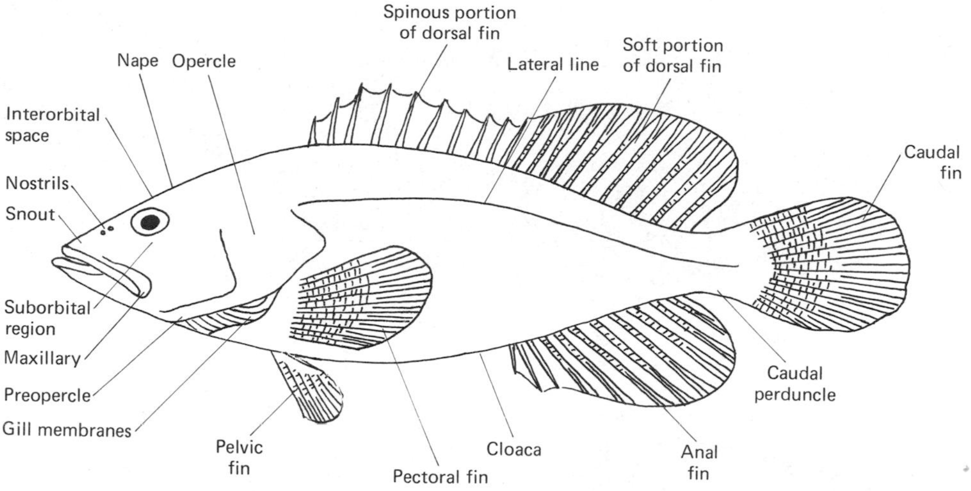

Fig. 14.1 External anatomical characters of bone fishes.

front and below the pectorals, in a **jugular position.** The unpaired fins are the **dorsal,** the **caudal,** and the **anal** (just behind the anus).

The mouth may be **terminal** or **inferior** (i.e., below the head, as in sharks and rays). The **nares** (nostrils) lie in front of the mouth, inferior in sharks and rays and on the upper side of the head in other fishes; the eyes lie on the upper side of the head; the jaw may be short, terminating at a level in front of the eyes, or long, terminating behind the level of the eyes. The maxillary bone of the upper jaw may slide under the cheek when the mouth is closed, or it may not be mobile in this way. Fishes that find their food in the bottom muds often have sensory tentacles hanging below the jaw, called **barbels.** (See Fig. 14.3.) The gills open at the sides of the head in most fishes; they have five to seven slitlike openings in the elasmobranchs (rays and sharks) or may be covered by a plate called the **operculum,** so that there is a single gill opening, as in teleosts. The skeleton may be **cartilaginous,** as in elasmobranchs, or **bony,** as in teleosts. Elasmobranchs have a tail in which the vertebral column extends into the upper or **epural** lobe, with a smaller **hypural** lobe below. Teleosts (bony fishes) usually have the vertebral column ending at the base of the tail. The former type of tail is termed **heterocercal,** the latter **homocercal.** If a homocercal tail is subdivided into similar epural and hypural lobes, it is said to be **diphycercal.**

Each fin may be supported wholly by flexible, segmented cartilaginous or **soft rays,** or the anterior rays may be rigid, bony **spines,** in which case a fin is said to comprise a spiny and a soft portion. (See Fig. 14.4.) In reporting the number of rays in any fin, the spines are numbered with Roman numerals, the soft rays with Arabic numbers. If the two parts are contiguous, then the two numbers are separated by a comma (for example, X, 9); but if the two parts are separated by an interval, then a plus sign is inserted between the numerals (IX + 9). These data are usually given for the dorsal, the anal, and the pectoral fins. Some spines may stand free of the fins, at the base of the tail, on the operculum, or on parts of the head; these variations characterize several families. In some families the dorsal, caudal, and anal fins may become continuous; or the pelvic fins may be lacking; or both pectoral and pelvic fins may be lacking. The teeth may be separate or fused into solid masses on the jaws. In teleosts the skin is usually covered by **scales.** The

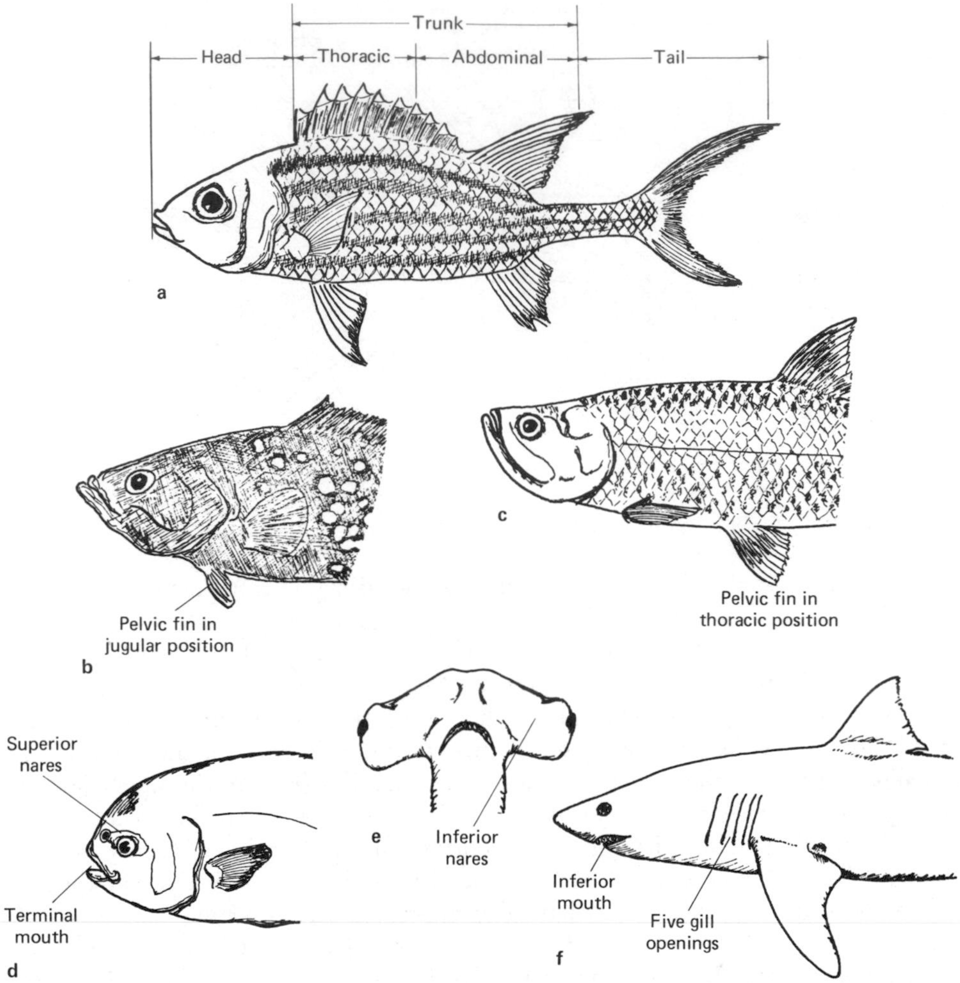

GERALD HESLINGA

Fig. 14.2 External diagnostic characters used in the identification keys in this book. (a) Squirrelfish. (b) Soapfish. (c) Tarpon. (d) Permit. (e) Hammerhead shark. (f) White shark.

scales may have concentric or **cycloid** growth lines, or the growth lines may form toothlike or comblike sets, as in **ctenoid** scales. In most teleosts the nostril is a double structure, but it is single in a few families. Behind the dorsal fin in some families small finlets may occur, like a fringe. The terms described are useful when identifying specimens of fishes in the field or in the laboratory.

How Fishes Swim

The structure of the fish body enables it to move rapidly, which is especially true of sharks. However, despite the robust tail, which provides the power for swimming, sharks suffer a constant tendency to sink because there is no means by which the specific gravity of the body can be matched to the surrounding medium. To overcome the sinking, the anterior part of the body, especially the snout and the head, is flattened horizontally. In addition, the pectoral fins are held in a horizontal position, slightly inclined downward on the trailing edge. These features, combined

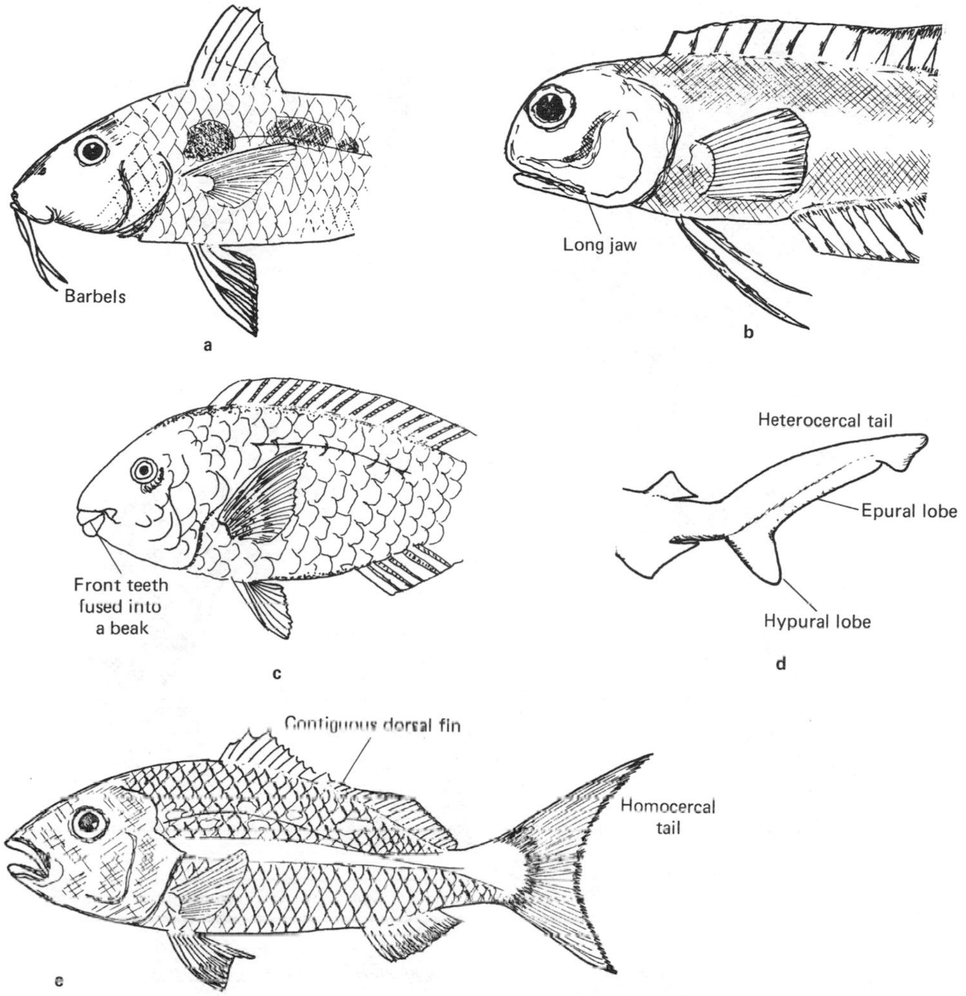

GERALD HESLINGA

Fig. 14.3 External diagnostic characters used in the identification keys in this book. (a) Goatfish. (b) Jawfish. (c) Parrotfish. (d) White-tipped shark. (e) Snapper.

with the motion imparted by the tail, maintain the fish's spatial relationship in much the same manner as for an airplane. Therefore, if the fish stops swimming, it will sink gently to rest on the seabed, just as an airplane will fall if it loses its forward motion. Rays have probably evolved from time to time from shark stocks as a response to the constant natural tendency of sharks to sink to the seafloor. Any fish that rests upon the bottom (or is imbedded in the sediments) is termed **demersal.** Demersal fishes, such as sand sharks, eagle rays, and sting rays, feed upon bottom-living organisms, especially upon mollusks, whose shells are crushed by their powerful jaws armed with flattened platelike teeth at the back of the mouth or all around the jaw.

In contrast to the sharks (which have a cartilaginous skeleton), the teleosts (i.e., fishes with a skeleton of bones) have evolved an organ called the swim bladder. This is an air-filled structure provided with blood vessels that is capable of secreting gas or absorbing it, and so adding to or subtracting from the amount of air in the bladder.

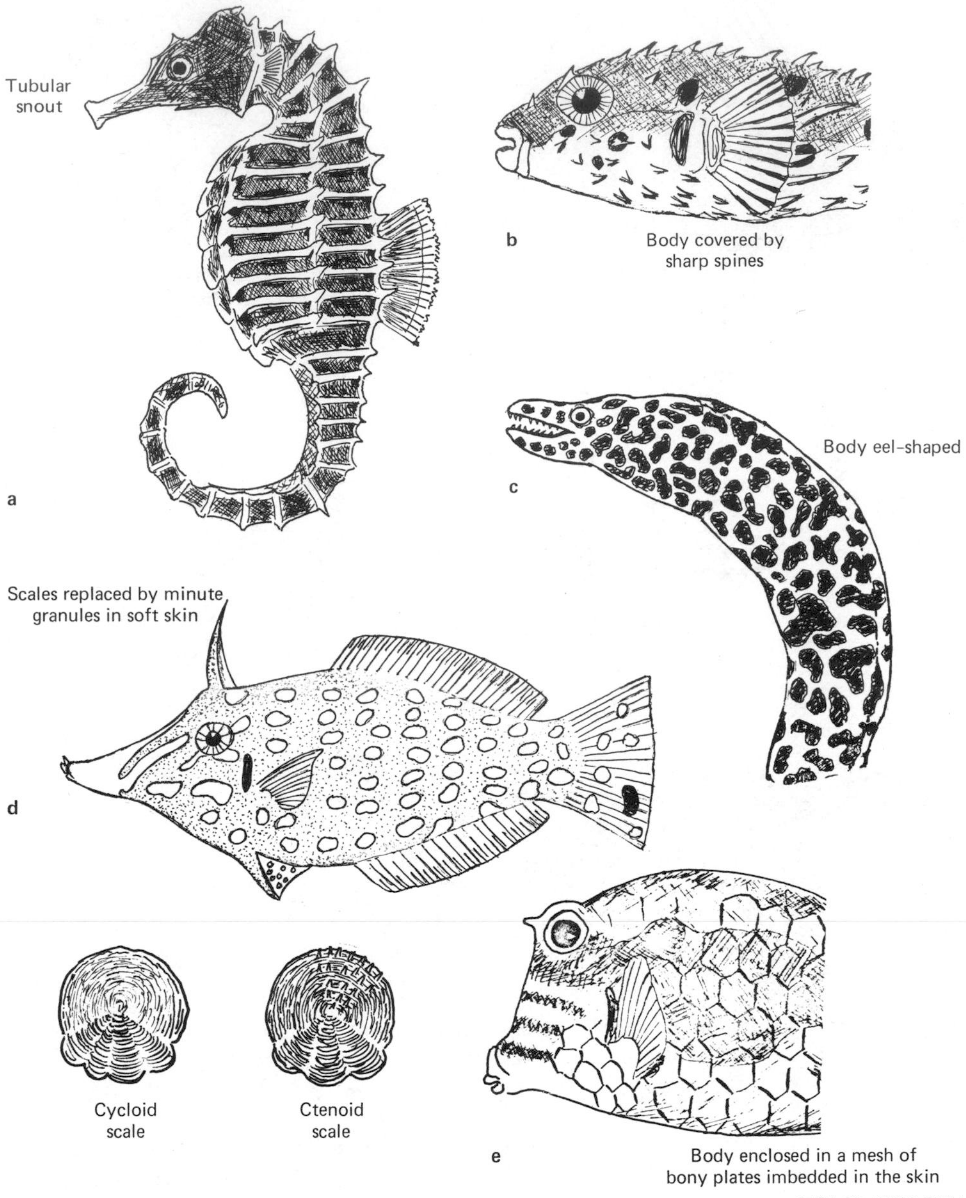

Fig. 14.4 External diagnostic characters used in the identification keys in this book. (a) Sea horse *Hippocampus* (family Syngnathidae, order Syngnathiformes). (b) Australian porcupinefish (see Chapter 27). (c) Spotted moray. (d) Australian pigfish (see Chapter 27). (e) Trunkfish.

The acquisition of the swim bladder by teleosts led to change in locomotor activity. The specific gravity of the fish body could now be adjusted to equal that of the surrounding medium, with a resultant disappearance of the tendency to sink to the seafloor whenever forward progression ceased. Thus the head region no longer had to be flattened in the horizontal plane, and the pectoral fins were liberated from the task of serving as stabilizers in the horizontal plane. Stabilization in the vertical plane could be achieved by the body's assumption of a deep laterally compressed form, with appropriate further extension in the vertical plane by means of the unpaired fins. Energy, as before, was provided for forward motion by the tail. The pectoral fins were now available to serve a new purpose, that of braking, and accordingly the attachment of the anterior edge of the pectorals rotates through 90° into the vertical plane and comes to lie just behind the operculum. The rapid braking, now possible, enables a teleost to reverse its direction in about its own length, a great increase in agility.

Certain teleosts have secondarily lost the swim bladder. These fall in two categories: (1) fishes that have adopted the demersal habit (e.g., flatfishes, anglerfishes, and their relatives) and (2) fishes that, through body armor (e.g., cofferfishes, porcupinefishes, cowfishes) or through venomous properties (e.g., puffers) are not susceptible to predator attack and therefore do not demand great agility.

Other variations in locomotor habit are seen in mudskippers (*Periophthalmus*), where the pectoral fins are converted into levers adapted to walking on damp or dry land and to ascending mangrove roots and branches in search of insects and crustaceans on which they feed while out of the water. Other modifications that may be cited briefly include pectoral fins adapted for gliding through the air from one wave crest to another, as in the pelagic flying fishes (Exocoetidae, Fig. 14.22) and the conversion of the dorsal fin into an attachment sucker in remoras, which adhere to sharks and other fishes and are thus transported passively. Morays among the tropical eels have lost both sets of paired fins, adopting instead a snakelike manner of progression on the seafloor, where much of the time is spent in concealment with only the head emergent, on watch for passing prey.

Rogue Sharks

Most sharks are nocturnally active, as is the case with most reef organisms. Pelagic species, however, that occasionally visit the reef, may include dominant carnivores of the open ocean; of these, particular individuals (rogue sharks) develop a taste for human prey. Nearly all attacks seem attributable to a few individual sharks, which once detected and caught, leave no impress on the behavior of other sharks. There is apparently a **diurnal periodicity** in shark attacks—most occurring between 2:00 P.M. and 6:00 P.M.; these, however, are the hours when most humans swim, so the cycle is probably an imposed one, dictated by the availability of prey.

Squirrelfishes (family Holocentridae), morays (family Muraenidae), and grunts (family Pomadasyidae) are nocturnally active, hiding near or on the bottom by day; however, a moray (Fig. 14.11) can be wakened and encouraged to feed if suitable prey is offered. The butterfly-fishes and angelfishes (Fig. 14.12) (family Chaetodontidae) are diurnal and become torpid, with color changes, at night. Wrasses (family Labridae) are also diurnal fishes, and at night young individuals sink to the soft bottom and conceal themselves as infauna. Squirrelfishes (Fig. 14.21) are related to mainly deep-water families and share with them the large eyes of animals that live in near darkness; hence the nocturnal habit is to be expected. Parrotfishes (Fig. 14.25) (family Scaridae) are diurnal and sleep on the bottom by night, sometimes secreting a nest of mucus about the body.

Protective Structures

Sharks, by their powerful locomotor organs and teeth require no additional protective structures; but other demersal forms, with the head, trunk and pectoral fins converted into the disk, are only feeble swimmers, and their teeth are suited more to crushing than to severing. These have become, for the most part, unaggressive animals, but many of them have developed a protective organ in the shape of a more or less erect and barbed dorsal spine at the base of the tail. An intruder, such as a human accidentally standing upon such a ray, may be severely injured by the animal as it makes its escape. Most wounds are in the ankle region; it is advisable, therefore, to shuffle if walking on the seabed in areas where rays abound, so that the animal may escape before being pinned down by the weight of the person standing upon its disk. The electric rays or torpedos (Fig. 14.18) can administer a sharp shock if trodden upon. The electric organs in this case are two lateral bands of modified pectoral muscle, reconstituted in the form of a charge accumulator of parallel plates separated by semipermeable membranes, and energized by part of the hind brain. Positive and negative ions, by separating on the upper and lower surfaces of the plates, may collectively build up a charge of several hundred volts, capable of sudden discharge through the body of the intruder. (See Fig. 14.5.)

Paleozoic fishes depend largely upon a strong, external armor plating of bony scales for protection against predators, which were chiefly giant invertebrates such as eurypterids and cephalopods. With the acquisition of the swim bladder and the agility imparted by the consequential elaboration of locomotor organs, the need for heavy external armor diminished. At this epoch, agility must be rated as a major protective characteristic of teleost fishes. However, other relevant features occur under this heading, and among those that are significant in the coral reef

PETER GARFALL

Fig. 14.5 Cow-nosed rays *Rhinoptera bonasus* (Rhinopteridae, Batoidei, tropical Atlantic) exhibiting courting behavior. The male is to the right. The young are born live and measure 35 cm across; the adult is 2 m across.

environment and in neighboring areas of the tropical shelf are sharp spines and the venom glands often associated with them, sharp anterior teeth, and protective coloration.

The surgeonfishes (Fig. 14.6) are essentially herbivorous and quite aggressive reef fishes that have evolved an effective organ of defense and also offense in the

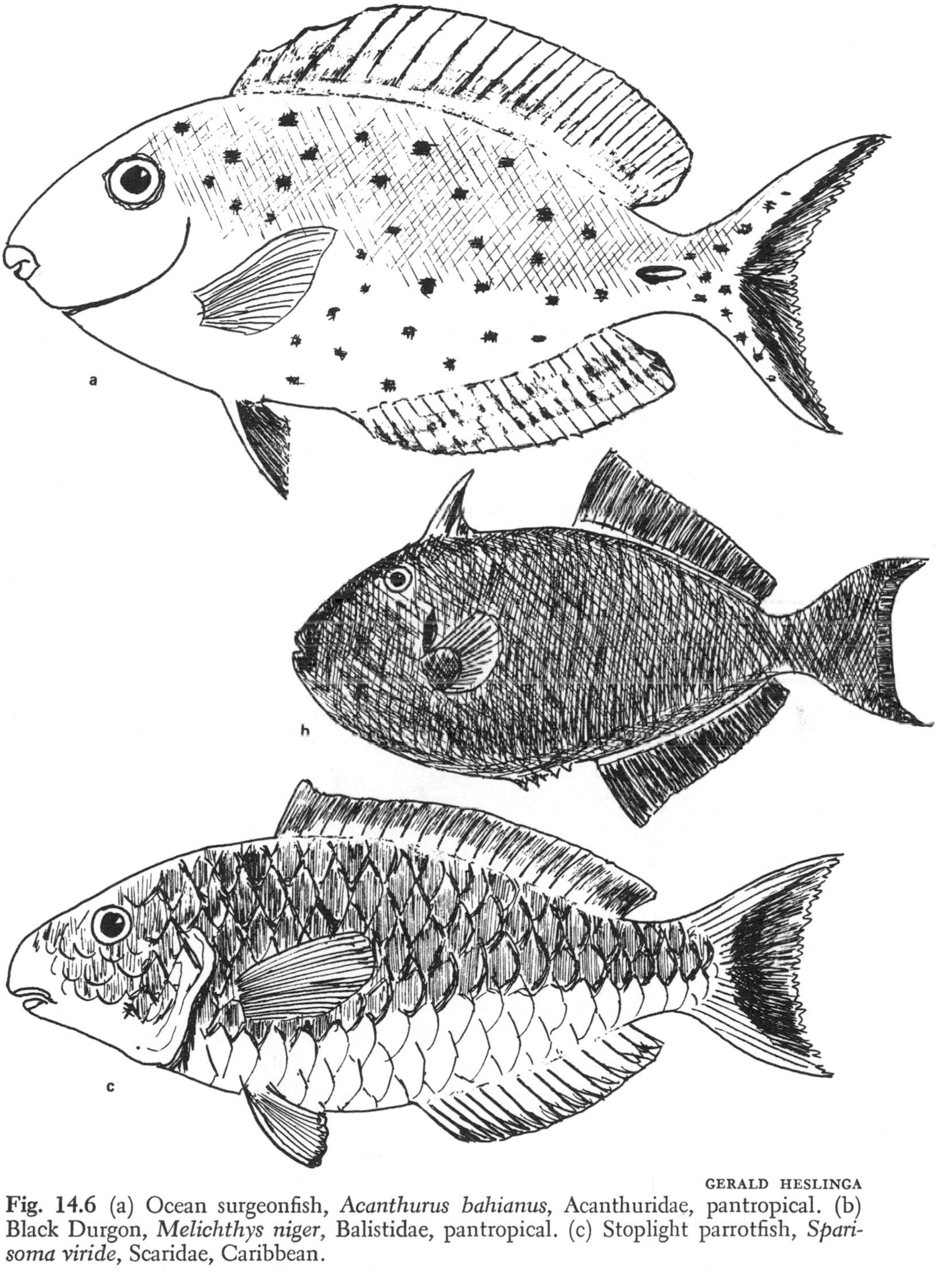

GERALD HESLINGA

Fig. 14.6 (a) Ocean surgeonfish, *Acanthurus bahianus,* Acanthuridae, pantropical. (b) Black Durgon, *Melichthys niger,* Balistidae, pantropical. (c) Stoplight parrotfish, *Sparisoma viride,* Scaridae, Caribbean.

shape of one or more erectile spines housed in a slot on either side of the base of the tail. According to Randall (1968), the mere threat of using the spine is enough to impart dominance to a surgeonfish forming part of a community in an aquarium tank. If provoked, the spine is erected and severe wounds inflicted upon other fishes by violent lashing of the armed tail. Among species listed by Halstead (1959) as capable of inflicting deep and painful wounds on man are: *Acanthurus xanthopterus* of the Indo-Pacific reefs, *Acanthurus bleekeri* of the Indo-Pacific exclusive of Hawaii, and *Naso lituratus* of the Indo-West Pacific, from Polynesia to east Africa and the Red Sea. *Acanthurus* has a single erectile spine, hinged at its posterior end, that rises like the blade of a pocketknife with the sharp edge facing anterior. *Naso* has on either side of the caudal peduncle two pairs of permanently erect spines, which are attached to plates imbedded in the skin. The Atlantic surgeonfishes all belong to the genus *Acanthurus*. It appears that no venom glands are associated with the spiny apparatus.

The scorpionfishes (Fig. 14.7) are mainly Indo-West Pacific reef inhabitants, although there are some representatives in the Caribbean and some also in temperate and Arctic seas. Their outstanding characteristic is the presence of erect spines derived from the unpaired and paired fins; each spine is associated with an investing venomous dermal gland. Halstead (1959) distinguishes three main groups according to the nature of the venom organs.

1. The *Pterois* type, exemplified by *Pterois volitans* (Fig. 14.8), is variously known as the lionfish, or dragonfish, zebrafish (on account of the contrasting bands of black and orange semantic coloration), or turkeyfish (from the habit of spreading the feathery fin spines in the manner of a turkey's tail) and is restricted to the tropical reefs of the Indo-West Pacific. In these fishes the venom spines are very long and slender, each invested by a thin venomous integumentary gland, with no venom duct. Several species range from the Red Sea, through the intervening tropical coasts, eastward to the reefs of Polynesia and northern Australia. They occupy shallow parts of the reef or crevices, or swim openly in unprotected shallows. They occur often in pairs. Their apparent fearlessness leads the incautious to be stung by the needle-sharp spines. Stings are also received when such a fish is hooked and is being taken from the line. The venom organs comprise 13 dorsal spines, 3 anal spines, and 2 pelvic spines. The biologist can readily recognize the nature of fishes of this genus, and is only likely to be stung if incautiously groping in crevices with unprotected hands. The effects of a sting include intense pain, nausea, and shock.

2. The *Scorpaena* type has a venom apparatus that is similar to *Pterois*, but the spines are shorter. Among fishes of this group are the waspfish *Centropogon australis* of northern and eastern Australia and the bullrout *Notesthes robusta*, with the same distribution; the scorpion fish *Scorpaena plumieri* (Brazil to Massachusetts), *Scorpaena guttata* of California, *Scorpaena porcus* of Africa and the Mediterranean; *Scorpaenopsis diabolus* of Australia, Polynesia and Indonesia; and the lup *Inimicus japonicus* of Japan. Effects of stings are like those of *Pterois*.

3. The stonefish type has short and robust stinging spines, each invested with a conspicuous dermal venom gland from which a duct runs along the spine to enter the puncture wound. These fishes are the most dangerous species and include the deadly stonefish *Synanceja horrida* (Fig. 14.9), of Indonesia, India, China, and Australia; the hime-okoze *Minous monodactylus* of Polynesia, China, and Japan; and *Chloridactylus multibarbis*, with the same distribution.

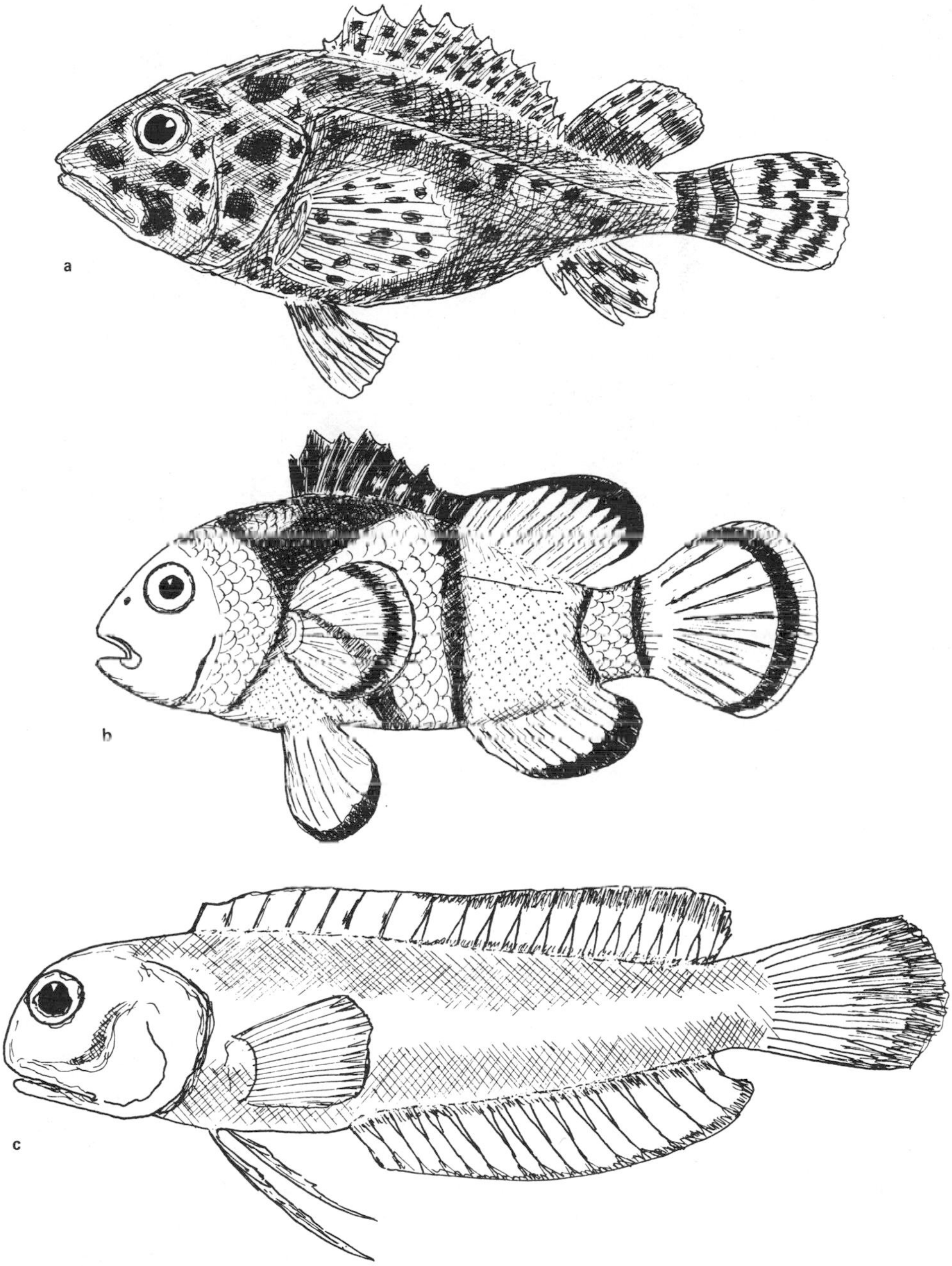

GERALD HESLINGA

Fig. 14.7 (a) Reef scorpion fish, *Scorpaenodes caribbaeus,* Scorpaenidae, Caribbean. (b) Anemone clownfish, *Amphiprion percula,* Pomacentridae, Indo-West Pacific. (c) Yellowhead jawfish, *Opisthognathus aurifrons,* Opisthognathidae, Caribbean.

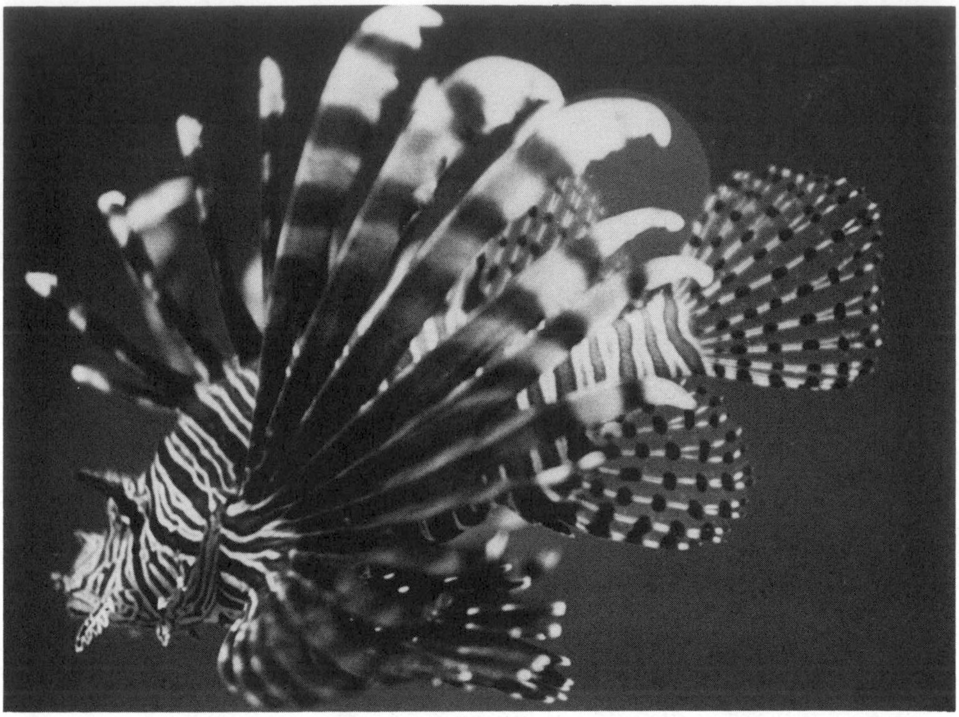

ISIDORO ZARCO

Fig. 14.8 Lionfish, *Pterois volitans,* Scorpaenidae, Indo-West Pacific.

Adaptive Coloration

Hugh Cott (1940) in the course of an exhaustive and scholarly study of animal coloration, isolated some 144 categories of what appear to be adaptive coloration of evolutionary value or significance in relationship to the mode of life or the environment. These he arranged in three broad groupings, namely concealment, advertisement, and disguise. Reef fishes were among the numerous assemblages of animals considered. The following is a resume of Cott's ideas, moderated by some more recent studies made possible in part by the development of free diving as an investigatory technique available to biologists.

1. Cryptic Coloration. Obliterative patterns include: **countershadings** as in offshore species, such as the Spanish mackerel *Scomberomorus,* which feeds at the surface and has its illuminated upper surface tinted bluish green and the undersurface a silvery color, to match the background according to the viewing angle (Fig. 14.10); **physiological color change,** as in the demersal pleuronectid, with their remarkable power to open or close chromatophore cells in the skin to match the background; **disruptive coloration,** as in most members of the family Chaetodontidae, in which contrasting bars of light and dark break up the outline of the fish; **coincident disruptive coloration,** in which vital organs such as the eye are concealed by disruptive means, again as in the Chaetodontidae; **general resemblance to medium,** in the blue chromis, which is common in blue water above the outer reefs, though highly conspicuous in another environment.

Demersal bottom-dwelling fishes may adopt a mottled color pattern, disrupting the outline of the animal and tending to match it to the variegated bottom.

ISIDORO ZARCO

Fig. 14.9 Deadly stonefish, *Synanceja horrida*, Scorpaenidae, Indo-West Pacific.

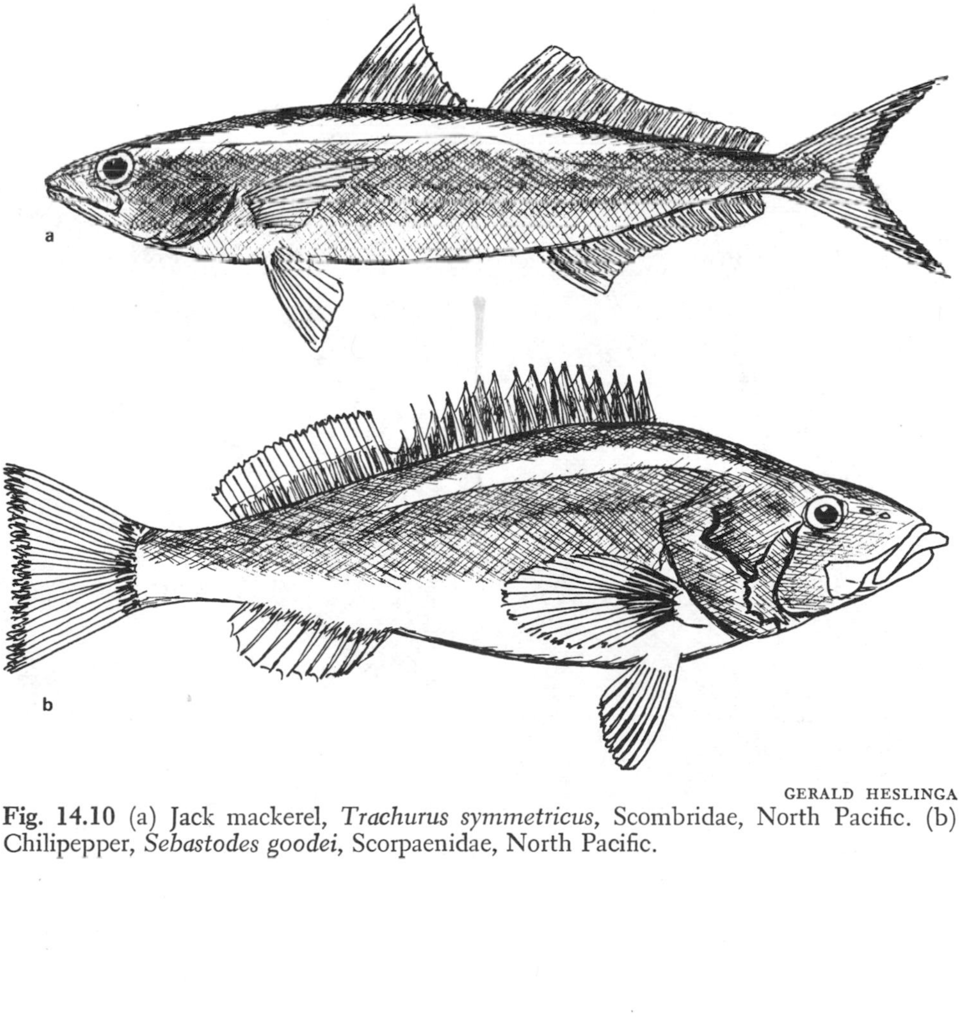

GERALD HESLINGA

Fig. 14.10 (a) Jack mackerel, *Trachurus symmetricus*, Scombridae, North Pacific. (b) Chilipepper, *Sebastodes goodei*, Scorpaenidae, North Pacific.

Examples are the torpedos, often with ring-shaped markings, and the mottled pattern of the upper surface of the spotted eagle ray (Fig. 14.18). These would appear to be protective rather than aggressive adaptations.

2. Sematic or Warning Coloration. This is adopted by many animals that are either poisonous to eat or can inflict a venomous wound. Examples of the former are spotted and chain morays (Fig. 14.11) (family Muraenidae), whose flesh is often toxic; and an example of the latter is the lionfish *Pterois* (family Scorpaenidae, Fig. 14.8). The various soft-skinned puffer-fishes (family Tetraodontidae, Fig. 14.29) frequently have patterns of conspicuous bars or stripes, possibly serving as sematic warnings. An example is the (somewhat tautologically named!) deadly death-puffer *Arothon hispidus* or Makimaki, which ranges the whole tropical Indo-Pacific reef region from the Red Sea to Panama.

3. Disguise Coloration. Deflective characters include **false eye marks** that deflect predator attack from vulnerable to less vulnerable parts of the body, as for example in the four-eye butterfly-fish (*Chaetodon capistratus*, Fig. 14.12), where the posterior end of the animal is likely to be mistaken as the head and **deflective** colors that divert a predator to less vulnerable members of a species, as for example, the dull colors of female parrotfishes and younger males, in contrast to the brilliant colors of terminal males. The sergeant major (*Abudefduf saxatilis*), a pantropical pomacentrid, in which the male is light yellowish to bluish, with several dark bars; but when guarding eggs it becomes a deep blue, and thus much less conspicuous.

Territorial Behavior

Some alternative (though not necessarily mutually exclusive) explanations for adaptive coloration have been offered in recent years. Among the chaetodonts, for example, where vertical contrasted stripes are of common occurrence, it has

DAVID F. MOYNAHAN

Fig. 14.11 Spotted moray, *Gymnothorax moringa,* Muraenidae, tropic amphiatlantic.

GERALD HESLINGA

Fig. 14.12 (a) French angelfish (young), *Pomacanthus paru*, Chaetodontidae, Caribbean. (b) Four-eye butterfly-fish, *Chaetodon capistratus*, Chaetodontidae, Caribbean.

been observed that these are apt not to be schooling species, but rather solitary or living in restricted reef situations as members of bonded pairs. It is probable that the distinctive species colors enable members of the same species to recognize one another, and when an intruder fortuitously enters the territory of another specimen of the same species, he either leaves after recognizing the occupier or is driven off when the occupier recognizes the intruder. Various progies, as for example *Calamus penna,* may adopt color patterns of dark vertical bars when near the bottom of the sea; this is suggestive of territorial advertisement, too.

Commensal Relationships

The advent of undersea observation by scuba-diving naturalists has brought to light many examples of cooperative coexistence (commensalism) or other interspecific relationships. Probably the most interesting of these have been the numerous cases of small species (or commonly the young stages of larger species) serving as **cleaners** (parasite removers) for other fishes, including very large species. A series of seven color photographs by Faulkner (1970) shows gobies and wrasses performing this function, and the young stages of some chaetodonts are also known to be cleaners. In some cases service stations are set up for visiting clientele, as for example the Caribbean neon goby (*Elacatinus oceanops*), a pair of 2-inch long gobies setting up a jointly operated clinic for fishes as large as a 60-pound sea bass (*Epinephelus itajara*). In Queensland waters off Australia, Faulkner found small wrasses (*Labroides dimidiatus*) similarly cleaning bass (*Plectropomus maculatus*), "fearlessly entering its mouth to pick parasites," later to "exist through the gill openings . . . like many fishes, the bass extends its gill rakers to facilitate cleaning." Faulkner found that other Pacific wrasse operate a cooperative clinic in the Scandinavian style, with more than one species of doctor attending to visiting patients. Still others, such as *Labroides bicolor*, operate a fixed clinic in the young stages then, as they grow older and larger, gradually extend the limits of the station until, at adulthood, they are on continuous housecall over a territory of up to 10,000 square feet. Schools of oceanic fishes, such as jacks (family Carangidae), apparently visit the reefs in order to be cleaned.

Fishes such as *Fierasfer* and *Carapus* shelter in the cloaca of large holothurians, backing in much the same manner as a jawfish (Fig. 14.7) backs into its burrow. Some fishes, such as the gobiesocid *Diademichthys* in Indonesia, take refuge among the long spines of *Diadema*, an urchin that presumably affords them protection.

Predator–Prey Relationships and Species Diversity

It follows that, because fishes are mostly predators, there is a great diversity of predators (and their patterns of predation) on coral reefs and in the associated shallow lagoons. As already noted, the invertebrates of coral reefs are the most diverse and richest of any marine ecosystem, a point stressed by Gunnar Thorson (1971). Predators and prey are intimately related in a complex food web. It might perhaps be supposed that the invertebrate biota would be even more diversified if they were not subject to the severe predation exerted by the rich predator community; but the case is the reverse, and the diversity of both predator and prey is mutually supported. A few years ago Paine (1966) artificially eradicated the dominant predators (mainly starfishes of the genus *Pisaster*) from a delimited stretch of coast in Washington. He then observed the consequent changes in the rather simple food web of the habitat and community. At first, as could be expected, there was a rapid increase in the prey organisms, especially a species of barnacle on which *Pisaster* feeds. But later certain other prey organisms, notably the bivalve *Mytilus*, became more common, eventually displacing much of the barnacle population.

The end result was a less diversified community than was initially the case. Evidently the removal of the predator permitted all the prey species to increase to the maximum carrying capacity of the habitat, followed by intense competition for food and space and eventual elimination by natural selection of the less vigorous or less efficient members. Thus it appears that the effect of predators is to apply selection to a prey community in such a way as to override the effects of competition based on individual efficiency or vigor, or in effect to randomize the predation pressure. So it seems likely that the more diversified the predator population, the more diversified the prey population. It is probable that the diversity of both predators and prey on coral reefs is an ancient feature and one in which the two elements are mutually interacting so that either party reinforces the diversity of the other. In the present rather puzzling population explosion of the starfish *Acanthaster*, certain reef communities have become saddled with a dominant uniform predator population that consistently destroys a fixed range of coral species. The corresponding depletion of reef populations would seem to be a predictable outcome of Paine's observations, as mentioned above. For example, since many of the butterfly-fishes and angelfishes (Fig. 14.12) feed largely upon the tentacles of tubeworms and the tube feet of echinoderms (sea urchins) no activity of *Acanthaster* in destroying living coral would affect them; but coral-eating ophiuroids would be adversely affected and would tend to be displaced from the community.

SYSTEMATIC REVIEW OF FISHES

Cyclostomes

These are archaic eellike animals with relatives in the lower Paleozoic. The living representatives are few and comprise the lampreys and hagfishes. The hagfishes are exclusively marine, but lampreys (Fig. 14.13) usually have at least the early stages of life in freshwater. All cyclostomes are predators, attacking larger vertebrates and using their rasping, plated tongue to bore through the body wall. Injured and dead whales are attacked by hagfishes in large numbers.

Marine Fishes

Sharks and Rays

Although more ancient types of sharks are known, the first sharks of modern aspect appeared at the end of the Triassic period, about 200 million years ago. The gills open to the exterior by a number of separate slitlike openings, a character described as **elasmobranch**, and used to form the name of the subclass to which true sharks and rays belong, as shown in Table 14.3.

By the late Jurassic the reef shark faunas had come to resemble those of modern seas, and in the following period, the Cretaceous, nearly all sharks prove to belong to the genera that still exist today. (See Figs. 14.14 and 14.15.) Thus shark genera have as great an antiquity as do most of the extant reef invertebrate genera. The pantropical distribution of such genera in modern seas can evidently be explained by the same mechanisms that have been used to explain the present dispersion of echinoderm genera, namely, the circumglobal continuity of tropical seas through most of the Tertiary.

The first modern types of rays differentiated from the general elasmobranch stock during the latter part of the Jurassic period. They were rare at first, but by

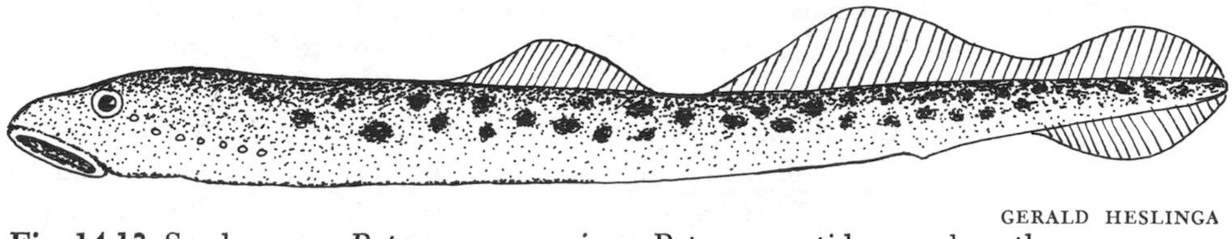

GERALD HESLINGA

Fig. 14.13 Sea lamprey, *Petromyzon marinus*, Petromyzontidae, cool northern seas.

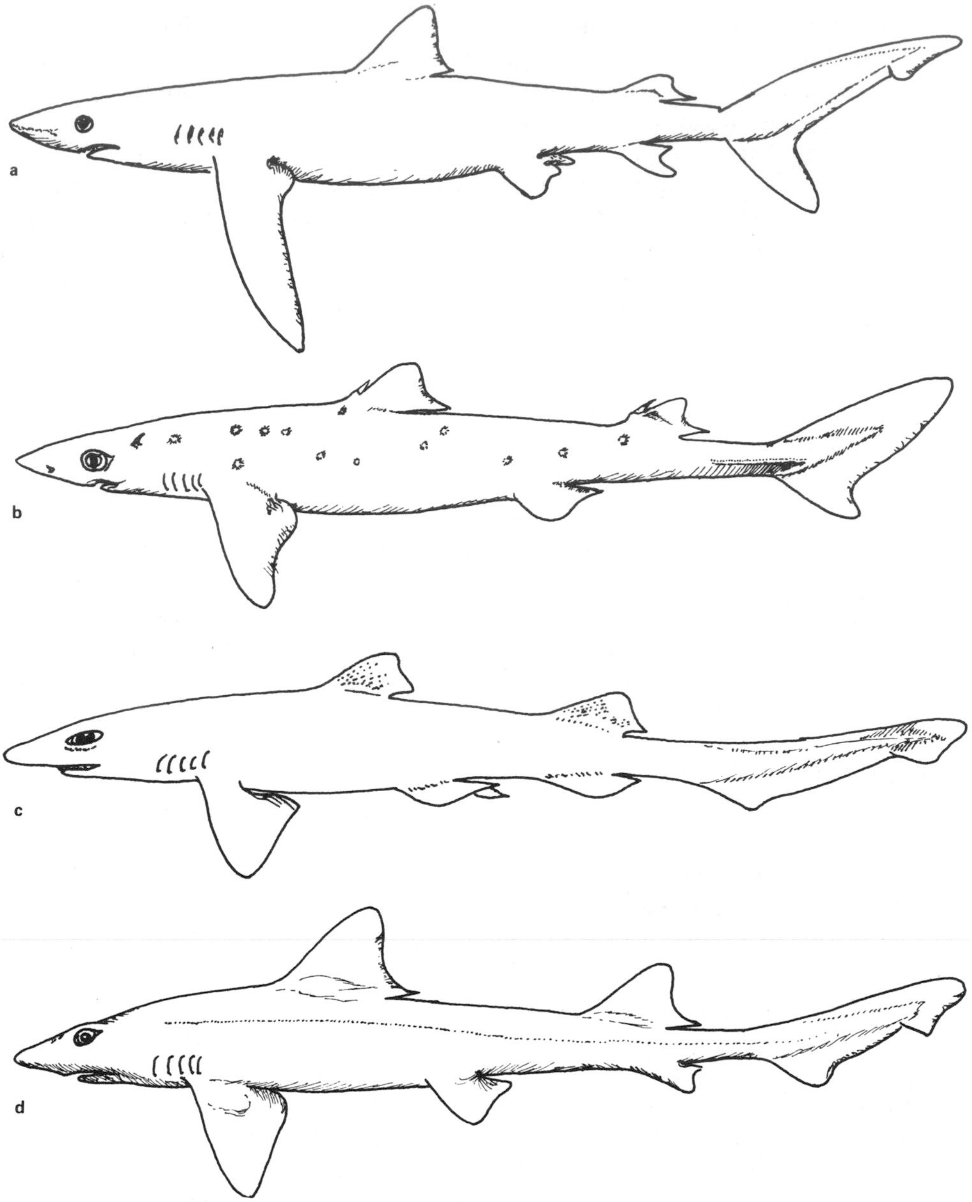

GERALD HESLINGA

Fig. 14.14 North Atlantic sharks and dogfish. (a) Great blue shark, *Prionace glauca,* Carcharinidae, 2 m. (b) Spiny dogfish, *Squalus acanthinus,* Squalidae, 1 m. (c) Smooth hound, *Triakis barnouri,* Triakidae, 1 m. (d) Dogfish, *Mustelus canis,* Triakidae, 1 m.

the late Cretaceous (about 80 million years ago) rays had assumed as conspicuous a position in reef faunas as is the case today, with all major existing families present except for eagle rays and electric rays (Fig. 14.18). These are Cenozoic differentiates.

In present-day seas about 500 species of elasmobranchs are known, most of them confined to shoal waters, although some of the best-known sharks and the largest rays are largely pelagic species of the open ocean. A few inhabit the deepest parts of the ocean. Apart from the diagnostic features already mentioned, elasmobranchs possess other characters. They have no swim bladder; their skin is not

Table 14.3 Key Characters of the Orders of Cartilaginous Fishes (Class Chondrichthyes)

Class characters: vertebrates breathing by means of gills in water; the body of fishlike shape, with paired pectoral and pelvic fins, upper and lower jaws; the skeleton of cartilage, without bone elements. Three extant orders, as follows:

Several (5 to 7) separate slitlike gill openings on each side of the neck region (subclass Elasmobranchia)		
Body elongate, tapering evenly into tail	Selachii	Sharks
Body flattened, abruptly narrowed behind trunk	Batoidei	Rays
Single gill opening on either side of neck region	Holocephali	Ghost sharks

covered by scales, but small, spinous dermal **denticles** are imbedded in it and impart a rough texture to most species; copulation occurs in all species, the male having clasping organs developed on the pelvic fin for attaching to and holding open the female genital aperture; the development of the embryo may either occur in the uterus, in which case live birth follows, or else fertilized encapsulated eggs may be laid and subsequently hatch independently of the parent. (See Table 14.4.)

Man as Prey to Sharks. Of the various, large carnivores that are capable of preying upon man, sharks alone remain uncontrolled. Although rigorous programs of selective trapping have been instituted on the Australian coasts (where most shark attacks occur) and have brought the problem under control, many fatalities still occur in exceptional circumstances, particularly after shipwrecks and plane crashes. Commercial shipwrecks and passenger-plane crashes are usually reported discreetly by the news media for the sake of good relations with the corporations concerned. Not so, however, is the case in time of war, when most losses involve noncommercial vessels and aircraft. During World War II it became apparent that severe losses of life following wrecks were due to attacks by swarms of sharks attracted to the scene by the presence of men in the water. Injured people are always attacked first, presumably because blood attracts a shark. People floating in life jackets are usually attacked below water first. Sharks that are the size of the tiger shark (*Galeocerdo*, Fig. 14.16), 3 to 5 m long, attack by dismembering the limbs one by one. The white shark *Carcharodon*, above 6 m in length, can swallow

Table 14.4 Key to the Principal Families of Tropical Elasmobranchs

Body tapering evenly to tail (order Selachii—sharks)	
Eyes on lateral projections of head (hammerhead sharks)	Sphyrnidae
Eyes normally placed	
A barbel before each nostril (nurse sharks)	Orectolobidae
No barbel	
Epural lobe of tail elevated, elongate (requiem sharks)	Carcharinidae
Epural lobe short, horizontal (leopard sharks)	Triakidae
Body flattened, tail abruptly narrowed behind trunk (order Batoidei—Rays)	
Head, trunk, and pectoral fins fused to form a flattened disk	
Tail bearing an erect dorsal spine (stingrays)	Dasyatidae
Tail bearing 1–2 dorsal fins (electric rays)	Torpedinidae Orectolobidae
Head partly distinct from kite-shaped body; tail whiplike, with spine at its base (eagle rays)	Myliobatidae

GERALD HESLINGA

Fig. 14.15 Hammerhead sharks. (a) Bonnet shark (or shovelhead), *Sphyrna tiburo,* pantropical, turbid shelf waters. (b) Hammerhead, *Sphyrna bigelowi.* (c) Great hammerhead, *Sphyrna makarran.* (d) *Sphyrna diplana.* All ca. 2 m.

GERALD HESLINGA

Fig. 14.16 Tropical sharks. (a) White-tipped shark, *Carcharinus longimanus,* 3 m, Carcharinidae. (b) White shark (or maneater), *Carcharodon carcarias,* 3 m, Carcharinidae. (c) Reef shark, *Carcharinus springeri,* Carcharinidae. (d) Sand shark, *Carcharias taurus,* Carcharinidae, 1 m. (e) Tiger shark, *Galeocerdo cuvieri,* Carcharinidae, 2 m.

a man whole; but the younger specimens usually bite a man in half first. About 20 species are believed to be potential maneaters. Other smaller species take an occasional bite if given the opportunity. During World War II tens of thousands of sailors are believed to have been killed by sharks, most while clinging to ropes on the sides of lifeboats or to flotsam, or while swimming or supported by life jackets. Of the few civilian cases on record in detail, an example is an airliner that crash landed apparently intact on the surface of the sea on November 16, 1959, 120 miles southwest of New Orleans. The 42 passengers and the crew all died, either from sharks or by drowning, 10 half-eaten bodies being recovered. The extensive research on so-called shark repellants has not thus far produced a remedy. The supposedly effective "repellant" issued to crewmen by naval authorities had, in fact, no more than a psychological value. Active research is in progress, with periodic international conferences by shark experts to review results, thus far mostly producing a confusing mass of seemingly contradictory data.

Sharks are widely distributed, mainly in warm waters, though not exclusively so. Many sharks are very large fishes, none are very small. The largest species now surviving are the basking shark, *Halsydrus maximus*, and the whale shark, *Rhineodon typicus*; both are known to reach lengths of 40 feet (12 m), but unconfirmed records of up to twice that size exist. However, extinct species reached 24 m. The largest sharks are unaggressive if unmolested, feeding on plankton and on invertebrates on kelp. The white shark *Carcharodon carcharias* (Fig. 14.16) occurs in warm seas throughout the oceans. (See Fig. 14.17.) A large specimen measures 36 feet (11 m), and one of that size at Sydney, Australia had teeth 2 inches long. Fossil teeth of the same genus are known about 6 inches across, suggesting a shark of 80 to 100 feet in length. A 20-foot white pointer swallows a man whole, whereas a 12-foot specimen first bites him in half. One Californian specimen contained a 100-pound sea lion. The tiger shark *Galeocerdo cuvier* is another maneater and ranges so far north as Iceland. A specimen taken at Durban was found to contain the anterior half of a crocodile, the hind leg of a sheep, three seagulls, two 2-pound cans of peas, and a can of cigarettes; a New Zealand specimen contained a large collie dog, one blue penguin, and two large lobsters. The hammerhead shark *Sphyrna lewini* (Fig. 14.15) ranges all warm seas and is often seen in mid-ocean. The young stages feed on mackerel. A specimen from Riverhead, New York, contained many portions of a man together with his clothes. Sharks enter rivers and swim far inland, some of the most serious Australian attacks having occurred in these circumstances. Nonetheless, the chances of death by shark bite are much less than the chances of being run over on a street. Sharks are valuable scavangers, and there can be no doubt that many of the land animals found in their stomachs were ingested after their cadavers were washed to sea by rivers.

The thorny skate *Raja radiata* ranges the continental shelf around the northern margins of the North Atlantic, and on the American side it appears as far south as Cape Cod. It feeds on shrimps, crabs, annelids, and small fishes. South of Cape Cod the species ranges on the continental slope at least so far south as North Carolina, where it descends to depth of 300 fathoms. Skates or rays occur all over the world (Fig. 14.18). All have a demersal habit. The largest rays are the devilfishes or mantas, mostly found swimming in open ocean waters near the surface. A large devil ray, *Manta birostris*, reaches 20 feet (6 m) across and weighs 3500 pounds (1.6 metric tons).

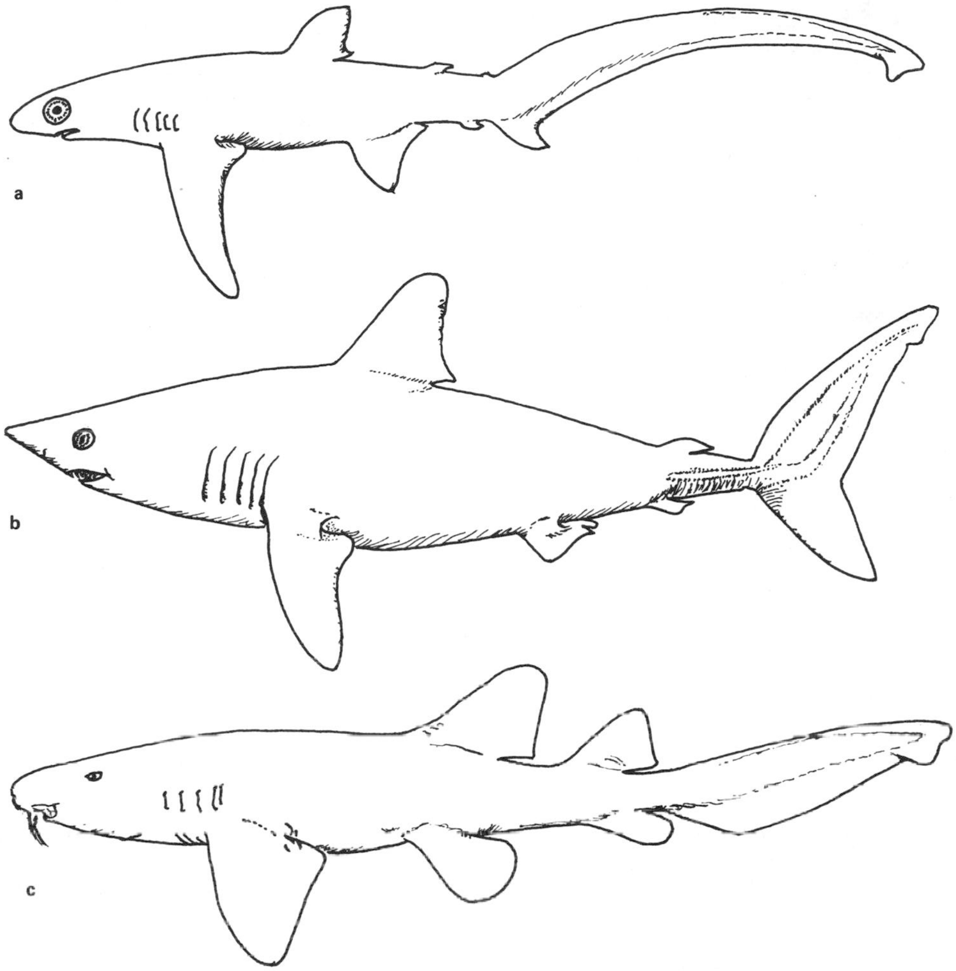

GERALD HESLINGA

Fig. 14.17 Tropical sharks. (a) Thresher shark, *Alopias superciliosus*, 2 m, Alopiidae. (b) Mackerel shark, *Lamna nasus*, Isuridae, 2 m. (c) Nurse shark, *Ginglymostoma cirratum*, Orectolobidae, 1 m.

Bone Fishes

Sturgeons (Fig. 14.19). Another ancient group of fishes are sturgeons. *Acipenser oxyrhynchus* ranges neritic waters from Hudson Bay to the Gulf of Mexico. It reaches a length of 18 feet (5.5 m). Sturgeons are bottom feeders. The barbels are probably the sensory organs with which they detect worms of the infauna; the snout is used to disturb the bottom in search of food. When mature sturgeons ascend rivers to spawn in freshwater, the young may remain in rivers and lakes for 1 to 3 years, leaving them for the sea when they reach 1 to 3 feet long. A sturgeon 8 feet long is estimated to be about 12 years old; it weighs about 85 kg (190 pounds). Sturgeon were common in American waters when the first colonial settlements were established. They are now rare, and the sturgeon fishery has disappeared. They are captured only occasionally as incidental species associated with other fisheries.

Table 14.5 Key Characters of the More Conspicuous Orders of Bone Fishes (Class Osteichthyes)

Class characters: vertebrates breathing by means of gills in water; the body of typical fishlike shape, with paired pectoral and pelvic fins, upper and lower jaws; the skeleton principally of bone, cartilage also occurring. Numerous orders, of which the following are important in marine environments:

Tail fin with long upper lobe containing tip of vertebral column (heterocercal); Mouth below the snout; bone plates form rows along flanks	Acipenseriformes	Sturgeons
Tail fin not heterocercal, the two lobes of fin usually similar (homocercal)		
Both eyes on one side of head; body flattened, demersal	Pleuronectiformes	Flatfishes
One eye on each side of head		
Dorsal and tail fins continuous; no pelvic fins	Anguilliformes	Eels
Dorsal fin separate from tail fin		
Snout tubular; body enclosed in bone plates	Syngnathiformes	Sea horses
Snout not tubular; body not enclosed in bone plates		
Fin rays of separate segments, not fused into stiff spines		
Chin barbel present; 2 or 3 dorsal fins	Gadiformes	Codfishes
No barbels; pelvic fins set far back	Clupeiformes	Herrings
Similar, but pectoral fins set high on body	Beloniformes	Flying fishes and halfbeaks
Fin rays not segmented, fused to form stiff or flexible spines		
Anterior dorsal fin placed on head	Lophiiformes	Angler fishes
Not so	Perciformes (and so on)	Perchlike fishes

Herrings. The clupeoid or herringlike fishes are also ancient. Fossils of the oldest fishes with clupeoid structures occur in Triassic rocks; and more herringlike members arose in the lower Jurassic, about 150 million years ago. Their characters are indicated in Table 14.6. The most distinctive feature, and one by which a clupeoid marine fish may be recognized at sight, is the low insertion and wide separation of the pectoral and pelvic fins, the latter lying well back on the body as is the case in sharks. About 30 families are recognized, best known being the Clupeidae, or herrings proper, and sardines. Other families include the salmon and trout, a few of which are marine, most however being freshwater fishes. Several other clupeiform families are encountered in freshwater environments. One deep-sea family, the Stomiatoid viperfishes and a few other deep-sea families, complete the roster of clupeiform fishes. (See Figs. 14.20 and 14.21.)

Table 14.6 Key to the Principal Families of Tropical Teleosts

Pelvic fins abdominal, without spines, and with 5+ soft rays (order Clupeiformes)	
Snout overhangs lower jaw, which does not reach behind eye (bonefish)	Albulidae
Mouth terminal, jaw opening extends behind eye (tarpons)	Elopidae
Pelvic fins thoracic or jugular	
No pelvic spines, 5+ pelvic soft rays; 1 or more anal spines	Holocentridae
Pelvic fins with 1 spine and fewer than 5 soft rays (order Perciformes)	
Dorsal fin subdivided into 2 or more separate fins	
Fringe of finlets extends from second dorsal fin to the forked tail	
2 spines precede anal fin (jacks)	Carangidae
No free preanal spines (mackerel and tuna)	Scombridae
No fringe of finlets follows the second dorsal fin	
Pair of barbels on chin (goatfishes)	Mullidae
Dorsal fin continuous, not subdivided into 2 or more parts	
Pelvic fins jugular, anterior to pectoral fin	
Pelvic spine very long; burrowers (jawfishes)	Opisthognathidae
Pelvic spine short; lower jaw projects (soapfishes)	Grammistidae
Pelvic fins thoracic, below or just behind base of pectoral fin	
Spiny ridge crosses cheek below eye (scorpion fishes)	Scorpaenidae
No spiny cheek-ridge	
11 or 12 branched caudal rays. Scales cycloid.	
Single series of teeth along front of jaw (wrasses)	Labridae
Front teeth crowded or fused into a beak (parrotfishes)	Scaridae
15+ branched caudal rays. Scales ctenoid.	
Two anal spines. Nostril a simple port (damselfish)	Pomacentridae
3 or more anal spines; 2 nostrils on either side	
No spines at tail base (angels and butterfly-fish)	Chaetodontidae
1 or more sharp spines on tail base (surgeonfish)	Acanthuridae
Mouth large, extending as far back as nostrils	
Maxillary not sliding under cheek (groupers)	Serranidae
Maxillary slides up under cheek when mouth closes	
Teeth feeble (grunts)	Pomadasyidae
Strong canine teeth	
Snout flattened; teeth on palate (snappers)	Lutjanidae
Snout convex; palate toothless (porgies)	Sparidae
Pelvic fins lacking	
Body eel-shaped. No pectoral fins (moray eels)	Muraenidae
Snout tubular; body wall encased in bony plates (sea horses, etc.)	Syngnathidae
Body more less of typical fish shape (order Tetraodontiformes)	
Dorsal fin divided into anterior erect spine(s) and posterior soft fin	
Scales replaced by minute granules in soft skin (filefishes)	Monacanthidae
Scales platelike forming stout armor (triggerfishes)	Balistidae
Single dorsal fin of soft rays only without spines	
Body covered by sharp spines. Teeth on each jaw fused into a beaklike structure (porcupine fishes)	Diodontidae
Body not covered by erect spines, though skin may have a prickly texture; median division separates fused teeth of either side of jaw (puffer fishes)	Tetraodontidae
Body enclosed in an interlocking mesh of bony plates imbedded in the skin. Frontal horns may occur (box- and cow-fishes)	Ostracionotidae

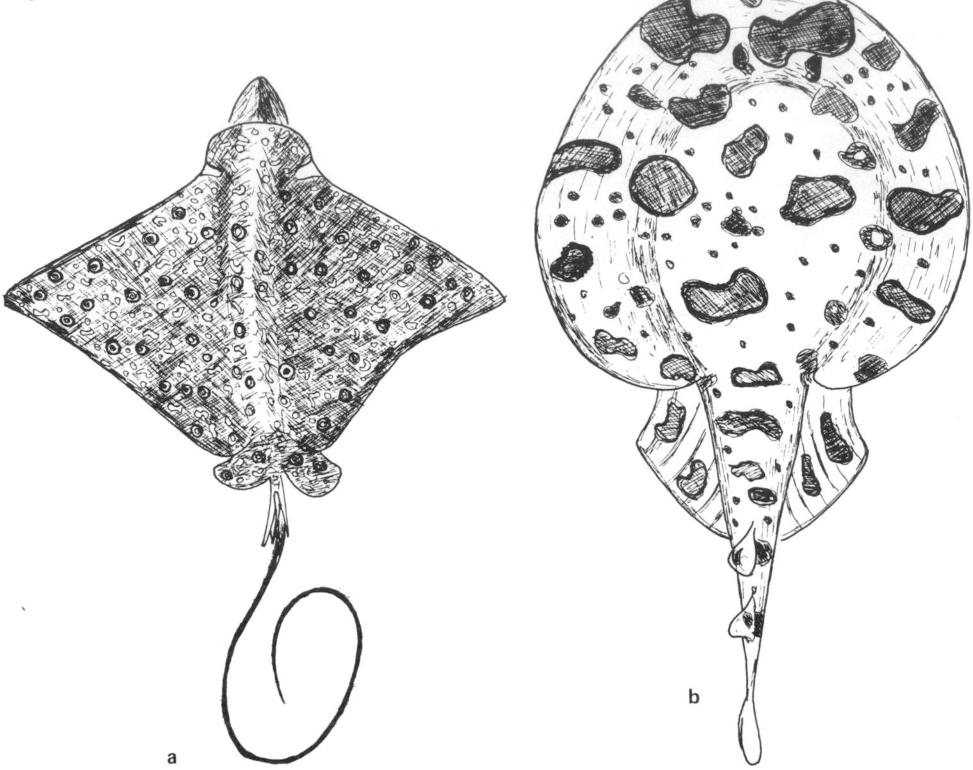

GERALD HESLINGA

Fig. 14.18 (a) Eagle ray, *Aetobatis narinari*, Myliobatidae, pantropical, 1 m. (b) Electric ray, *Narcine brasiliensis*, Torpedinidae, 100 volts, Caribbean.

The herrings and the salmonids are both essentially northern groups. They are valuable food fishes in the Northern Hemisphere, where extensive fisheries are located. The set herring *Clupea harengus* occurs in large schools, especially in open coastal waters. It feeds on plankton and has a life span of about 10 years, reaching a maximum length of about 37 cm (15 inches). It is eaten by larger fishes and by man.

Flying Fishes and Halfbeaks (Fig. 14.22). A small group of herringlike fishes resemble the clupeoids in the wide separation of the pelvic and pectoral fins, but differ in having the pectoral fins placed high on the body, with the pectoral insertion lying at or above the horizontal axis. There are no fin spines; the swim bladder, which usually has a duct in clupeiform fishes, has no duct in the beloniform species. Best known members of this small order are the flying fishes. Other members include the long-snouted billfish, halfbeaks, lizard fishes with elongate body, and

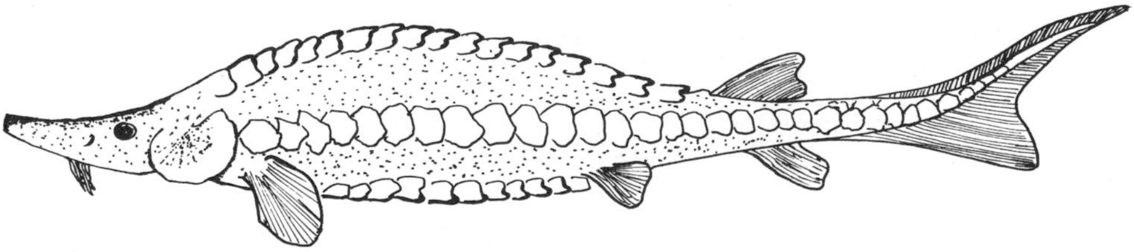

GERALD HESLINGA

Fig. 14.19 Shovelnose sturgeon, *Scaphirhynchus elatorhynchus*, North Pacific.

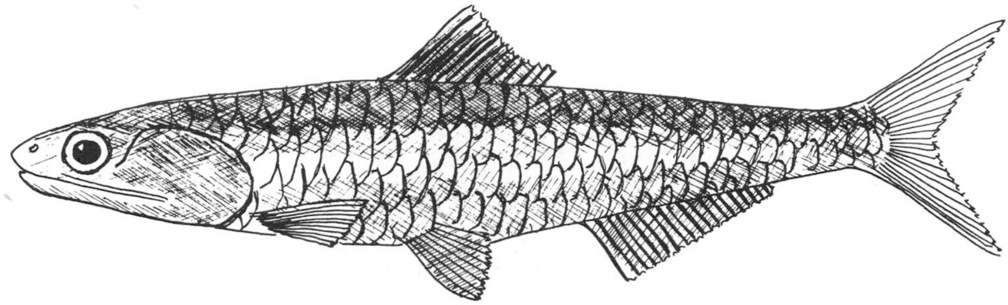

GERALD HESLINGA

Fig. 14.20 Northern anchovy, *Engraulis mordax*, Clupeiformes, North Pacific.

various other fishes, which are mostly inhabitants of warm surface waters. Flying fishes have enlarged pectoral fins that they use to plane from wave crest to wave crest, gliding rather than flying. They break the surface when pursued by larger pelagic fishes and are equally disturbed by the passage of a ship's hull through the water. Flying fishes most often are seen taking to the air in shoals from the bow waves of ships. With a strong wind the fish may be carried 5 or 6 m in the air and fall on the deck. The oldest known beloniform fishes date from the Eocene period, ca. 70 million years ago.

Eels. The most ancient eels seem to date from the Cretaceous period, about 100 million years ago. Best known are the species that enter freshwater, although these are marine at the breeding stage. Among the strictly marine members of the order are the moray eels of shallow tropical seas, particularly common on coral reefs, and the conger eels. The American Atlantic conger, *Conger oceanica,* ranges from Massachusetts Bay southward, perhaps as far south as Brazil, ranging the whole shelf region from inshore shallows to beyond the edge of the shelf (to 150 fathoms). The adults feed on smaller fishes, crustaceans, and mollusks. They do not enter freshwater. In the breeding season the conger migrate into deep oceanic waters, where the young stages (called leptocephali) are found.

Sea Horses and Pipefishes (Fig. 14.4). Tubular-snouted fishes are mostly found in warm shallow waters, but the American Atlantic species or Northern sea horse ranges from the Carolinas northward to Nova Scotia. *Hippocampus hudsonicus* conceals itself by clinging to vegetation of any kind in shallow water. It feeds on planktonic crustaceans, as do also the pipefishes. The male carries the developing eggs in a brood pouch from which a hundred or more seahorses may be born simultaneously. Pipefishes (*Syngnathus* spp.) have elongate tubular bodies and they lack the prehensile tail, whereas the sea horses have a more inflated trunk and a slender prehensile tail.

Codfishes (Fig. 14.23). These are structurally intermediate between soft-rayed fishes and the spiny-rayed (or perchlike) fishes. One character they share with the latter is an internal swim bladder that lacks any connection with the gut. This means that deep-sea members of the order, if trapped and hauled too rapidly to the surface, suffer evisceration through the mouth on account of the sudden expansion of gases in the swim bladder under the reduced pressure near the surface. The oldest known members of the order date from the Eocene period, about 70 million years ago. There are five families of Gadiformes, the best-known being the Gadidae, or codfishes. Cod and their relatives (pollock, haddock, hake) all tend to be rather shallow-water fishes, distributed in northern seas. *Gadus callarias* occurs

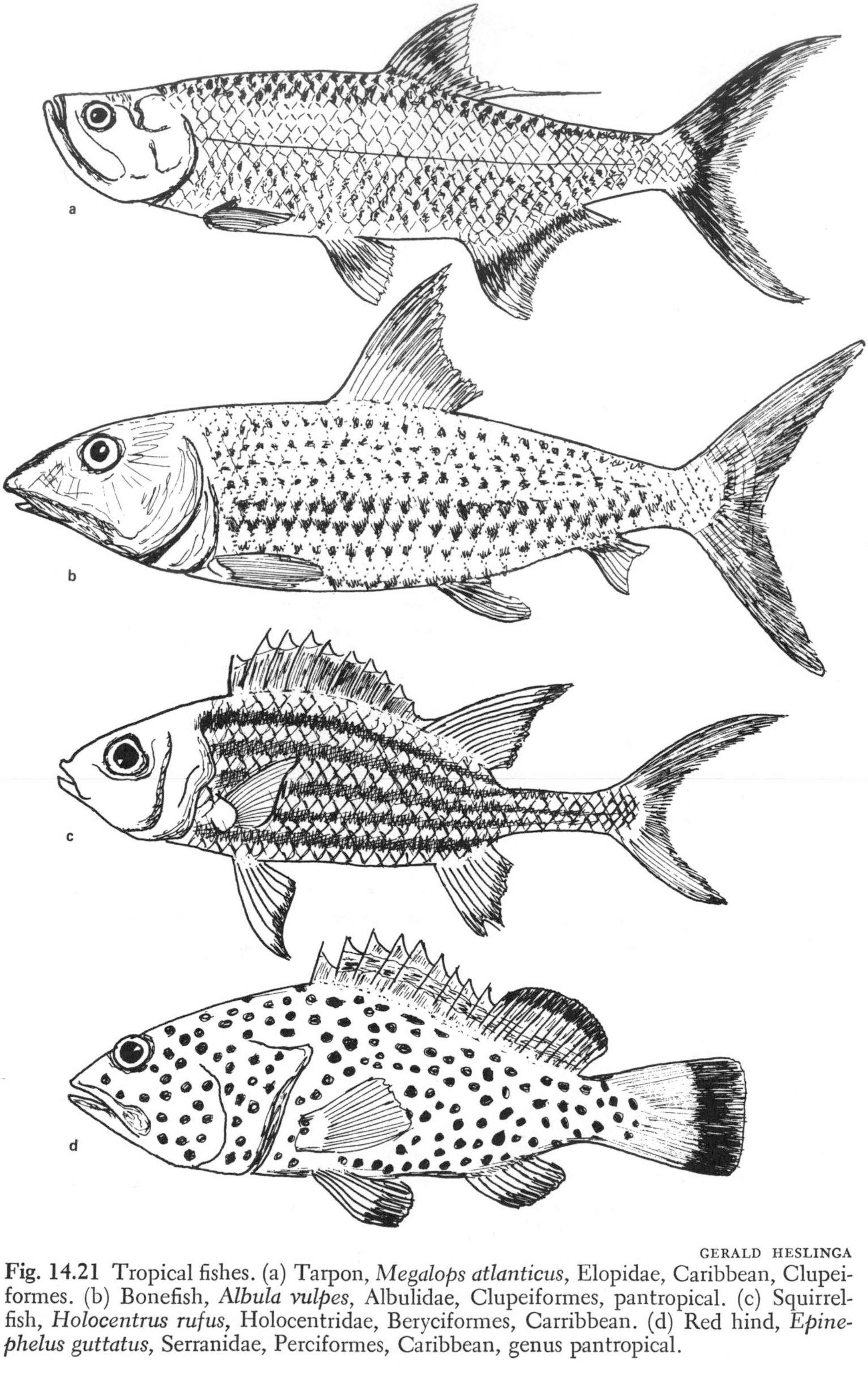

GERALD HESLINGA

Fig. 14.21 Tropical fishes. (a) Tarpon, *Megalops atlanticus,* Elopidae, Caribbean, Clupeiformes. (b) Bonefish, *Albula vulpes,* Albulidae, Clupeiformes, pantropical. (c) Squirrelfish, *Holocentrus rufus,* Holocentridae, Beryciformes, Carribbean. (d) Red hind, *Epinephelus guttatus,* Serranidae, Perciformes, Caribbean, genus pantropical.

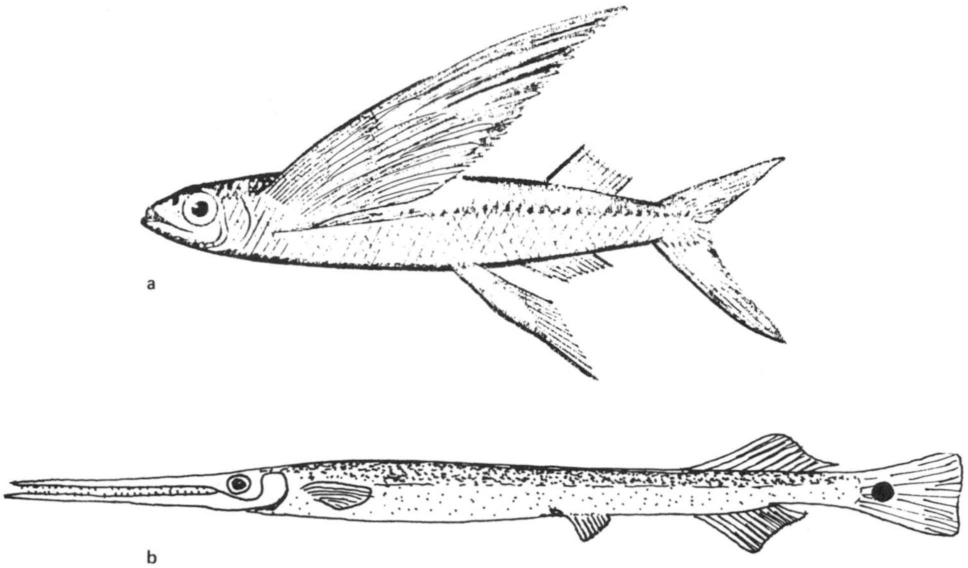

GERALD HESLINGA

Fig. 14.22 (a) Flying fish, *Exocoetus volitans*, Exocoetidae, Beloniformes, pantropical. (b) Blackspot longtom, *Tylosaurus strongylurus*, Belonidae, Beloniformes, Southern Ocean.

on both sides of the North Atlantic, ranging from western Greenland southward to Virginia; but it is most common from Labrador to Cape Cod, where the main fishery occurs. Cod reach a length of about 2 m and a weight of nearly 100 kg; however, most specimens are much smaller. At an age of 7 or 8 years, a cod measures about 1 m. Cod feed on bottom organisms, particularly crabs, lobster, and mollusks; some squid and smaller fishes are also taken. They swallow bright or unusual objects lying on the seabed, such as stones, rope, and bottle caps; jewelry has been found in their stomachs. The cod are fished throughout the year in the more northern waters. The other families of Gadiformes are mainly deep-water fishes.

Flatfishes. This order of fishes is one of the most easily recognizable, for the included members have partially lost the bilateral symmetry that characterizes vertebrates. Young specimens resemble other fishes, but during the course of growth the eye of one side of the head migrates across the top of the skull, and comes to lie beside the eye of the other side of the head. The fishes of this order may be separated into right-eyed and left-eyed groups, depending upon which eye migrates. Any one species is normally always a member of one group, but occasional reversals occur. The purpose of the eye-migration is to permit the fish to lie on the bottom of the sea, on the blind side, without any loss of vision. Fishes that habitually rest on the bottom are termed demersal. Demersal fishes thus become temporary members of the benthos, although they retain the ability to swim and may still be considered as nekton. There are about 500 species of flatfishes, mostly inhabitants of the continental shelves of the world. They generally prefer temperate or warm waters, although a few occur in the polar regions. Of the six families, among the best-known are the Pleuronectidae, or right-eyed flatfishes (Fig. 14.23). These forms, which rest upon the left side, include the North Sea plaice (*Pleuronectes*); Pacific flounders, such as *Pleuronichthys*; and west Atlantic flounders, such as the

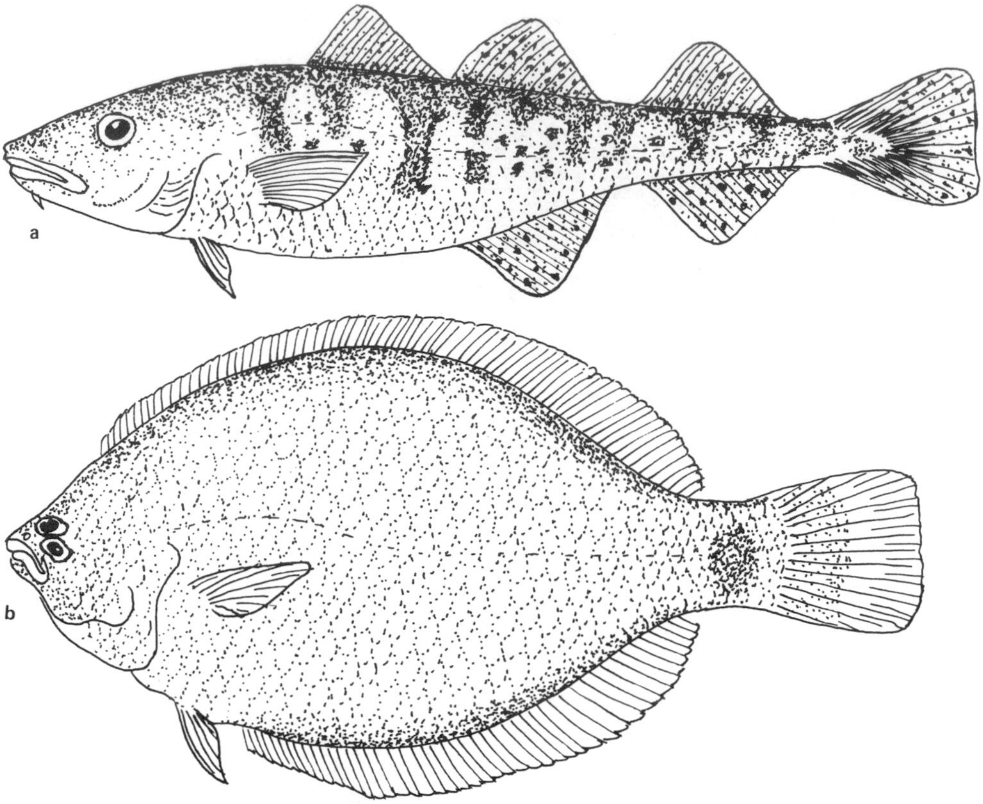

GERALD HESLINGA

Fig. 14.23 (a) Tomcod, *Microgadus tomcod,* Gadidae, Gadiformes, North Atlantic. (b) Bay-whiff flounder, *Citharichthys spilopterus,* Bothidae, Pleuronectiformes, North Atlantic.

blackback (*Pseudopleuronectes americanus*) and the yellowtail (*Limanda ferruginea*)—both species caught in considerable numbers on the New England coast. Flounder occupy quiet waters, especially in sheltered inlets at depths from a few fathoms down to about 50 fathoms. Some species enter rivers. The oldest known flatfishes date from the middle Eocene, about 50 million years ago.

Anglerfishes. Members of the order Lophiiformes are recognizable by the characteristic translocation of the anterior dorsal fin rays to a position on the head, where they are converted to form a fishing lure in many of the species. Most anglers are deep-sea fishes, often of small size, but some species are large demersal forms that inhabit the continental shelf. The best-known species are the east Atlantic anglerfish, *Lophius piscatorius,* and the west Atlantic congener, known in America as the goosefish, *Lophius americanus.* The latter species ranges from Newfoundland south to the Carolinas, reaching a length of over a meter and a weight of over 20 kg. There is no air bladder in anglers. The goosefish is extremely tolerant of wide variation in depth of habitat, probably because of its immunity to changes of pressure. The species is known to occur in water as shallow as the low-tide zone, and as deep as about 350 fathoms on the continental slope. It lies quietly on the bottom, but when a potential food animal approaches it darts forward very rapidly to engulf the prey in its oversized mouth and stomach. In shallow water it is reputed to seize seabirds.

The Perchlike Fishes. As Table 14.6 shows, these are spiny-rayed fishes. There are several orders, differing only in respect to certain internal characters (especially in respect to bones of the head); so it is not feasible to make key to these orders using only external features. However, the great majority of the perchlike fishes belong to the order Perciformes, the largest known order of vertebrates. The Perciformes are the dominant fishes of existing seas, therefore, and they have been so since Eocene times, about 70 million years ago. Although the precise ordinal characters of perchlike fishes are difficult to state without undue technicalities, the various families, on the other hand, are often easy to recognize on account of their external familial characters. There are about 1200 genera arranged in about 140 families. A few of these are noted here.

Examples of the Perciformes. **Jacks** are mainly tropical, but they often drift into temperate regions; mainly fishes of the open sea, but some enter reef environments to hunt other fishes. They include the lookdown, *Selene vomer*, a strangely flattened North Atlantic example, and also the permit and pompano, both species of *Trachinotus* (Fig. 14.24), important Caribbean food fishes. **Tunas and mackerel** are swift-swimming pelagic fishes, often in schools, mainly in offshore blue water. Mackerel, *Scomber scombrus*, feed on plankton, especially copepods. Tuna may reach a length of over 4 m and a weight of 800 kg (1800 pounds). Large tuna do not school; they feed on herring and mackerel. **Snappers** are mostly benthic nocturnal carnivores, frequenting reefs or mud bottom, or in deeper water beyond the edge of the shelf. Yellowtail snapper, *Ocyurus chrysurus* (Fig. 14.24), occurs on the Caribbean reefs. The fishes of this family feed mostly on benthic crustaceans, and the adults also capture smaller fishes. **Angelfishes and butterfly-fishes** (Fig. 14.12) frequent tropical coral reefs. They are usually brightly colored, bold, and active fishes. They are diurnal and feed chiefly on sponges and algae. An example is the French angel, *Pomacanthus parus* (Fig. 14.12), found in the Caribbean. **Parrotfishes** are colorful, herbivorous reef fishes. Their teeth have been modified into strong plates for scraping algae from hard substrates and also for grazing on eelgrass. An example is the blue parrotfish, *Scarus coeruleus* (Fig. 14.25), which is found in the Caribbean. **Scorpion fishes,** mainly inhabit the tropics. They are found over hard bottom and have venomous spines. An example is *Pterois volitans*, the deadly lionfish of the Indo-West Pacific, that ranges from the Red Sea to Polynesia. The wounds from this fish may be lethal. No species of *Pterois* occurs in the Atlantic. The genus *Scorpaena* ranges all warm seas. **Sculpins** are north temperate representatives of the scorpion fishes, lacking the venomous character of their tropical relatives. They have conspicuous fin spines. An example is *Myxocephalus octodecimspinosus*, a sculpin found off New England coasts. **Lumpfishes** have a ventral suctorial disk by which they attach to the seabed. An example is *Cyclopterus lumpus* (Fig. 14.26), which is found off the Massachusetts coast.

Other Perchlike Fishes. Squirrelfishes frequent tropical reefs and also deeper water. They are nocturnal, probably because they are related to deep-sea forms, and have large eyes; by day they hide in crevices. They feed mainly on crustaceans. An example is the red squirrelfish, *Holocentrus rufus* (Fig. 14.21), which frequents the Caribbean. **Triggerfishes** have the first fin spine of the dorsal fin enlarged. Their body is compressed, their eyes high on the head, and their scales are very strong. They are found in outer tropical reefs and tropical open ocean and graze in shallow water on benthic plants, rising to the surface to feed on plankton. An example is black durgon, *Melichthys niger* (Fig. 14.6), which is pantropical. (See Fig. 14.27.)

GERALD HESLINGA

Fig. 14.24 (a) Soapfish, *Rypticus saponaceus*, Grammistidae, Perciformes, Caribbean. (b) Palometa, *Trachinotus goodei*, Carangidae, Perciformes, Tropical Atlantic. (c) Permit, *Trachinotus falcatus*, Carangidae, Perciformes, Tropical Atlantic. (d) Yellowtail snapper, *Ocyurus chrysurus*, Lutjanidae, Perciformes, Tropical Atlantic. (e) Spanish mackerel, *Scomberomorus maculatus*, Scombridae, Perciformes, genus pantropical.

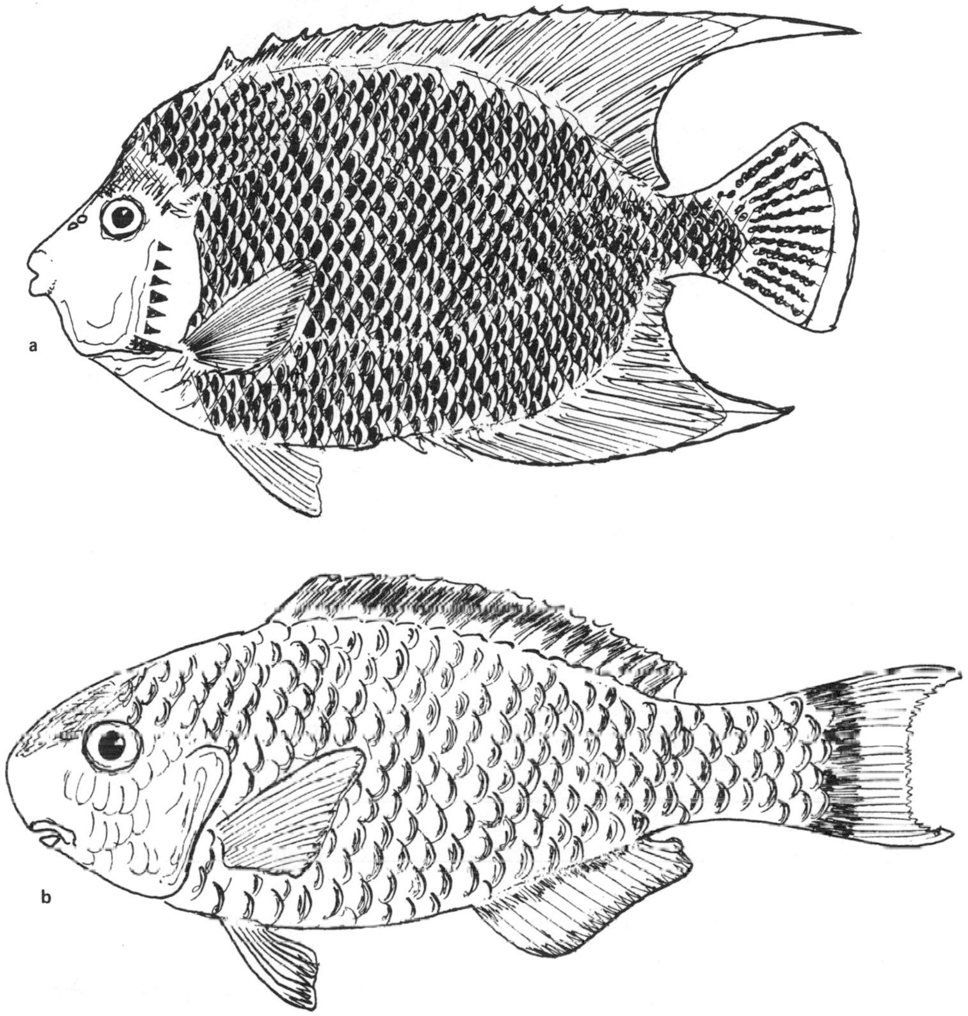

GERALD HESLINGA

Fig. 14.25 (a) Blue angelfish, *Holocanthus isabelita,* Chaetodontidae, Perciformes, Caribbean, genus pantropical. (b) Blue parrotfish, *Scarus coeruleus,* Scaridae, Perciformes, Caribbean.

How to Recognize Fishes

In addition to the anatomical features given in this chapter, you will find numerous illustrations of typical fishes scattered through the book; these will help you to become familiar with the names and classification. A selection of drawings appears in this chapter, and others will be found in the ecological chapters dealing with various regions (Chapters 23 to 28). Study the drawings and match them with the anatomical characters given in related tables found in the chapters mentioned. Soon you will be able to name many fishes at sight.

ISIDORO ZARCO
Fig. 14.26 Lumpfish, *Cyclopterus lumpus*, Cyclopteridae, Perciformes, North Atlantic.

FISHES OF TROPICAL REEFS

> The coasts are free from rocks, except that they all have some reefs near the land under water, on which account it is necessary to keep a sharp look out, and not to anchor very near the shore. . . . There are here fish, so unlike ours that it is a marvel; there are some shaped like Dories, of the finest colors in the world—blue, yellow, red, and of all colors, and others painted in a thousand ways.

By these entries in the log of the *Santa Maria* for October 15 and 16, 1492, and later incorporated into a letter to Ferdinand and Isabella, Columbus gave the world the first hints as to the extraordinary nature of the coral-reef fauna he had discovered in the Caribbean. The coloration and variety of reef fishes in the tropics provide a thought-provoking feature that never fails to fascinate the visitor from higher latitudes—biologist and layman alike. (See Figs. 14.28 and 14.29.)

The rich benthic communities of reef invertebrates yield an abundant source of prey for the vertebrate secondary heterotrophs of tropical seacoasts, of which fishes are by far the most important element. Thus, **reef nekton** must be viewed as an integral part of the tropical seafloor community; fishes occupy most of the upper trophic levels in the pyramid of energy flow through the reef ecosystem. However, in addition to the secondary heterotrophs and dominant carnivores contributed by reef fishes, some of the families operate at the primary heterotrophic level, for example, the essentially herbivorous parrotfishes and the surgeonfishes. On account of their greater mobility, most fishes are much less intimately tied to particular substrates than is the case with most invertebrate groups. In this review, therefore, the fishes of tropical reefs and lagoons are treated in a single systematic sequence.

Faunal Content. Families of fishes that are conspicuous in coral reef communities include: sharks (especially of the families Triakidae, Orectolobidae, also

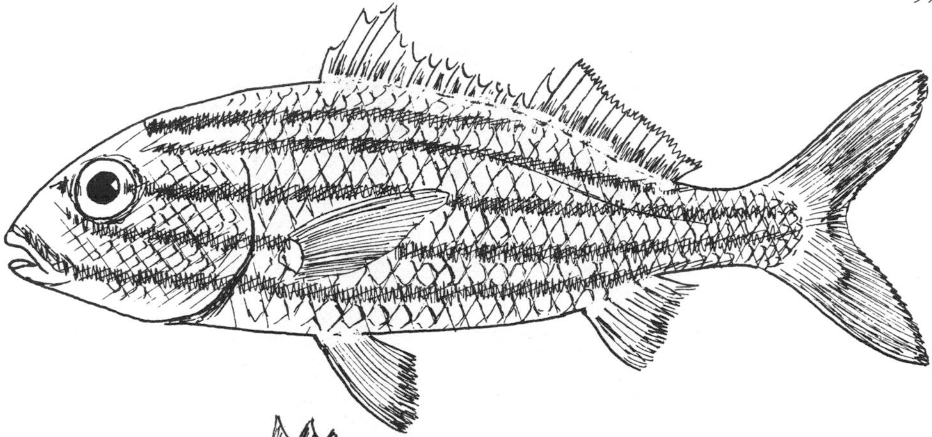

a

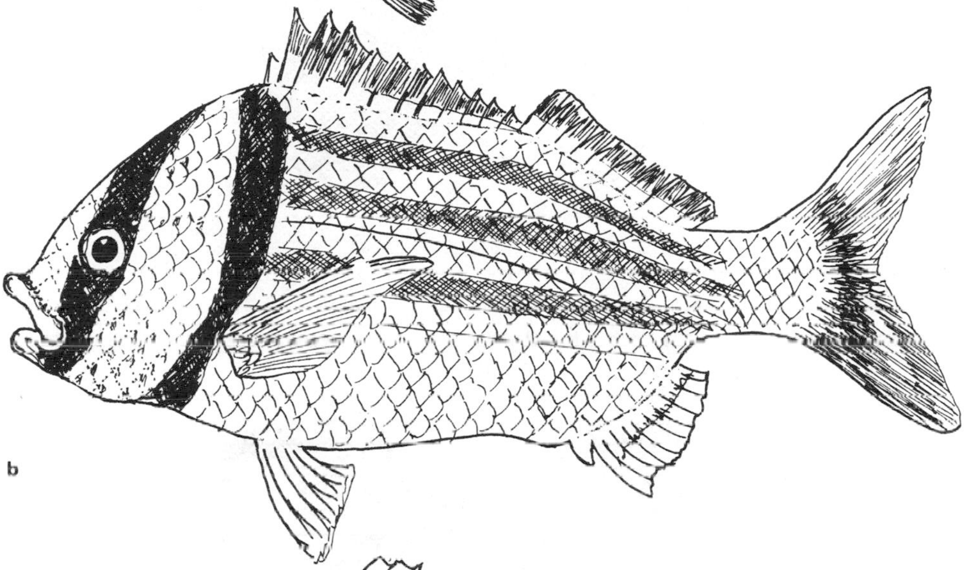

b

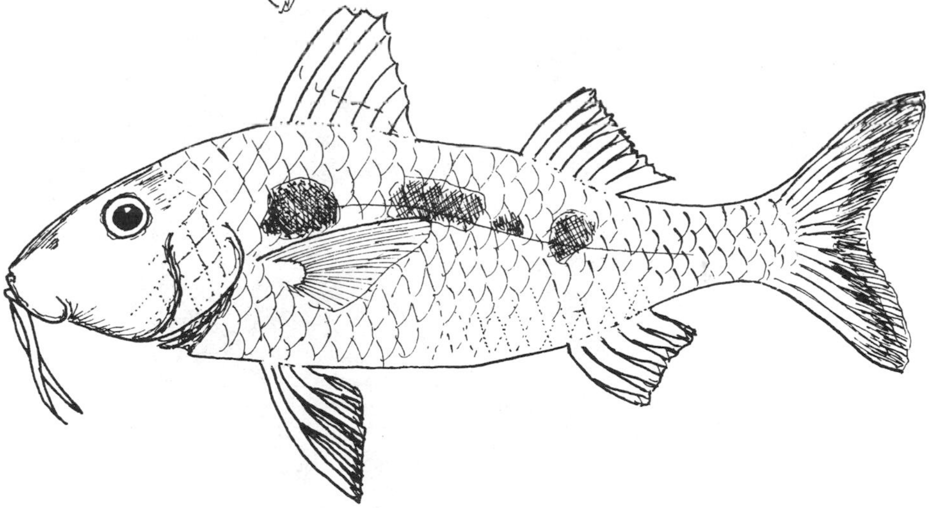

c

GERALD HESLINGA

Fig. 14.27 (a) Smallmouth grunt, *Haemulon chrysargyreum*, Pomadasyiidae, Perciformes, Caribbean. (b) Porkfish, *Anisotremus virginicus*, Pomadasyiidae, Perciformes, Caribbean. (c) Spotted goatfish, *Pseudupeneus maculatus*, Mullidae, Perciformes, Caribbean, genus pantropical.

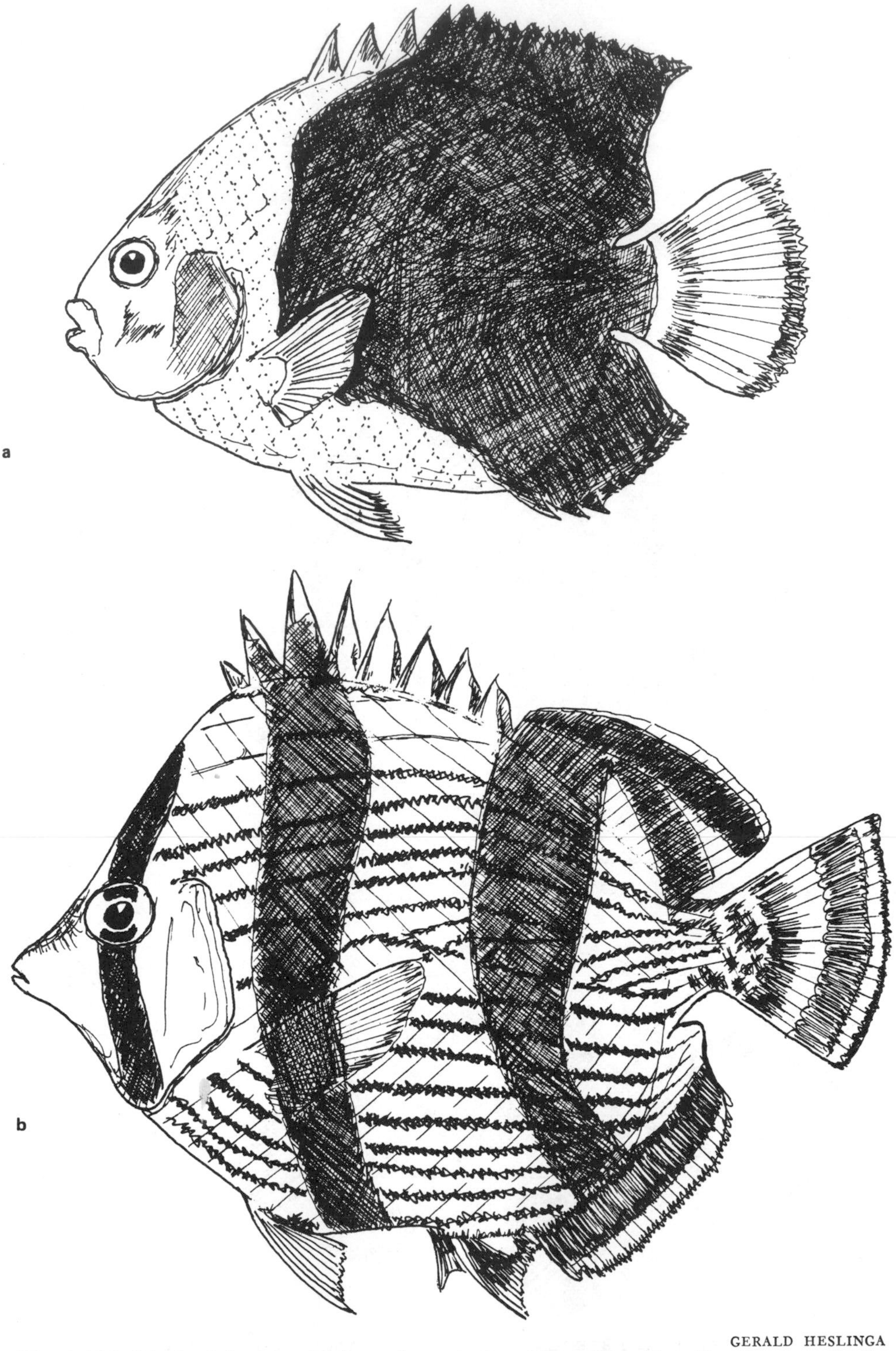

a

b

GERALD HESLINGA

Fig. 14.28 (a) Rock beauty, *Holacanthus tricolor,* Chaetodontidae, Perciformes, Caribbean, genus pantropical. (b) Banded butterfly-fish, *Chaetodon striatus,* Chaetodontidae, Perciformes, tropical Atlantic.

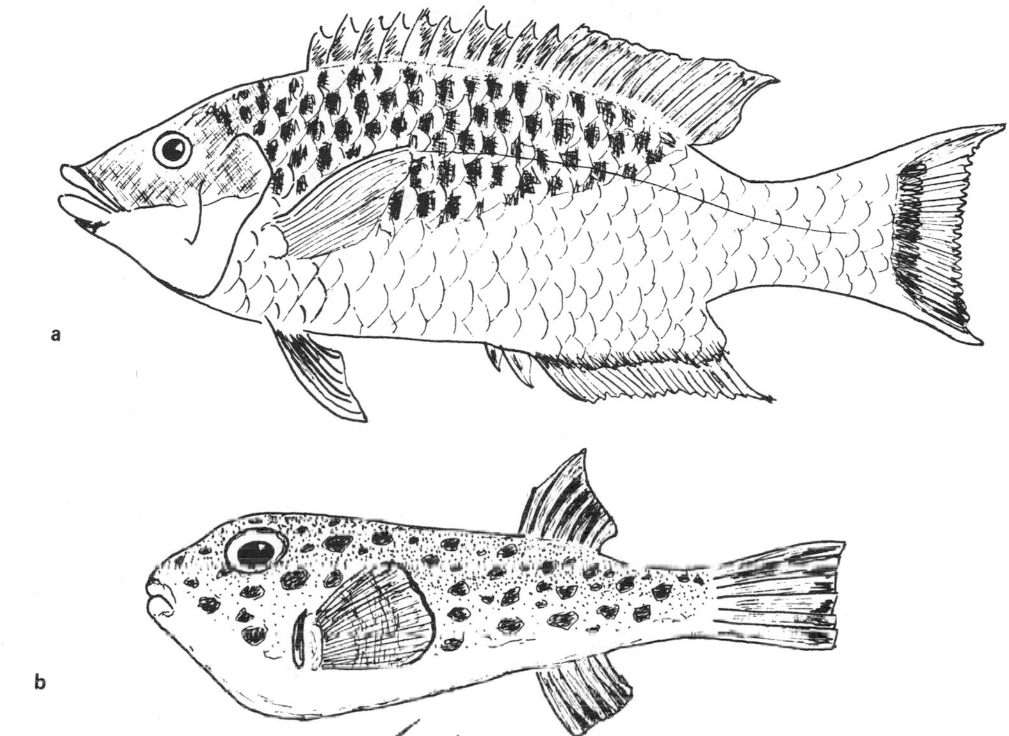

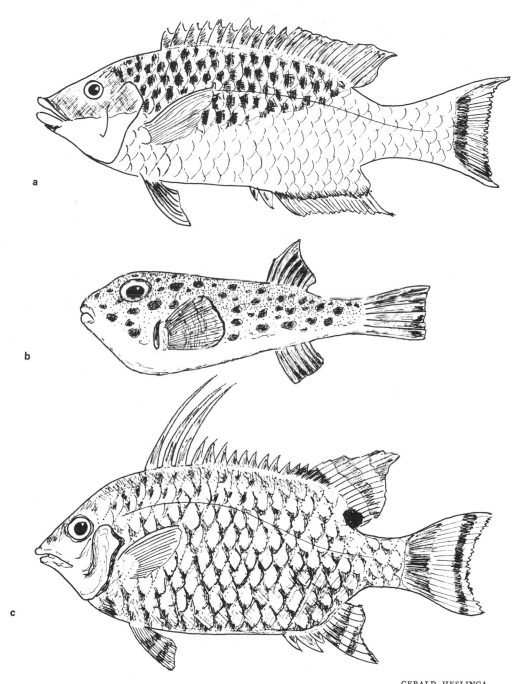

GERALD HESLINGA

Fig. 14.29 (a) Spanish hogfish, *Bodianus rufus*, Labridae, Perciformes, Caribbean. (b) Puffer, *Sphaeroides greeleyi*, Tetraodontidae, Tetraodontiformes, Caribbean. (c) Hogfish, *Lachnolainus maximus*, Labridae, Perciformes, Caribbean.

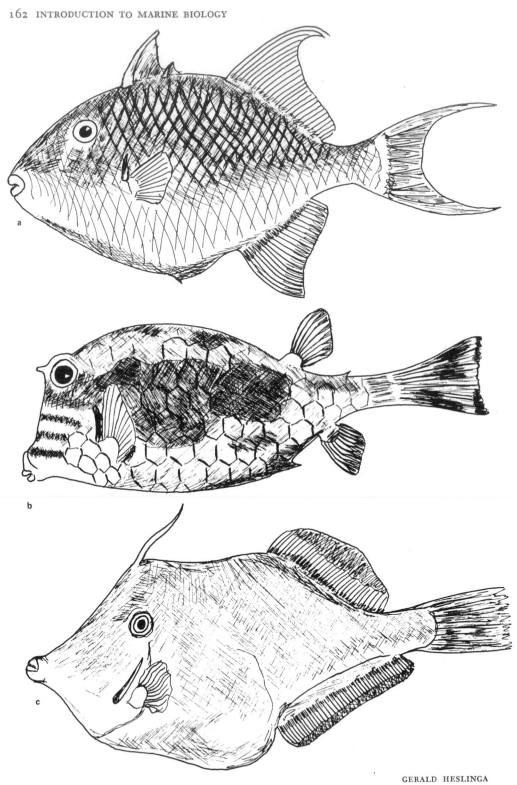

GERALD HESLINGA

Fig. 14.30 (a) Queen triggerfish, *Balistes vetula*, Balistidae, Tetraodontiformes, tropical Atlantic. (b) Cowfish, *Acanthostracion quadricornis*, Ostracionotidae, Tetraodontiformes. (c) Filefish, *Alutera schoepfii*, Monacanthidae, Tetraodontiformes.

Carcharinidae and Sphyrnidae); rays (families Dasyatidae, Myliobatidae, and Torpedinidae); tarpon (Elopidae) and bonefish (Albulidae); morays (Muraenidae); squirrelfishes (Holocentridae); groupers (Serranidae); jacks (Carangidae); Spanish mackerel (Scombridae); snapper (Lutjanidae); grunts (Pomadasyidae); porgies (Sparidae); goatfishes (Mullidae); jawfishes (Opisthognathidae); scorpion fishes, lionfishes, waspfishes, and stonefishes of the Indo-Pacific reefs (Scorpaenidae); angelfishes and butterfly-fishes (Chaetodontidae); Moorish idol (Zanclidae); damselfishes or demoiselles (Pomacentridae); wrasses (Labridae); parrotfishes (Scaridae); surgeonfishes (Acanthuridae); triggerfishes (Balistidae); filefishes (Monacanthidae); puffers (Tetraodontidae); trunkfishes and cowfishes (Fig. 14.30) (Ostracionotidae); porcupine fishes (Diodontidae).

Representative genera of these families should be examined in the laboratory and sketched or photographed to show the diagnostic features. Live materials may be studied at most public aquariums or in the field.

Among the families of fishes notably absent from coral reef communities may be noted: herrings (family Clupeidae, northern temperate and boreal seas); sea trout and salmon (Salmonidae, northern temperate and boreal seas); eels and congers (Anguillidae and Congridae, replaced here by moray eels); flying fishes (Exocoetidae, pelagic); cod, haddock, hake, and pollock (Gadidae, northern temperate and boreal seas); swordfishes and marlin (Xiphiidae and Istiophoridae, pelagic); flatfishes (order Pleuronectiformes have only minor representation).

Identification of Living Materials

The distinctive color patterns of many of the tropical fishes makes it possible to identify many species in the field by comparing them with color illustrations in a field guide such as *Caribbean Reef Fishes* by John E. Randall. Color photographs made in the field can similarly be identified in most cases. Unfortunately the colors fade on preservation.

Additional illustrations and text relating to fishes are given in Chapters 24 North Atlantic Shores, 25 North Pacific Shores, 27 Marine Biology of the Southern Oceans, and 28 The Deep-Sea Fauna.

15

MARINE REPTILES

Marine turtles / Marine crocodile / Sea snakes

Reptiles are air-breathing animals primarily adapted for life on land. However, a few of the surviving groups have become secondarily adapted to living in tropical and subtropical seas. These are the marine turtles, one species of crocodile, and the sea snakes.

Marine Turtles (Order Chelonia). There are five living genera, all of which occur in all warm seas of the world. The oldest known fossil marine turtles date from the Cretaceous, about 100 million years ago, and they seem to have arisen from marsh-inhabiting stocks. Marine turtles come ashore to lay their eggs; the young make their way back to the sea after hatching, and at this stage in their life they are extremely vulnerable to predator attacks. Turtles are threatened species and should be protected, for they are very ancient animals of great scientific interest. The leatherback, *Dermochelys coriacea,* is the only known representative of its family. On the other hand, the family Cheloniidae is represented by four genera: the green turtle, *Chelonia mydas* (Fig. 15.1); the hawksbill turtle, *Eretmochelys imbricata;* the loggerheads, *Caretta caretta* and *C. gigas* (Fig. 15.2); and the ridleys, *Lepidochelys olivacea* and *L. kempii.* Green turtles are mainly herbivorous; the hawksbill is omnivorous; the loggerhead is mainly carnivorous (on fish, crabs, conchs); the ridleys eat crabs, sea urchins, and mollusks; and the leatherback is omnivorous.

Marine Crocodile (Order Crocodylia). The only marine crocodile is *Crocodylus porosus* (note the spelling), of the Indo-West Pacific. It ranges from India to Northern Australia and Fiji. It inhabits estuaries but also is found at sea, far from land, and it has reached many oceanic islands. The meter-long skull at Harvard was taken from a 9-m-long specimen killed in the Philippines by a Boston merchant nearly 150 years ago; this animal was reputed to have eaten many people. Specimens today seldom exceed 6 m, but the species is dangerous when 3 m long. Adults feed mainly on mammals approaching water to drink; the newly hatched

DAVID F. MOYNAHAN

Fig. 15.1 Green turtle, *Chelonia mydas*, all warm seas.

young crocodiles feed on water beetles, changing to fish as they grow older. They live to about 20 years of age. The males fight during the breeding period. The females come about 60 m inland to lay 25 to 60 eggs in a dome-shaped nest of leaves. The eggs are eaten by jackals, otters, and monitor lizards as well as mongooses. Man is the only enemy of the adult.

Marine Snakes and Lizards (Order Squamata). About 16 genera of sea snakes are recognized, all confined to the Indo-Pacific. The most widely distributed species is the yellow-belly, *Pelamis platurus* (Fig. 15.2), which ranges from Madagascar to Panama and from Japan to New Zealand. Although it is less than a meter long, it is able to cross wide ocean gaps, as its range also shows. But this fact seems to have escaped the notice of some paleontologists who argue that the presence of comparably sized fossil reptiles in South America and South Africa in Carboniferous times must mean that these two continents were joined since supposedly the reptile concerned (*Mesosaurus*) would be too small to cross the Atlantic. Sea snakes are often gregarious; Lowe (1932) records swarms numbering millions crossing the sea between Sumatra and Malaya, scattered over a region some 60 miles across. Specimens that reach New Zealand (the extreme southern range) are solitary females, probably swept southward by currents; they penetrate several miles inland and have been known to enter houses and even get into bed—to the great surprise of New Zealanders who are taught at school that no snakes occur in that country. Most species are venomous. Like all water snakes, sea snakes have a compressed, flattened tail, adapted to eellike swimming movements. They feed on fishes. Some sea snakes have become adapted to life in freshwater lakes on oceanic islands.

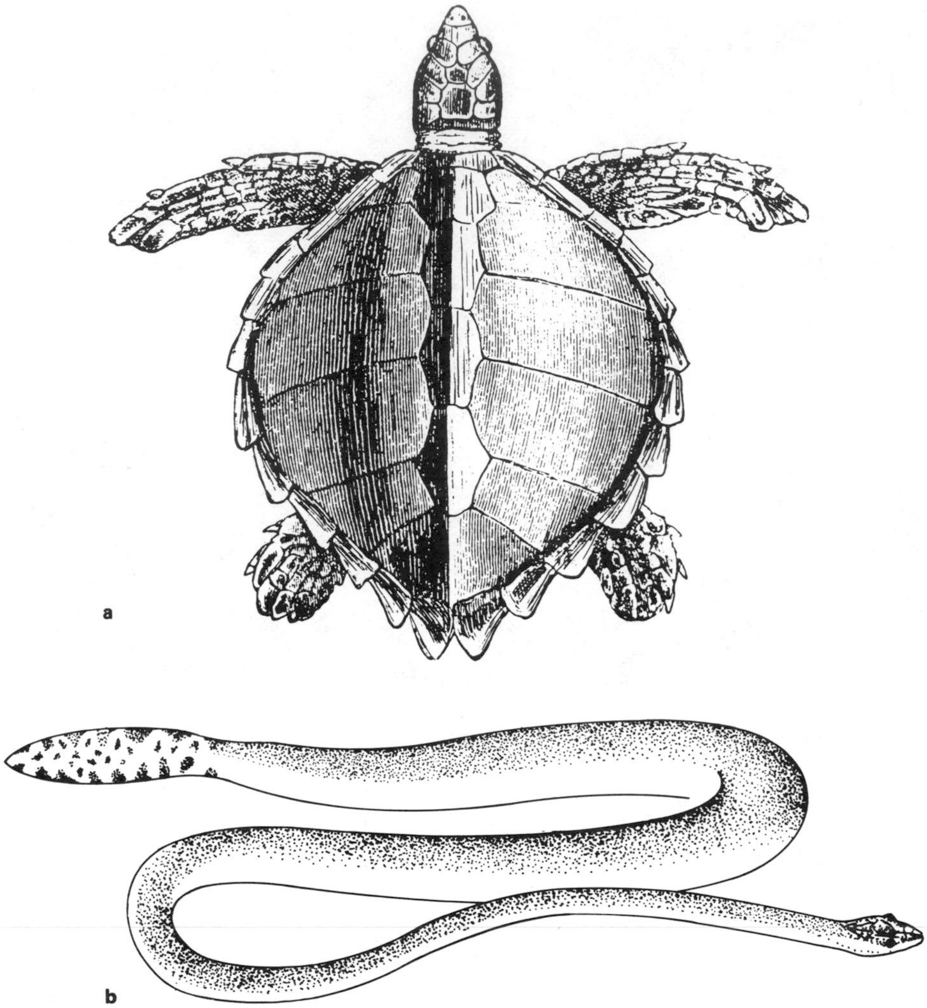

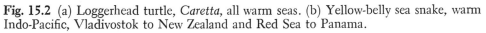

Fig. 15.2 (a) Loggerhead turtle, *Caretta,* all warm seas. (b) Yellow-belly sea snake, warm Indo-Pacific, Vladivostok to New Zealand and Red Sea to Panama.

In contrast to snakes, very few lizards have adopted a marine life. In the Galapagos archipelago there is a marine iguana. In the island of Komodo in Indonesia a giant monitor lizard, *Varanus komodoensis,* frequents the seashore, and often has been observed swimming at sea. Neither of these species forsake the coast for the open sea, however.

16

MARINE BIRDS

Penguins / Loons / Albatrosses, petrels, and shearwaters / Transoceanic migration / Ducks, geese, and swans / Pelicanlike birds / Herons / Gulls, skuas, and waders

Birds, although creatures of the air, are known to have frequented aquatic (including marine) environments ever since their first appearance on earth. The Jurassic *Archaeopteryx* occurs as fossils of specimens that fell into the lagoons of coral reefs, and the toothed birds of the Cretaceous period resembled sea gulls and loons.

Seven existing orders of birds include members that are specially adapted for life in or near water, having, for example, webbed feet, or long legs adapted to wading in shallow marginal water, and analogous modifications enabling them to find food in water environments (Fig. 16.1). Adaptations of this kind have enabled birds to occupy not only inland waters, but also marine habitats. The systematic characters of the orders of birds are set out in Table 16.1.

Penguins (Order Sphenisciformes). Penguins are exclusively marine birds of the far southern oceans, one species alone reaching the equator at the Galapagos archipelago. They occupy the southern islands and occur on the Antarctic continent itself (Fig. 16.2). They prey upon fish and are themselves eaten by marine mammals such as sea leopards and toothed whales.

The single surviving family of penguins, Spheniscidae, comprises the most completely aquatic of all living birds. The largest living species is the emperor penguin, *Aptenodytes forsteri* (Fig. 16.3) of Antarctica. However, a much larger fossil species, *Pachydyptes ponderosus*, lived in New Zealand during the Oligocene (ca. 35 million years ago) and reached a weight of about 90 kg, or 200 pounds (for comparison, the emperor reaches 40 kg, or 90 pounds). Emperors, and the other species that live on surface ice or that nest on exposed sheets of Antarctic tundra or rock or the remote islands of the southern ocean, seem to have highly developed communal instincts and form very large rookeries, or penguin cities. In these

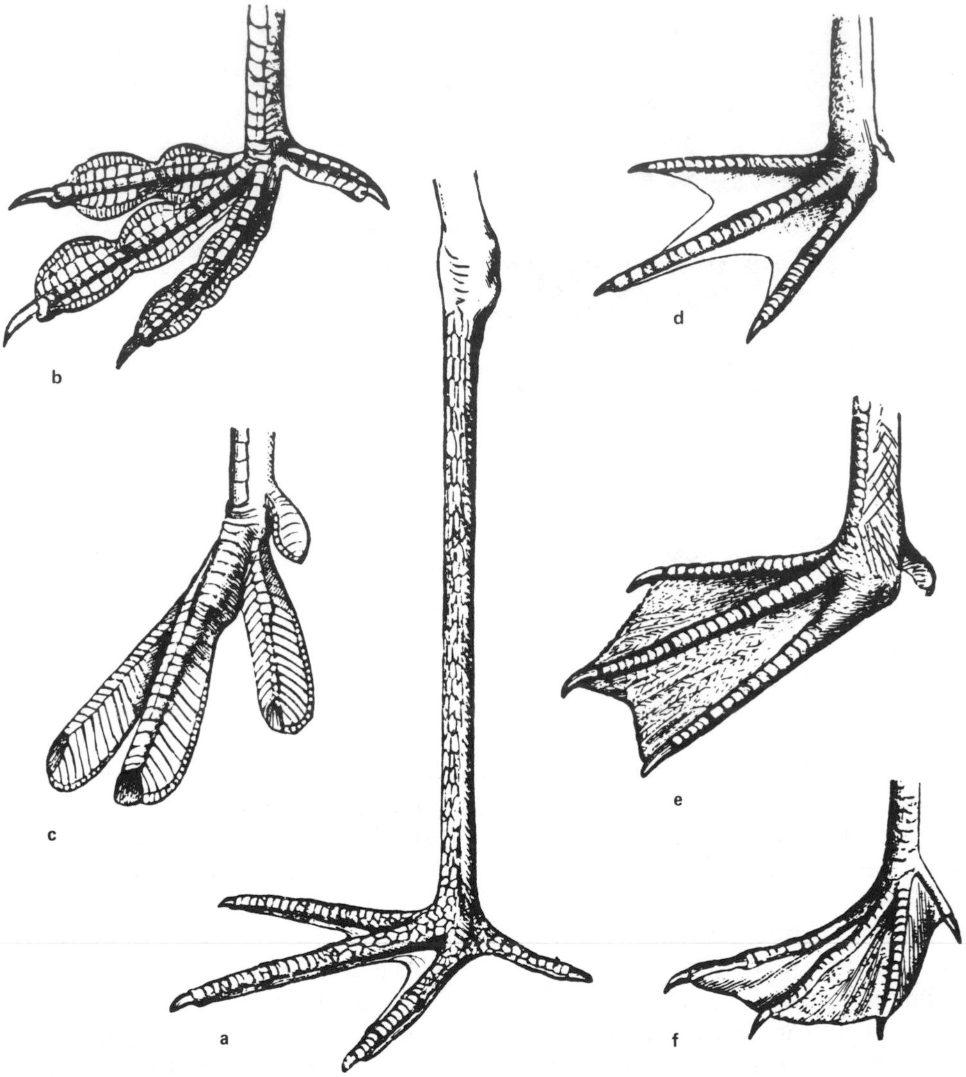

Fig. 16.1 Key characters of oceanic and shore birds. (a) Order Ciconiiformes, family Ardeidae, herons. (b) Lobed webs of some marsh birds occasionally found on coast. (c) Fringed webs and flattened nails of grebes, order Podicipiformes (not in Table 16.1), essentially freshwater birds, but migrating across oceans. (d) Foot as in wading birds, order Charadriiformes, and (e) duck, Amseriformes. (f) Order Pelecaniformes.

communities there are regulated egg-laying territories and organized traffic routes for birds going to or coming from the fishing grounds. Newly emerging penguins coming fresh and clean from the sea show an evident distaste for penguins that have been ashore long enough to become fouled with excrement; thus the two lines of outgoing and incoming traffic are kept apart. Penguins of the more temperate islands do not form such clearly organized communities and, although they may fish in groups, they tend to nest alone in burrows (or under beach houses of *Homo sapiens*) and the accent is on family rather than community ties. An example of the latter group is *Eudyptula minor*, the little blue penguin of New Zealand and Australia.

Table 16.1 Key Characters for Recognizing the Orders of Marine Birds (Class Aves)

Class characters: warm-blooded vertebrates in which wings are normally developed, only the two hind limbs used for walking; feathers normally occur, and eggs laid. Seven orders, as follows:

Wings converted to flippers used for swimming; Southern Hemisphere only	Sphenisciformes	Penguins
Wings normally developed, used for flying		
Hind toe joined by web to the front toes	Pelecaniformes	Pelicanlike birds
Only front toes connected by a web		
Nostrils open to exterior through raised tubes on bill	Procellariiformes	Albatrosses, petrels, and shear-waters
Nostrils open directly to exterior, not by way of tubes		
Ankle joint (tarsus) flattened from side to side; legs placed far back on body, diving birds	Gaviiformes	Loons
Ankle joint not obviously flattened; legs not set far back		
Hind toe very long; bill, neck, and legs long	Ciconiiformes	Herons
Hind toe small or absent		
Outer edge of bill serrated or fringed	Anseriformes	Ducks, geese, and swans
Outer edge of bill straight	Charadriiformes	Gulls, skuas, and waders

JOHN H. DEARBORN

Fig. 16.2 Adelie penguins, *Pygoscelis adeliae*, order Sphenisciformes, at Cape Crozier, Ross Island, in the Ross Sea, East Antarctica.

JOHN H. DEARBORN

Fig. 16.3 Young and adult emperor penguins, *Aptenodytes forsteri,* order Sphenisciformes, at Coulman Island, Ross Sea.

Loons (Order Gaviiformes). Loons are diving and fishing birds of the northern part of the Northern Hemisphere, about the size of a goose. The legs are placed far back on the trunk to favor propulsion in water and to make walking on land difficult. The birds can fly, but only if they take off from water. Other characters are given in Table 16.1. There is a single living family, containing four species. The best known is the species distributed over the southernmost part of the range, namely the great northern diver or common loon (*Gavia immer*). Certain diving birds of the Cretaceous period, the Hesperornithiformes, somewhat resemble the loon, differing in having vestigial wings and teeth in the jaws. Whether there is any genetic relationship between those two stocks is doubtful, but it is clear that birds with a similar role in the ecosystem to that of loons today had already developed about 100 million years ago.

Albatrosses, Petrels, and Shearwaters (Order Procellariiformes). This order encompasses oceanic fish-eating birds and is readily recognizable by the peculiar tube-shaped nostrils (Table 16.2). The group apparently is ancient, for an albatross bone that is 60 million years old has been recognized in the Eocene sediments of New Zealand. The largest members are the existing albatrosses, especially the wandering albatross, *Diomedea exulans* (Fig. 16.4), of the Southern Ocean (south of 30°S), where the wingspread reaches 2.5 m, or 8 feet. There is a great range of size within the order, the smallest members being the petrels, which skim over the wave crests like sparrows; the smallest species is a petrel with the rather long name of *Holocyptena microsoma.* All members of the order are able to spend months at sea, sleeping afloat or on the wing. One southern albatross specimen is known to have circumnavigated the world at least 3½ times—it was observed on successive

Table 16.2 Key Characters for Recognizing the Families of Petrels (Order Procellariformes)

Nasal tubes formed of horny material, placed one on either side of the upper ridge of the bill (culmen), so as to leave a clear median strip along bill	Diomedeidae	Albatrosses
Nasal tubes in contact, placed along the median part of the culmen of bill		
No hind toe; Southern Hemisphere only	Pelecanoididae	Diving petrels
Hind toe present		
Two distinct nasal tubes, fused together along culmen	Procellariidae	Prions and shear-waters
Nasal tubes fused together to make a single tube	Hydrobatidae	Storm petrels

occasions in Australia, New Zealand, and South America, each time recognized by a distinctive birdband placed on one leg. The larger species of the order feed on fishes and squid, the smaller species take plankton.

Transoceanic Migration

Members of the Procellariiformes are famed for their long-distance flights. Since the eighteenth century, mariners have reported the wandering albatross, *Diomedea exulans*, as following the wake of a ship for days at a time, apparently for the sake of garbage thrown overboard. Banded birds of this species are known to have made long flights. Over 90 cases are on record of birds banded in New South Wales, Australia, turning up later in South Georgia, in the South American quadrant of Antarctica; and 12 of these returned to Australia. A royal albatross, *Diomedea epomophora*, banded at Campbell Island, south of New Zealand, was found near Santiago, Chile, one year later, and others made the same passage to Valdivia, Chile, flights over open ocean of 5000 miles being involved. The white capped mollymawk, *Diomedea cauta*, has been banded in South Africa and subsequently taken at Cape Campbell, New Zealand, while two others are reported to have made the same flight but in the opposite direction. The giant petrel, *Macronectes giganteus*, is another ocean voyager. Many specimens banded in various island localities on the New Zealand plateau have subsequently been retaken in Chile, in Uruguay, and in Argentina, after intervals of 6 to 7 months; one reached Rio de Janeiro after 8 months; and one was taken in southwest Africa after only 3 months. A sooty shearwater, *Puffinus griseus*, banded on Macquarie Island, south of New Zealand, was taken 31 months later on the California coast, nearly 8000 miles away. Most of the southern seabirds, however, remain in the southern part of the Southern Hemisphere, circling the pole, probably aided by the planetary west-wind system.

Duck, Geese, Swans (Order Anseriformes). Members of this order are mainly found on or near freshwater, but some species may be reckoned as shore birds. Best-known of these are the eider, *Somateria mollissima*, of Arctic shores. In the Auckland Islands group to the southeast of New Zealand there occurs a sub-Antarctic species of very feeble flight, *Nesonetta aucklandica*; this duck frequents

JOHN H. DEARBORN

Fig. 16.4 Wandering albatross, *Diomedea exulans*, order Procellariiformes, photographed in mid-ocean between New Zealand and Antarctica. The body length in an adult bird is about 4 feet, the wing span 8 feet.

sheltered coves, finding food among the kelp beds. The black swan of Western Australia, *Cygnus atrata*, was introduced into New Zealand and soon became much more numerous (on saltwater lagoons and in sheltered harbors) than in its original home.

Pelicanlike Birds (Order Pelecaniformes). Members of this order are aquatic birds characterized by having all four toes joined by a web. The included families may be distinguished by Table 16.3. Most members of the order are found in

Table 16.3 Key Characters for Recognizing the Families of the Pelicanlike Sea Birds (Order Pelecaniformes)

Ordinal character: hind toe joined by a web to the front toes. Six families, as follows:		
Bill hooked at the tip		
The horny plate forming the upper part of the bill (culmen) about the same length as the ankle joint (tarsus) of the leg	Phalacrocoracidae	Shags, or cormorants
The culmen is two or three times longer than tarsus		
Tail short, not forked	Pelecanidae	Pelicans
Tail long, deeply forked	Fregatidae	Frigate birds
Bill straight, not hooked at tip		
Chin region covered by feathers	Phaethontidae	Tropic birds
Chin region naked		
Tail rounded	Anhingidae	Darters
Tail pointed at tip, or wedge-shaped	Sulidae	Gannets and boobies

tropical and subtropical waters, in both fresh and saltwater. Some members, particularly the gannets and the shags, occur in cool northern and southern seas. All are fish eaters. Group nesting behavior is highly developed, particularly among the gannets, which have large colonies on various rocky headlands and scattered rocky islets in far northern and southern seas. Among the better-known genera may be noted *Sula* and *Moris* of the Sulidae; *Phalacrocorax* among the shags; *Fregata* or frigate birds, tropical and subtropical long-winged species, which characteristically snatch food in flight from other birds; *Phaethon*, the tropic birds, with long tail plumes, colored bright red in one pantropical species; and *Pelecanus*, the well-known pelicans, with a large gular pouch for holding fish below the large bill.

Herons (Order Ciconiiformes). Seven families comprise this order, but only two of them frequent marine environments. These are distinguished in Table 16.4. The order is made up of long-necked and long-legged birds. Herons, which include reef and coastal species, feed mainly on fishes and other smaller aquatic animals. Flamingos have a strainer type of bill adapted for eating small crustaceans and other animals from lagoon and lake floors. Some species of flamingo inhabit salt estuaries and mangrove swamps.

Gulls, Skuas, and Waders (Order Charadriiformes). A large order of about 16 families, of which 6 families are restricted to nonmarine habitats. The other families may be distinguished by Table 16.5.

The order is a very large one, and its members range widely. Remarkably long annual migrations are performed across the oceans by some species, such as the golden plover (*Pluvialis dominicus*) and arctic tern (*Sterna paradisaea*). Some of the better-known species and genera: Gulls, genus *Larus*, range the shorelines of the world (and also occur on lakes and rivers); in recent years in many countries gulls have been changing their habitats to occupy, more or less permanently, city garbage dumps, where they feed on refuse; some flocks now occur far inland. Terns are related to gulls, but have forked tails like swallows. Oystercatchers frequent rocky shores of the world; plover, knots, and sanderling favor sandy shores. Quiet lagoons are the main haunts of the avocet sandpipers. The Alcidae, comprising auks, puffins, and their relatives, are mainly found on far northern coasts and rocky islets. Skuas or sea hawks are predators of Arctic and Antarctic seas. One species, *Stercorarius parasiticus*, breeds in Greenland and other Arctic lands during the northern summer and appears in New Zealand during the southern summer, when it is not a breeding species. Skuas prey upon the eggs and young of other birds. The Antarctic species parasitizes the adelie penguin in this way. Skuas attack land birds when opportunity offers, and in the farmlands of the Southern Hemisphere they even attack lambs and sickly sheep, as gulls occasionally also do. Skuas are birds that even naturalists do not defend, for their depredations on other wildlife are so severe as to rival those of man.

Table 16.4 Conspicuous Marine Families of the Heronlike Birds (Order Ciconiiformes)

Ordinal characters: aquatic birds; the wings developed normally for flying; the front toes connected by a web; the hind toe well developed; the bill, neck, and legs all elongate. Seven families, but only the following are conspicuous:

Bill curved, front toes fully webbed	Phoenicopteridae	Flamingoes
Bill straight, front toes incompletely webbed	Ardeidae	Herons

Table 16.5 Key Characters for Recognizing the Families of Wading Birds (Order Charadriiformes)

Ordinal characters: front toes alone connected by web; nostrils not enclosed in tubes; hind toes small or absent; outer edge of bill straight, not serrated. Ten families, as follows:

The bill flattened in vertical plane (compressed), bladelike	Rhynchopodidae	Skimmers
The bill is not obviously compressed		
Front toes completely webbed		
No hind toe	Alcidae	Auks
Hind toes normally present		
Upper bill formed of 3 horny plates	Stercorariidae	Skuas
Upper bill formed of a single horny element	Laridae	Gulls
Front toes webbed only at the base of toes, or without webs		
Claw of middle toe less than half as long as ankle joint	Recurvirostridae	Stilts and avocets
Claw of middle toe longer than half the length of ankle joint		
Toes with fleshy lobes along their sides	Phalaropodidae	Phalaropes
Toes without such lobes		
Bill long, not curved downward at tip		
Bill flattened in vertical plane at tip	Haematopodidae	Oystercatchers
Bill not flattened in vertical plane at tip		
Bill stout, not slender	Charadriidae	Plovers
Bill slender	Scolopacidae	Snipe, curlews, and sandpipers
Bill short, turned downward at tip	Glareolidae	Pratincoles

17

MARINE MAMMALS

Dolphins and whales / Extinct Archaeoceti / Odontoceti or toothed whales / Dolphins and porpoises / Killer whale / Sperm whales / Mysticeti or baleen whales / Rorqual or finback whales / Right whales / Blue whale / Manatee and dugong / Carnivora Pinnipedia / Eared seals / Walrus / True seals / Carnivora fissipedia / Polar bear / Sea otter

Three orders of mammals have become adapted in greater or lesser degree to life in the sea. All retain the characteristic features of mammals, having regulated body temperature, a viviparous habit, and the ability to breathe air. This last feature requires marine mammals to return to the surface of the sea at regular intervals. Table 17.1 sets out the diagnostic features.

Dolphins and Whales (Order Cetacea)

With the exception of one family of freshwater dolphins all Cetacea are marine. The included families are contrasted in Table 17.2.

Archaeoceti. The most ancient known members of the order Cetacea are known as the *Archaeoceti.* Fossils of these extinct, carnivorous-toothed forms occur in Eocene sediments that are about 60 million years old. The last-known survivors inhabited New Zealand seas in the Miocene, about 25 million years ago. The ancestor of these early whales is thought to have been some carnivore, but this is uncertain. Two other extant suborders of whales are known, both still extant: the Odontoceti or toothed whales, with a simplified dentition derived evidently from the Archaeoceti; and the Mysticeti or baleen whales, which have lost the dentition.

Odontoceti. The Odontoceti are carnivorous, and they comprise most surviving cetaceans. They are active predators, feeding on large animals such as fishes and squid, including apparently giant squid. The common dolphin *Delphinus delphis* (Fig. 17.1) has been known in written records since the time of the

Table 17.1 Key Characters for Recognizing the Orders and Suborders of Marine Mammals (Class Mammalia)

Class characters: warm-blooded vertebrates in which hair is developed and, in the aquatic orders, the young are born after placental development in the uterus. Three orders, as follows:

Mammals in which a full dentition occurs, with prominent canine teeth for seizing prey; sharp chisellike incisors for cutting flesh; and large cusped carnassial molars for tearing prey animals apart	Carnivora	
Forelimbs not modified into swimming paddles	Fissipedia (suborder)	Polar bears and sea otters
Forelimbs paddle-shaped; hind limbs present	Pinnipedia (suborder)	Seals
Mammals in which canine teeth lacking and hind limbs lacking		
Teeth either all uniform or else lacking altogether	Cetacea	Dolphins and whales
Teeth of 2 kinds, incisors and molars; tropics only	Sirenia	Manatees and dugongs

Table 17.2 Key Characters for Recognizing the Families of Aquatic Mammals (Order Cetacea)

Ordinal characters: aquatic mammals in which the forelimbs are modified to assume the form of swimming paddles (flippers); the hind limbs are lacking; the teeth are either all similar or else lacking and replaced by horny baleen plates. Eight families, as follows:

Teeth developed, at least in the lower jaw; no horn baleen plates (suborder Odontoceti)		
Numerous teeth in both jaws		
Teeth spade-shaped; snout short and blunt	Phocaenidae	Porpoises
Teeth conical; snout usually pointed in front	Delphinidae	Dolphins
One very long twisted tuck in the upper jaw of male	Monodontidae	Norwhals
No functional teeth in upper jaws		
Numerous teeth in the lower jaw	Physeteridae	Sperm whales
Only 2 or 4 teeth in the lower jaw	Ziphiidae	Beaked whales
No teeth developed at all; instead, a fringe-filter of fibrous horn plates (baleen) developed around the margin of the upper jaw (suborder Mysticeti)		
Throat region marked by conspicuous longitudinal grooves		
Numerous grooves developed on the throat; dorsal fin present	Balaenopteridae	Finback whales
Only 2 or 4 throat grooves; no dorsal fin	Eschrichtidae	Gray whales
Throat region without any longitudinal grooves	Balaenidae	Right whales

NOTE: A few genera are not identifiable without recourse to anatomical detail beyond the scope of this book.

Fig. 17.1 Dolphin, *Delphinus delphis*, Cetacea.

Phoenicians. It grows to a length of 2.5 m (8.5 feet). Dolphins travel in groups, called schools. They accompany ships for considerable distances, performing intricate leaps and entering the sea head first with scarcely a splash. Their interlocking teeth and crocodilelike jaws permit seizure of fishes, which are taken whole and tossed into the air to facilitate swallowing them headfirst. The harbor porpoise of the North Atlantic, *Phocaena phocaena,* also travels in schools; its favored prey include herring, mackerel, and squid. Similar species occur in the South Atlantic and in the South and North Pacific. Most porpoises have a relatively blunt snout. The name dolphin is applied to a fish, and some ichthyologists claim that all cetaceans should be called porpoises. However, historical tradition and the meaning of the ancient Greek word *delphis* clearly dictate the priority of the cetacean to the name. The grampus or killer whale, *Grampus orca* (Fig. 17.2), is a member of the Delphinidae, ranging all seas nearly from pole to pole. The sperm whale, *Physeter catodon,* is one of the largest whales (males reach 18 m or 60 feet, females are half as long) and has a relatively large head, though small jaws; it ranges most seas, feeding mainly on squid. The pigmy sperm whale, *Kogia breviceps,* is a related cosmopolitan species, around 3 m long.

Mysticeti. The rorqual or finback whales belong in the Mysticeti, having baleen plates and a planktonic diet, combined with a taste for cod and herring. Right whales are also baleen whales, but they lack the longitudinal throat grooves of the rorquals and gray whales. They were formerly very common, but were nearly extinguished by the whalers a century ago. Right whales are placed in the genus *Eubalaena,* also *Balaena.* The largest animal in the world is the blue whale, *Sibbaldus musculus,* which is a member of the rorquals reaching a length of 30 m (100 feet); like other baleen whales, it feeds on plankton.

Whales are rapidly becoming extinct, owing to persistent overhunting and the failure of international agencies to implement recommended conservation measures. The probable economic collapse of the industries based on whale products may occur early enough to save some stocks for future generations. The only effective conservation measures thus far have been temporary circumstances dictated by accident or human perversity, for example, the almost total destruction by fire of the Union whaling vessels in the Bering Sea by the *Shenandoah* brought respite to the northern whales. In 1871 and again in 1877 the Arctic whaling fleets were destroyed by pack-ice, once more delaying the final extinction by a few years. The New Zealand whaling industry, based on the southern humpback whale collapsed abruptly a few years ago when the last remnants of the regional herd failed to put in the expected annual appearance on migration. This pattern of abrupt disappearance is likely to be seen elsewhere.

DAVID F. MOYNAHAN

Fig. 17.2 (a) Dolphins, *Delphinus*. (b) Killer whale, *Orca*.

Manatees, Dugongs (Order Sirenia)

The Sirenia are herbivorous aquatic mammals of tropical coasts, coastal seas, and estuaries, having the characters given in Table 17.1. Two families are distinguished: (1) Dugongidae, dugongs of the Indo-West Pacific, the tail fluke deeply notched; and (2) Trichechidae, the Atlantic manatees, the tail fluke without any twofold division.

The dugong, *Dugong dugon,* occurs along the coasts of Northern Australia, Asia and East Africa. The manatee, *Trichechus* spp., range the Atlantic tropical coasts of America and West Africa. Fossils show that sirenians were already as modified for equatic life in the Eocene, 60 million years ago, as are modern forms. The closest related mammals are believed to be the elephants. Some of the fossil sirenians preferred habitats such as coral reefs and areas of the continental shelf; the extant forms prefer sheltered estuaries and lagoons. *Hydrodamalis,* exterminated within recent times, inhabited the Bering Sea; it was discovered in 1741, when it occurred in large herds. By 1770 it had almost been extinguished by excessive hunting; the last specimens finally succumbing in 1854.

Seals (Order Carnivora, Suborder Pinnipedia)

The Pinnipedia include the seals, walrus, and sea lions—all aquatic carnivores largely marine in habitat. They are strongly modified for aquatic life, but other features of their morphology imply a rather clear relationship with the terrestrial carnivores, such as cats, bears, dogs. Therefore, the pinnipedes are usually treated as a suborder of a larger order, Carnivora, to include all the types mentioned (Figs. 17.3 and 17.4). The oldest-known fossil seals date only from the Miocene, about 25 million years ago; and by that time they had already assumed virtually all the characters of extant pinnipedes. Three families of pinnipedes are recognized, as shown in Table 17.3.

The Otariidae. These comprise the sea lions, sea leopards, and sea bears. They occur in both the Northern and Southern hemispheres, usually near coasts. They spend a good deal of time on land, where they are able to move about rather freely, for the hind flippers can also be used as limbs for walking, galloping, and climbing rocks. The largest species are the sea lions, which reach a length of 4 m (13 feet). The males arrive at the breeding grounds in the spring and fight fiercely for territory near the seashore; defeated males are compelled to settle further inland, where

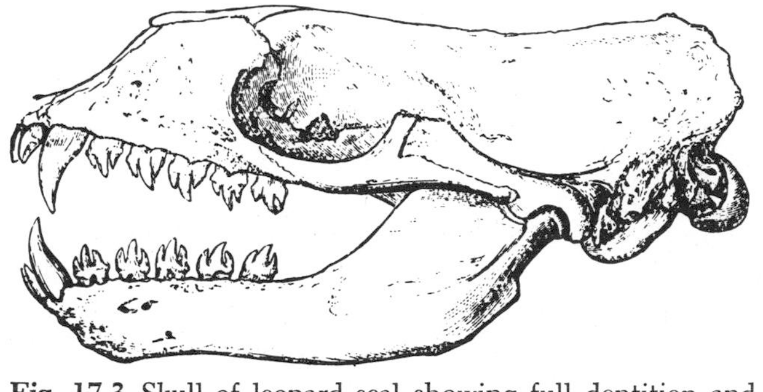

Fig. 17.3 Skull of leopard seal showing full dentition and enlarged canine teeth, characteristic of order Carnivora.

F. JULIAN FELL

Fig. 17.4 Sea leopard, *Hydrurga leptonyx*, on ice floe in the Antarctic Ocean.

the property values are depressed on account of the long journey needed to reach the fishing grounds offshore. Most of these settlements are on remote islands of the far northern and far southern oceans. The females arrive after the initial settlements have been established. They are then rounded up into harems of up to 15, the largest dominant bulls having the biggest harems. Excessively jealous of the attentions of unattached males, the harem bulls neither eat nor sleep during the breeding season. This lasts several weeks, and they are enfeebled at the end of the mating sessions. The cubs are born on land and weigh a few pounds.

Odobaenidae. Walrus occur only in the Arctic Ocean. There are only two species, *Odobaenus rosmar* in the Arctic Atlantic and *O. divergens* in the Arctic Pacific. Large males grow to about 3.5 m long, weighing 1400 kg or 1.5 tons. The long tusks are used for excavating the seabed because they feed almost exclusively on infaunal invertebrates such as bivalve mollusks and some crustaceans. Walrus live in herds. They may gather on drifting ice floes, huddled together in vast biomasses. They are unaggressive animals, but have close ties and defend their herds fiercely when attacked.

Phocidae. The true seals have a more extensive modification of the hind flippers for life in the water; therefore they are not active on land. They have thick, close fur and thick, protective, heat-insulating layers of fat (blubber) beneath the skin. This permits them to occupy extremely cold regions, such as Antarctica. The largest species are the elephant seals, also the largest of all Pinnipedia, growing to 6 m (20 feet). There are two species, *Miroungia leonina* (Fig. 17.5) of the Southern Ocean and *M. angustirostris* of Southern California. The name is derived from

Table 17.3 The Families of Seals (Suborder Pinnipedia)

Subordinal characters: Carnivora with a complete dentition, including conspicuous canine teeth; the forelimbs modified to form paddles for swimming. Three families, as follows:

External ears present	Otariidae	Eared seals
No external ears visible		
Upper canines developed as long tusks	Odobaenidae	Walruses
Upper canines not tusklike	Phocidae	True seals

a short, hollow, inflatable trunk on the males, who use it as an amplifier for their vocalizations. The well-known common seals of various species are much smaller members of this family. Some are easy to tame and have become attached to their owners, as terrestrial carnivores also do.

Some species of Pinnipedia are severely hunted either for skins or oil. The joint Russian-American hunting treaty permits 12,000 walrus to be killed each year in the North Pacific. Far more are killed than can be used, and the floating carcasses of about 6,000 are abandoned every season. Two hundred such carcasses washed ashore at the town of Kotzebue in Alaska on August 12, 1971. About 50,000 seals are clubbed to death each year by hunters operating in the waters of the Bering Sea, under joint Russian-American agreements. Nearer at hand, Canadian, Norwegian, and other hunters operate in the eastern seaboards of Canada, their activities often involving intolerable brutality; the Canadian-Norwegian catch for harp seal for 1973 was 160,000.

F. JULIAN FELL

Fig. 17.5 Sea elephants, *Miroungia leonina*, in West Antarctica.

Sea Otter (Order Carnivora, Suborder Fissipedia)

Enhydra lutris. Once very common on North Pacific coasts, the species was nearly extinguished after its pelt was made fashionable by Empress Catherine the Great. Successful conservation measures are now bringing it back to its former habitats, and it has spread as far south as Monterey, California. Sea otters feed on sea urchins, which they eat while floating on their back—sometimes cracking open the shell with a stone, at other times holding the urchin in the two forefeet like a hamburger and taking bites out of the uninviting object. The return of the sea otter to California will reduce the excessive sea urchin populations that developed after the otter's original destruction. This change will stop the constant wastage of kelp, which had been drifting to sea following urchin attack on its seabed habitats. The conservation of the kelp will restore the abalone populations, provided that man can stop inserting himself into the food chain.

Polar Bear (*Thalarctos maritimus*). Polar bears are now approaching the status of an endangered species on account of excessive hunting for trophies and pelts. The species is thought to have originated in Eurasia during glacial conditions and to have spread to North America. It now has a general Arctic distribution. Occasional specimens drift as far south as Newfoundland each year, and as many as 50 per year once used to drift on ice floes across the Denmark Straits from Greenland to Iceland. The numbers now, however, are rapidly being reduced. The species is the only aquatic bear. Modifications for life in cold seas include hair on the sole of the foot, permitting a firm grip on the ice (a method adopted by skiers, too, when fitting skins beneath the runners), with the hair directed toward the rear. In winter polar bear eat seals and walrus and in coastal areas a great deal of fish. In summer they are mainly herbivores. This variety in diet, according to season, characterizes other species of bears. In fall the female retires to an ice cave to hibernate and to give birth to cubs in winter. The mothers care for the cubs assiduously, sharing her food if the supply is short. When alarmed, polar bear become extremely dangerous animals.

18

STARFISHES

General biology / Anatomical features / Platyasterida / Phanerozonida / Spinulosida / Euclasterida / Forcipulatida

Starfishes are among the most numerous and most conspicuous of the larger benthic invertebrates. They range all seas and are common at all depths from the tidal zone to the deep trenches. As already noted (Table 5.2), starfishes comprise a subclass of echinoderms, the *Asteroidae*.

GENERAL BIOLOGY

Starfishes usually creep about the seabed with the aid of their tube feet. No swimming forms have yet been recorded. Those species that have suctorial tube feet are able to climb and descend obstacles to capture food. A number of genera have peculiar peglike tube feet without suckers; they usually occur on submarine mud banks, where they seem to use their tube feet as oars or as stilts—I imagine them as tottering or paddling through the semifluid, slimy mud.

Although starfishes cannot see because they do not have eyes, they are able to detect light fluctuations with the aid of a photosensitive eyespot at the outer end of the ambulacral groove where it reaches the tip of the arm. Starfishes cannot be said to "hear," although they evidently detect the grosser vibrations often associated with sound. At the tip of the arm are some special tube feet in some starfishes, which are held erect when the animal is active and probably serve as taste organs (chemoreceptors).

The sex of a starfish is seldom obvious without dissection, but in those forms that carry or brood the young, the female can be recognized during the breeding season. Their breeding habits are very varied. Apart from brood-protecting forms, such as *Calvasterias* in littoral waters, there are deep-sea forms, such as *Pteraster* with a dorsal brood chamber developed on the upper side. This is provided with a water-conditioning system not unlike the water-circulatory system of a sponge and has a large central **osculum** for the outflow of the water current. One Greenland genus, *Leptychaster,* somehow contrives to hatch its eggs inside its stomach without accidentally digesting them. Many starfishes have free-swimming larval stages, others lay yolky eggs that undergo a direct development.

Starfishes are probably short-lived animals with a lifespan of only a few years. Some, at least, are known to be sexually mature after one year, although growth continues for about four years. Some kinds can regenerate lost arms. *Luidia* fragments upon even gentle handling and must spend much of its life regenerating its lost members, to judge by the unevenly matched arms specimens often exhibit. In the southern Pacific another genus, *Allostichaster,* regularly reproduces by asexual transverse fission. In the tropics *Linckia* can regenerate the whole animal from a single arm, producing curious "sea comets" in the process; species of *Scelerasterias* sometimes show the same habit.

The brilliant colors of starfishes are caused by biochromes, which usually fade on preservation and are chemically changed by alcohol. The colors include orange or vermilion (the two most common) crimson (deep-sea forms), yellow, green, blue, violet, brown, as well as variegated patterns.

There are usually five arms, but many species have more than five—extreme cases occurring among the deep-sea Brisingidae where as many as 44 long, fragile arms may be present. Square starfishes with only four arms are abnormalities but are by no means rare. The largest starfishes (*Linckia* spp.) reach a meter in diameter.

Starfishes are fast-growing animals and require a great deal of food. The carnivorous types will attack any animal that they can swallow whole and will feed upon pieces of larger animals if their stomach can be applied to the food. Some species with suctorial tube feet attack bivalve shellfish and sea urchins. The valves of shellfish are forced apart by long-sustained suction exerted by the tube feet and the arms; in this way members of the family Asteriidae do much damage to oyster fisheries.

Starfishes are not known to be dangerous to man, although tropical species occasionally cause septic wounds in the foot, as in the case of the Indo-Pacific genus *Acanthaster,* a coral-reef dweller with sharp spines. No American species is likely to cause injuries of any kind. Starfishes, provided they are cooked first, are edible, although they are said to be bitter and unattractive. Of the few animals that do use them as food, the walrus is perhaps most notable, being immune to the toxins of the raw venom glands of the pedicellariae.

A number of parasites attack starfishes. A sea snail, *Stylifer,* bores into their skin; a polychaete worm, *Achloe,* inhabits the ambulacral groove; a cirripede destroys the sex gland of *Coscinasterias* and others; and in the tropics a slender fish, *Fierasfer,* inhabits the body cavity (apparently entering the mouth of its host and boring through the stomach wall).

There are about 1500 known species of starfishes, inhabiting the coasts of all known seas but avoiding brackish and freshwater. Some live at low-tide level or float on seaweed at the surface of the sea (*Calvasterias*); others burrow into the seafloor. Twelve families are known to range into waters more than 2 miles deep, and one species (*Porcellanaster caeruleus*) has been recovered from the seabed at a depth of 4 miles. Next to holothurians, starfishes are the deepest-dwelling animals of the sea—probably because, like holothurians, they can extract nutriment from mud. Individual species usually have rather restricted bathymetric ranges; but, where the continental slope is relatively steep, deep-water forms often ascend the slope and shelf forms often tumble into deep water.

ANATOMICAL FEATURES

The well-known genus *Asterias* is widely used as a laboratory animal (Fig. 18.1).

Body. The body is star shaped, consisting of a central disk and five symmetrically arranged arms. It is enclosed in a strong, flexible integument that contains numerous calcareous ossicles. The mouth lies at the center of the ventral surface, guarded by five interradial jaws. Radiating from the mouth are five ambulacral furrows bordered by two or three rows of movable adambulacral spines. Lateral to the adambulacral spines are three rows of ventrolateral spines, which are fixed, and a marginal series that runs along the border separating the ventral from the dorsal surface.

Near the center of the dorsal surface lies the anus. On one of the dorsal interradii is the madreporite—a flat, circular plate. The two arms between which the madreporite lies are called the bivium, and the remaining three are called the trivium. On the dorsal surface are irregularly shaped ossicles that support short, stout spines arranged in irregular rows parallel to the long axes of the rays. In the spaces between the ossicles are minute dermal pores through each of which projects a small soft respiratory process or papula, which is retractile. Scattered on the body are a number of very small pedicellariae, each of which is supported on a long or short flexible stalk and consists of three calcareous pieces—a basilar piece at the

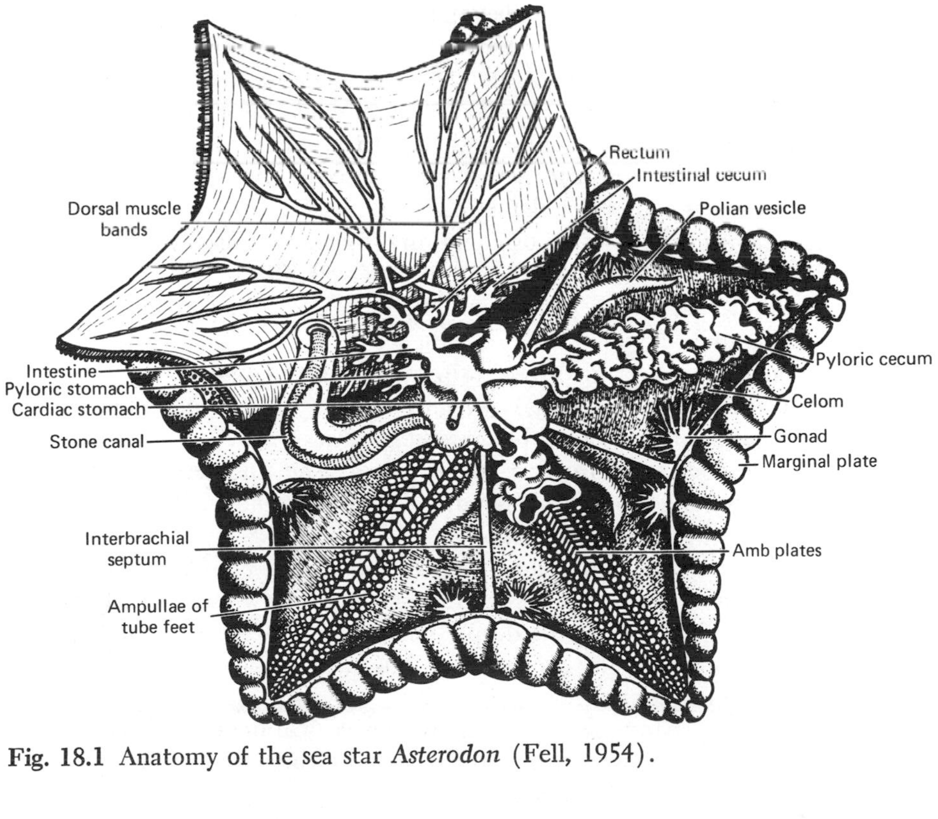

Fig. 18.1 Anatomy of the sea star *Asterodon* (Fell, 1954).

extremity of the stalk and two jaws which are articulated with the basilar piece and can be moved by certain sets of muscle fibers. In some pedicellariae the jaws are straight; in others they are crossed.

Locomotor Organs. Two double rows of tube feet lie in each ambulacral furrow. At the distal extremity of each furrow is a small, red eyespot; and above it is a sensory tentacle. The sensory tentacle is similar to a tube foot, but lacks a sucker; it is an olfactory organ. The ambulacral grooves are walled by two rows of ambulacral ossicles that meet at the apex of the groove. At each edge of the groove is a row of adambulacral ossicles; these carry the adambulacral spines.

Digestive System. The mouth leads, by a short esophagus, into the cardiac portion of the stomach, which is a broad, five-lobed sac with each lobe lying opposite one of the arms. The walls of the sac are greatly folded; and the sac can be extruded through the mouth, wrapped around food material, and then retracted with the aid of special retractor muscles. The cardiac stomach connects with a smaller pyloric portion, which in turn opens into a very short conical intestine. The intestine opens at the anus. The pyloric portion of the stomach gives rise to five pairs of pyloric caeca which extend into the arms and give off laterally two series of short branches connected with bladderlike pouches. The caeca are glandular and act as digestive glands. The short intestine also gives rise to five pairs of intestinal caeca, but these are short, interradially placed, and confined to the disk.

Water Vascular. Running downward from the madreporite to the area of the mouth is a cylinder, the stone canal, which is enclosed in a membranous sac. The walls of the stone canal are supported by a series of calcareous rings. The canal opens into a ring vessel that surrounds the mouth. The ring vessel gives off five ambulacral water vessels, each of which runs along the midline of the ambulacral furrow on the lower surface of the ambulacral ossicles. From each ambulacral vessel arise short side branches that form ampullae of the tube feet. The ampullae are internal, and connect with the external tube feet through ambulacral pores that lie between the ambulacral ossicles. Contraction of an ampulla injects contained fluid into the corresponding tube feet, causing their protrusion. In the interradii the ring vessel gives off five pairs of pear-shaped, thin-walled Polian vesicles. A pair of small, rounded, glandular bodies (Tiedemann's bodies) arises from the neck of each Polian vesicle (except in the interradius containing the stone canal, where there is one on one side only). The various parts of the water-vascular system have a muscular wall and an internal lining epithelium. The muscular layer is best developed on the tube feet, ampullae, and Polian vesicles.

The stone canal is accompanied by a fusiform axial gland; and surrounding the gland is a blood sinus (the axial sinus), which connects with an oral ring vessel. This ring vessel gives rise to five radial blood vessels, each of which is divided longitudinally by a vertical septum into two lateral halves. The ring vessel is similarly divided into two halves; the inner half communicates with the coelom, while the outer half connects with the axial sinus.

Nervous System. A pentagonal nerve ring surrounds the mouth and gives off five radial nerves, each of which traverses the ambulacral furrow immediately below the radial water vessel. Another set of deeper nerves parallels the path of the first set, and a third set, comprising the coelomic nervous system, extends along the roof of each arm superficial to the muscles.

Sex. The sexes are separate. The gonads are masses of rounded follicles, a pair of which lies in each interradius. The gonoducts open on the dorsal surface through several perforations on a pair of sievelike plates which are situated interradially close to the bases of the arms.

SYSTEMATIC REVIEW

The following summary will serve as a guide to recognizing the most important families of starfishes in world seas. In addition, all are mentioned in later chapters.

Several families have tube feet that lack a sucking disk, namely the lingthorns, the porcellanasterids, and the combstars. These are believed to be ancient groups.

Lingthorns are spiny starfishes with strap-shaped arms, bordered by a single row of marginal plates. The gut is blind because there is no anus. They live on sandy bottom, mainly in warm seas, and are active predators. The single surviving genus *Luidia* includes many species, mainly living in shallow water, and constitutes the family Luidiidae (Fig. 18.2).

Among the deep-water sea stars a notable family is the Porcellanasteridae, differing from lingthorns in having a double row of plates along the edge of the arm and distinguished by having what are called **cribriform organs**—ciliated grooves that occur between all or some of the marginal plates. These grooves are supposed to serve as respiratory or perhaps feeding organs. The family is characteristic of the deepest seafloors and is conspicuous in trench faunas. However, one genus, *Ctenodiscus,* occurs on the northeastern continental shelf of North America and is commonly dredged in shallow water off Maine (Fig. 24.2).

The **combstars,** family Astropectinidae, have pointed tube feet, lack cribriform organs, and have two rows of marginal plates around the edge of the arms. These sea stars constitute a large and widely distributed family with representatives in all seas at all depths. In the tropics the family is more diversified than in temperate seas. The most widely distributed genus is *Astropecten,* which occurs on the shelf of all continents except Antarctica. The North American species occur mainly to the south of about 40°N.

All the remaining families of starfishes have tube feet with suctorial disks.

The family Benthopectinidae are easily recognized because they have a double row of marginal plates around the edge of the arms. The upper plates alternate with the lower ones instead of lying one above the other. These are mainly found in deep water, on the continental slope of all continents, including Antarctica. *Benthopecten,* shown in Fig. 18.3, is typical of the family.

There are five more families of sea stars with conspicuous marginal plates defining the edge of the arms; in addition, they all have pincerlike cleaning or protective organs called **pedicellariae,** usually shaped like a miniature vise, located on various plates of the body. These five families are grouped as the order Valvatida.

In the first family of this group, the tooth stars (family Odontasteridae), there are paired or unpaired recurved spines on each jaw. The genera of this family are mainly Southern Hemisphere forms, inhabiting the continental shelf, especially of Antarctica, New Zealand, and South America. In temperate latitudes they tend to lie on rocky bottom in situations accessible only to scuba divers. In Antarctica they occupy the shelf and in summer ascend to the water's edge, retreating to deeper water in the winter. Examples are *Odontaster* (one oral spine), *Asterodon* (two oral spines). These are illustrated in Fig. 27.4.

Members of the family Archasteridae resemble *Astropecten,* but have suckers

GEORGE PUTNAM

Fig. 18.2 Shallow-water sea stars. (a) *Oreaster*, Caribbean. (b) *Luidia*, tropical and subtropical. (c) *Linckia*, pantropical. (d) *Archaster*, Indo-West Pacific. (e) *Coscinasterias*, Indo-West Pacific. (f) *Nidorellia*, tropical East Pacific. All approximately one-quarter actual size.

on the tube feet. It is found in the Indo-West Pacific on tropical sandy coasts, just below low tide, and is fairly common. There is only one genus, *Archaster*. It has male–female superposition during release of sex cells, the only known case of copulation in asteroids.

GEORGE PUTNAM

Fig. 18.3 Deep-water sea stars. (a) *Freyellaster*, Brisingidae. (b, c) *Benthopecten*, Benthopectinidae. (d, f) *Zoroaster*, Zoroasteridae. (e, g) *Porcellanaster*, Porcellanasteridae. First three genera abyssal and bathyal, *Porcellanaster hadal*. (a, d, f) One-quarter actual size; (b, c, e, g) half actual size.

Members of the family Goniasteridae resemble the tooth stars, but lack the recurved oral spines. These valvate pedicellariae, however, are normally well developed. There is a large and varied assemblage of forms, especially in the tropical Indo-West Pacific, but also occurring widely in deeper water in all oceans, mainly on soft bottom. Following are representative examples of the range of form encountered: *Pentagonaster* (Fig. 27.4), with enlarged penultimate marginals, chiefly Australia and New Zealand in recent seas, but widely distributed in Cretaceous times. *Ceramaster*, like *Pentagonaster* but without enlarged plates; cosmopolitan on outer shelves and continental slopes, including North America. *Hippasteria*, large and small ossicles scattered over dorsal surface, usually with conspicuous spines and pedicellariae on them. Deep-slope bottom, but occasionally ascending to the shelf, even into water of only 10 fathoms on very rare occasions (once taken by a scuba diver). Never common. Cretaceous species much like existing ones. Seem to favor cold water. *Pseudarchaster* mainly inhabits warm shelf seas, pedicellariae spiniform rather than valvate. *Mediaster* is similar but different in details of oral structure. Examples are shown in Fig. 25.1.

The family Oreasteridae includes some large sea stars. The disk is large, high, or swollen, usually with a meshed skeleton. They are essentially tropical shelf forms that are common on coral reefs of Indo-West Pacific and Caribbean. *Culcita* is shown in Fig. 18.4. *Oreaster* has a large body with pentagonal outline and a reticulate dorsal skeleton well exposed (Fig. 18.2). *Protoreaster* is like *Oreaster* but with large tubercles. Indo-West Pacific. *Nidorellia* (Fig. 18.2), body more flattened, strongly tuberculated, East Pacific only.

The serpent stars, family Ophidiasteridae, are readily recognized by the small, central-disk region and long arms that are slender and, more or less, cylindrical. Tropical. *Linckia*, (Fig. 18.2) common on shallow coral reefs in full sunlight. The plates of its arms are not arranged in regular longitudinal series, often covered by leathery skin, blue or purple. In *Ophidiaster* the arm plates occur in regular longitudinal series, usually brown or red. These forms can reproduce from broken arms.

In the remaining families the marginal plates are either lacking or are very small and inconspicuous.

The sun stars, family Solasteridae, have a large disk, from 5 to 15 arms, a dorsal reticulate skeleton with imbricating plates. Mainly cool or cold seas, continental shelf and deeper seafloors. *Solaster* and *Crossaster* are examples of this family.

The Pterasteridae are viviparous. Their dorsal surface is covered with cross-shaped or lobed plates supporting a protective marsupium from which young escape via a central aperture. Star-shaped deep-water forms, cosmopolitan. *Pteraster* is shown in Fig. 28.1.

The starlets, family Asterinidae, have the dorsal surface covered by minute imbricating plates bearing single or grouped spinules. Their body outline is usually pentagonal or nearly so. Mainly intertidal or shallow-water infratidal forms on continental and island coasts in all seas. The major species of this family are *Asterina* and *Patiria* (California) (Fig. 25.1).

In far northern and far southern seas some puzzling sea stars occur that have been grouped to form the family Poraniidae. The dorsal surface is covered by soft skin. The systematic position is in doubt, as some specimens show conspicuous marginal plates (such being referred to the supposed genus *Poraniomorpha*), whereas others show no marginal plates (*Porania*), and still others show varying degrees of loss of almost all plates. Each version comprises a separate genus. Perhaps all these are local forms of one genus, *Porania*, with the loss of calcareous material apparently related in some way to influx of fresh ice-melt water in the fjords where many occur.

Long-armed forms with a very narrow ambulacral groove comprise the cosmo-

GEORGE PUTNAM

Fig. 18.4 Tropical shallow-water sea stars. (a, b) *Acanthaster,* Indo-Pacific. (c) *Linckia,* pantropical. (d, e) *Culcita,* Indo-West Pacific. All one-fifth actual size.

politan family Echinasteridae, occurring mainly in shallow water. *Henricia* and *Echinaster* are the best-known members (Fig. 24.2).

The crown-of-thorns stars, family Acanthasteridae, are multiarmed forms with a large disk, with the disk and arms covered by very large tapering spines. There is only one genus, *Acanthaster,* in the Indo-Pacific, known to attack coral organisms on coral reefs (Fig. 18.4).

Certain deep-water sea stars with a very small disk and with numerous slender, easily detached cylindrical arms that lack a dorsal skeleton in the outer part of the arm, comprise the family Brisingidae (Fig. 18.3).

The remaining families have pedicellariae that resemble miniature forceps.

The family Heliasteridae are sun stars with numerous arms (up to 50), disk large. *Heliaster*, East Pacific (Fig. 27.4).

The zoroasters, family Zoroasteridae, have a small disk, and five slender arms, tapering and subcylindrical, with ossicles in close longiseries. Deep-water forms. An example is *Zoroaster* (Fig. 18.3).

In the last family, the Asteriidae, the valves of the pedicellariae are crossed, like scissors. They have a minimum of five arms ambulacral groove with four rows of tube feet. Crossed pedicellariae usually occur in wreath formation at base of dorsal spines. World-wide predators, in deep and shallow water. Examples: *Asterias, Allostichaster, Calvasterias* (southern oceans, brood-protecting, rafting), *Coscinasterias* (tropical and warm temperate, especially Indo-West Pacific), *Leptasterias* (boreal and north polar), *Pisaster* (found primarily in California in shallow water), *Pycnopodia* (west coast sunflower starfish, Fig. 25.1). Classification of this family is presently rather chaotic and the literature is very hard to follow. Extensive research is needed on a world-wide scale.

CLASSIFICATION OF SEA-STAR FAMILIES MENTIONED ABOVE

Following is the grouping in orders of the families mentioned here. Although detailed information has been omitted, if you examine the various illustrations in this chapter and in the chapters on regional faunas, you will notice that sea stars of similar appearance are grouped in the same orders. In time you will be able to determine the relationship of many species encountered in field or laboratory study. A good reference work is W. K. Fisher's volumes of *North Pacific Asteroidea* (Memoirs, 1911–1930, U.S. National Museum, Washington, D.C., USA). There is no comprehensive monograph on the sea stars of the world.

Order Platyasterida
  Family Luidiidae.
Order Phanerozonida
  Families Porcellanasteridae, Astropectinidae, Benthopectinidae, Odontasteridae, Archasteridae, Goniasteridae, Oreasteridae, Ophidiasteridae.
Order Spinulosida
  Families Solasteridae, Pterasteridae, Asterinidae, Poraniidae, Echinasteridae, Acanthasteridae.
Order Forcipulatida
  Families Brisingidae, Heliasteridae, Zoroasteridae, Asteriidae.

19

BRITTLESTARS

General biology / Anatomical features / Oegophiurida / Phrynophiurida / Ophiurida

Brittlestars are the most conspicuous bottom invertebrates on the deep seafloor; they also range the continental shelf. They are slender relatives of the starfishes and, like them, comprise a subclass of echinoderms. Their almost universal occurrence in marine environments and their enormous numbers have led to their extensive use as **dominants** and **subdominants** in the nomenclature and definition of marine communities. By these terms a given community can be named. For example, the *Amphiura-Echinocardium* community is a soft-bottom assemblage of species in which the brittlestar *Amphiura* is the most conspicuous member, the dominant, and the heart urchin *Echinocardium* is the next most prevalent (subdominant). These animals are also called ophiuroids.

From the earliest days of deep-sea oceanography, the abundance of ophiuroids in bottom samples has been the subject of comment. More recently the perfection of seafloor photography has placed in our hands truly astonishing pictures of the bed of the ocean, which is often carpeted with regularly spaced brittlestars extending into seemingly limitless distance (Fig. 19.1). Besides this, they are the most numerous of echinoderms in species, with about 1900 nominal extant forms referred to 220 genera and grouped in three orders. Their fossil history ranges back to the earliest Ordovician.

GENERAL BIOLOGY

Although no ophiuroids are used by man for food, they are of indirect economic importance, partly through their participation in food chains involving commercially significant fishes and partly through their function as scavengers disposing of the by-products of life and death of other marine organisms. Ophiuroids have probably

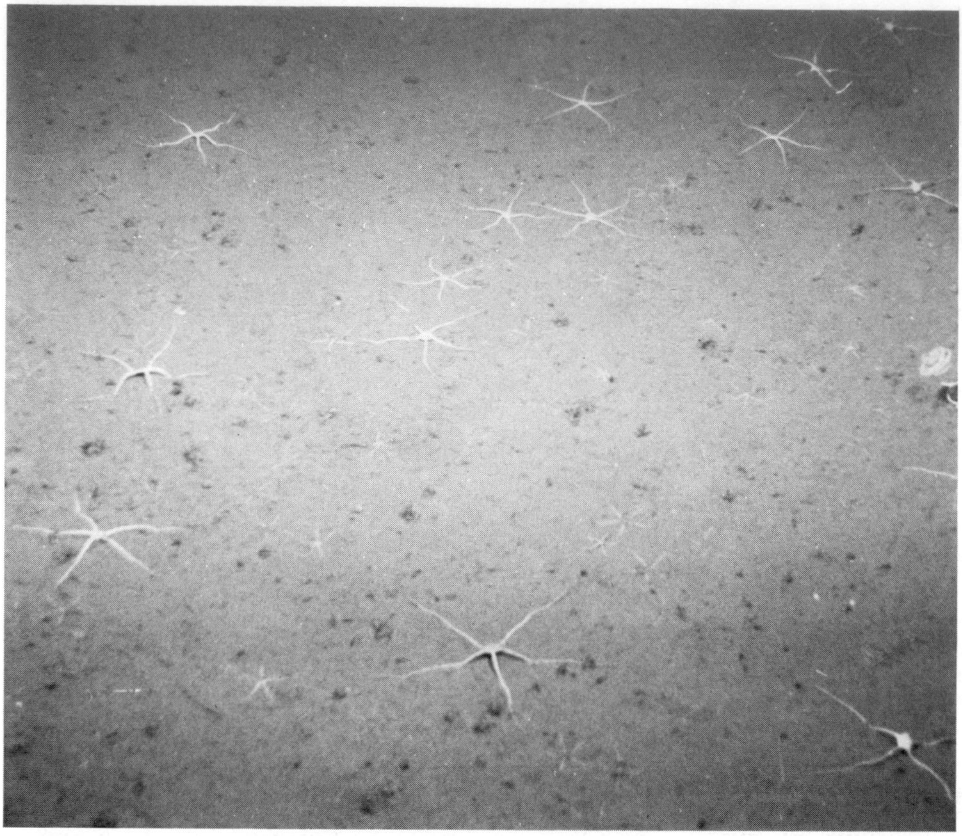

U.S. NAVY

Fig. 19.1 Part of an *Ophiomusium lymani* community on the deep-sea muds that cover the floor of the San Diego Trench at a depth of 2100 m. This view was obtained from the port of the bathyscaphe *Trieste*.

been very significant elements in the economy of the seas ever since they first arose in the lower Paleozoic.

At maturity most ophiuroids achieve a maximum size on the order of 10 to 20-mm disk diameter, with an armspread of around 100 mm. In small forms, such as species of *Ophiomisidium*, the disk and arms together measure only 3 to 5 mm across. The largest ophiuroids are all members of the suborder Euryalina, in some of which the arms branch many times, thus considerably adding to the total mass. The most massive species are basket stars of the genus *Gorgonocephalus*, where the disk sometimes measures about 100 mm across, and the armspread is around 350 mm. There are also some large euryaline ophiuroids that have unbranched, elongate arms.

Ophiuroids are generally rather active animals that move in a very characteristic manner by means of the arms, which are thrown forwards in pair, pushing the animal forward in jerks. One of the arms (any one of them) is directed in front, or is trailed behind, and does not actively assist locomotion, but probably serves as a sensor. This way of moving is especially characteristic of species living on hard bottom (*Ophiura*, for example). Some species, especially *Ophiocomina nigra*, *Ophiura robusta*, and *Amphipholis squamata*, are able to climb up the walls of aquariums by means of their tube feet (Mortensen, 1927).

In accordance with their retiring habits, amphiurids as a rule are less active than other ophiuroids; and when placed on a hard substrate they do not push themselves forward like other ophiuroids. Instead, they move by extending an arm, then contracting it in a series of sinusoidal curves, while the distal portion is fastened to the substrate, so that the body is dragged forward and the other four arms are dragged along with it. *Ophiocomina nigra* often climbs the sides of aquariums, attaching itself by a pair of arms, while the other arms are stretched out under the surface of the water. Some sea stars, for example, *Stegnaster,* will also do this when first introduced into an aquarium, as if responding to a reduced oxygen content of the environment, later becoming adapted and descending the sides to the floor of the tank. The tube feet of *Ophiocomina* are strongly papillate, and the slime secreted by these papillae serves as a means of attachment.

Sexual Reproduction. Most species are egg-laying (oviparous), the spawning occurring in the early fall; one New Zealand ophiuroid *Ophiomyxa,* however, spawns in the early spring. Viviparous species (that hatch the eggs within the body) are numerous. In the southern oceans (Antarctica, South America, New Zealand), where they may comprise up to 50 percent of the fauna, brood protection, however, is not necessarily associated with a cold environment, for the New Zealand shelf waters are temperate, and the genera involved there include *Pectinura,* which is of tropical Indo-West Pacific derivation. A general review has been given by Fell (1948). When extruding the genital products, some ophiuroids (amphiurids) assume a peculiar position. Their disk is raised some centimeters above the bottom, resting on the proximal part of the arms as if on five props, and the distal part of the arms remains buried in the substrate. When the sperm or ova are shed, the animal relapses back into its normal buried position. *Ophiura texturata* takes 2 years to reach sexual maturity, but probably at least 3 or 4 years to reach full size. Grieg (1928) believed that the deep-water species *Ophiomusium lymani* takes about 3 years to reach full size, and he thought this might also apply to most other deep-water species of ophiuroid in the northern Atlantic. However, species such as *Ophiopholis aculeata, Ophiocten sericeum,* and *Ophiura sarsi* barely reach half their maximum size over a 2 to 4 year period, so it would seem likely that most ophiuroids continue to grow for periods of 8 years at least, and probably for 10 to 15 years.

The ecology and feeding habits of *Astrotoma agassizii* (Fig. 19.2) have been studied on the basis of seafloor photographs from stations in the Ross Sea where specimens were dredged. The ophiuroid lies on its dorsal side with its mouth directed upward and its long arms sweeping through the overlying water, apparently collecting small nekton and plankton fall, which is conveyed to the mouth from the tube feet by drawing the arms across the oral opening (Fell, 1961). A conspicuously different manner of feeding in the same environment is exhibited by *Ophiurolepis gelida* (Fig. 19.2), an individual that can be seen only a short distance away from a specimen of *Astrotoma agassizii,* which is behaving in the manner indicated above. *Ophiurolepis gelida,* unlike *Astrotoma agassizii,* practices detritus feeding, thrusting one arm up a small mound of the substrate as it searches for organic material. The associated fauna in photographs and samples where *Astrotoma* is present include tubicolous polychaetes and bryozoans, often oriented directionally and suggesting the presence of a bottom current. This would confirm the inference that plankton and small nekton are available for ophiuroids that fish the overlying water as *Astrotoma* must do. (See Fig. 19.3.)

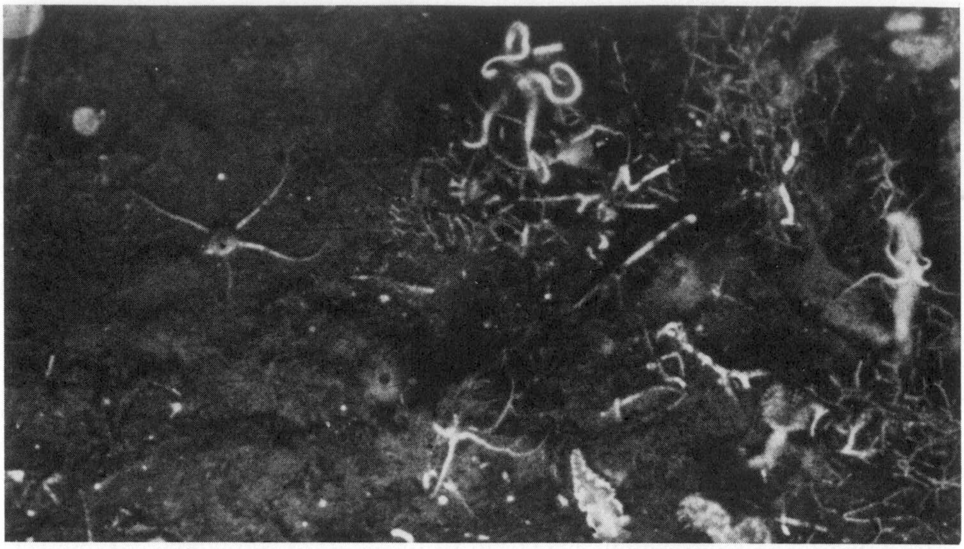

NEW ZEALAND OCEANOGRAPHIC INSTITUTE

Fig. 19.2 Community at 260 m in McMurdo Sound, Antarctica, photographed by remotely controlled camera (John Bullivant). To left, *Ophiurolepis gelida* sensing the substrate; above center, *Astrotoma agassizii* in feeding position for taking plankton fall; to right, *Ophioceres incipiens* "fishing" from an erect sponge. See text for further explanation. (Fell, 1962)

Ophioceres incipiens mounts corals, tubicolous polychaetes, and sponges, to which it clings by one or more of the arms while the other arms are held out for fishing like the rods of an angler. In the original negatives of some of the photographs obtained of the Antarctic seafloor by Bullivant, it is possible to distinguish the tube feet; it appears that these organs are probably sticky and catch small plankton and nekton, which are then transferred to the mouth (Fig. 19.2). Nichols (1963) notes that the ophiuroid tube-foot system is concerned with feeding and in some cases with burrowing. Although the power of protraction of the tube foot is generally limited, in a few species the tube feet can be extended at some length.

The food of *Ophiothrix fragilis* consists mainly of worms and crustaceans (Mortensen, 1927), but it also eats small clams, echinoderms, compound ascidians, forams, and so forth. It appears that ciliary currents also play some role in catching food and detritus. Vevers (1952) reporting on the English Channel and Hurley (1959) on Cook Strait, New Zealand, both record large, dense populations relying upon a steady flow of suspended matter, with evidence that seafloor currents supply this.

ANATOMICAL FEATURES

The body comprises a circular, central disk and five elongate, snakelike arms—all enclosed in an integument of calcareous plates. The mouth lies at the center of the ventral surface, guarded by five interradial jaws. A pair of radial shields lies at the base of each arm on the aboral side of the disk; and on the oral side there are five conspicuous oral shields, one in each interradius. One of these oral shields functions as the madreporite. Also on the oral side at the base of each arm are two

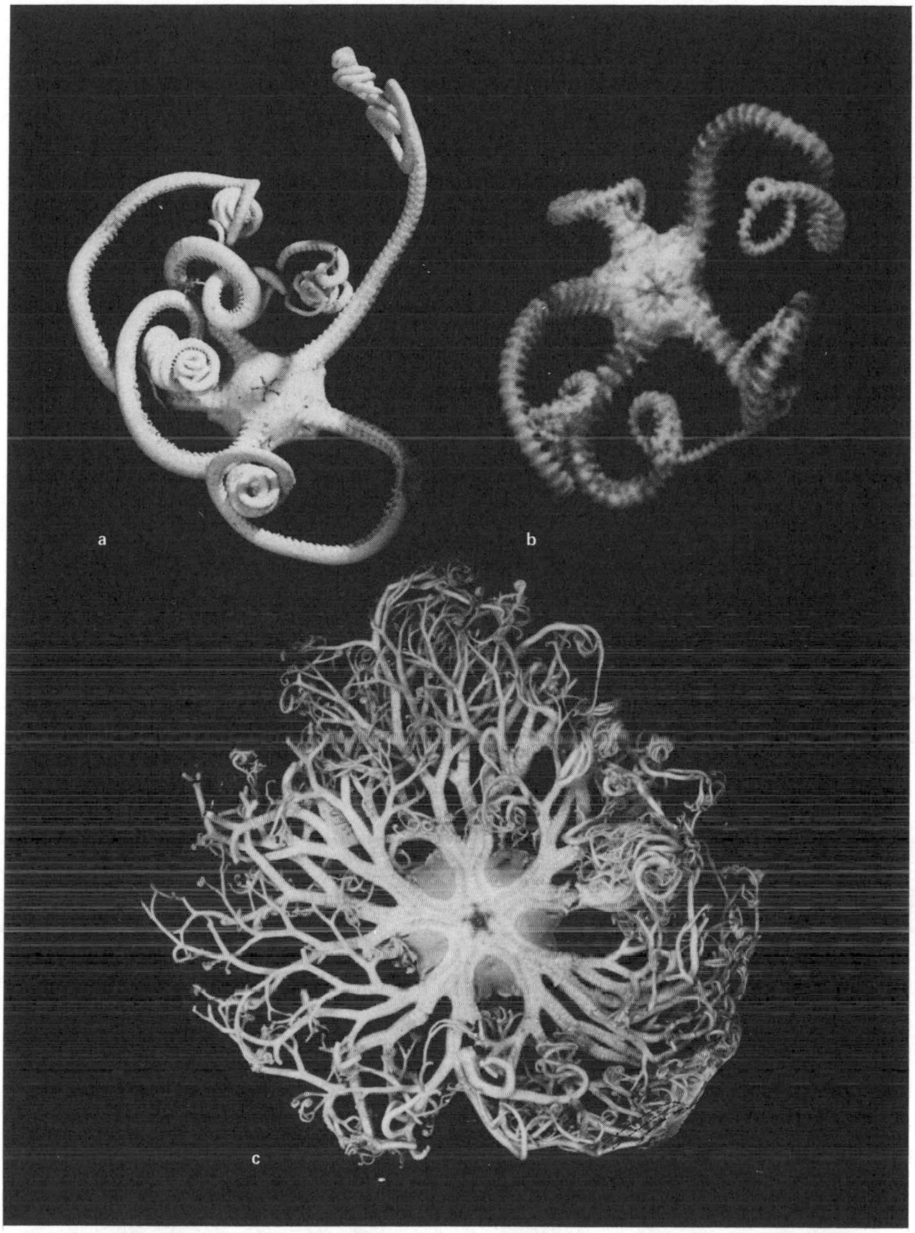

MERVYN D. KING

Fig. 19.3 (a) *Astrotoma agassizii.* (b) *Astrochlamys bruneus.* (c) *Gorgonocephalus chilensis.* All members of the Euryalae, reduced several diameters. (a, b) Antarctica. (c) West-wind-drift distribution in Southern Hemisphere.

adoral plates and two oral plates. The oral plates contribute to the jaws and carry oral papillae. On each side of the base of each arm is a genital slit, which opens into a genital bursa. The arms are solid, comprising a series of ossicles, termed vertebrae, which are held together by well-developed longitudinal muscles. The vertebrae are covered externally by dorsal, ventral, and left and right lateral plates, of which the latter two carry spines. Tube feet project between the ventral and lateral plates, and modified spines, or tentacle scales, protect them. Pedicellariae are absent.

Because the arms are solid, the alimentary and reproductive systems are confined to the disk. The alimentary system is simple. The mouth opens into a blind sac, the stomach, which has 10 pouches; there is no anus. Food material is digested within the stomach and egesta are voided through the mouth. The water vascular system is similar to that of asteroids, except that here the tube feet lack suckers and ampullae. The nervous system is typical of the phylum; no organs of special sense are known. The genital bursae are blind sacs. The gonads are attached to the coelomic walls of the bursae and consist of small sacs. The sexes are usually separate. (See Fig. 19.4.)

The coelomic cavity is greatly reduced in ophiuroids, owing to the presence of both gut and gonads in the small disk.

IDENTIFYING BRITTLESTARS

The classification of the numerous genera of brittlestars is too complex to cover in an introductory text, but keys have been separately published (Fell, 1960). Following are notes on some of the more conspicuous families in marine communities.

Order Oegophiurida. These are small ophiuroids that have branches of the gut and reproductive organs extending into the arms. Only one family survives in existing seas, comprising the single genus *Ophiocanops*. It is a small predator that clings to corals in Indonesian seas.

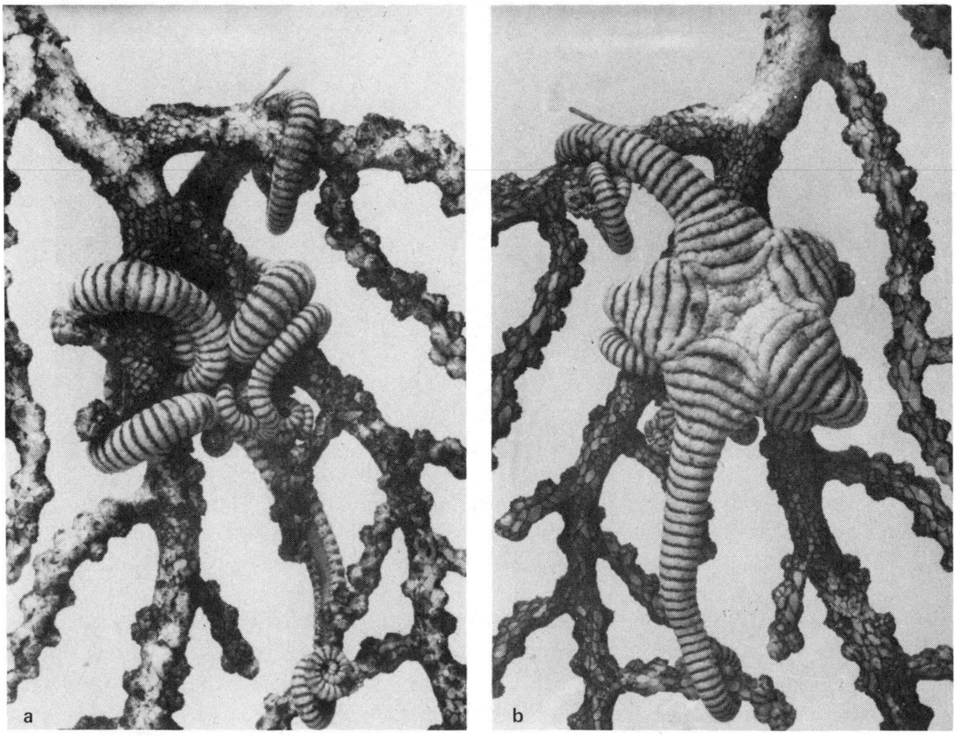

PATRICIA M. RALPH

Fig. 19.4 *Astroporpa wilsoni*, a deep-water euryalid from northern New Zealand clinging to coral sea-tree. Continental slope.

Order Phrynophiurida. In these brittlestars the arms usually coil spirally in the vertical plane, and they may branch by repeated forking. The skin of the arms is either soft or granulated and is not covered by distinct rows of plates. The gut is restricted to the disk. If the arms fork many times, the animals are called basket stars. *Gorgonocephalus,* with large species (armspread up to 35 cm), frequenting cold Northern and Southern Hemisphere seas, is an example. In some forms the

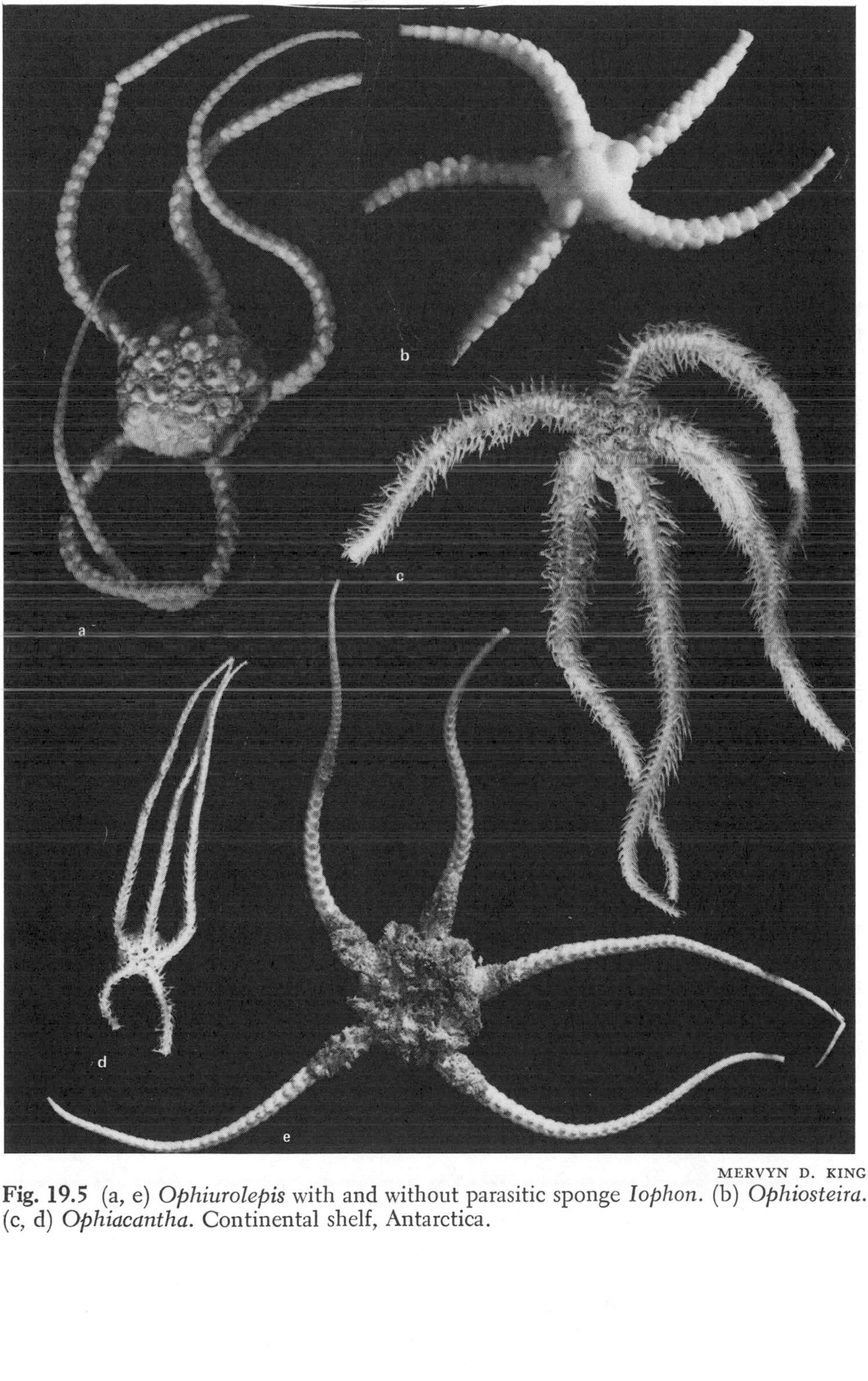

MERVYN D. KING

Fig. 19.5 (a, e) *Ophiurolepis* with and without parasitic sponge *Iophon.* (b) *Ophiosteira.* (c, d) *Ophiacantha.* Continental shelf, Antarctica.

arms are unforked (i.e., "simple") and are covered above by soft skin, flexing in the horizontal plane, not coiling spirally. *Ophiomyxa* and related genera occur in tropical and warm temperate seas in shallow water. (See Fig. 19.5.)

Order Ophiurida. In this large group the disk and arms are covered by a mosaic of plates; the arms flex in the horizontal plane and are not branched and do not usually coil spirally. There are many families, only a few of which can be mentioned here. In the Ophiocomidae the jaws are bordered by small plates called **oral papillae,** and at the inner tip of each jaw there is a cluster of spinelike **tooth papillae.** *Ophiocoma,* with species in tropical and warm temperate seas, sometimes forms dense communities on seafloors where currents deliver suspended particulate nutrients. The family Ophiothricidae do not have oral papillae, but spinelike tooth papillae are present at the tip of the jaw. Arm spines are usually conspicuously long. *Ophiothrix,* with spines or thorns on the disk, is sometimes a conspicuous member of faunas. The Amphiuridae comprises delicate species, mostly found on soft substrates. The jaws have a pair of papillae at the inner end, and the sides of the jaw lack papillae. *Amphiura* occurs in all seas, and is one of the most common brittlestars. The Ophiactidae are brittlestars having jaws with a single papilla at the tip. *Ophiopholis,* with small platelets around each arm plate, is a very common littoral genus in northern seas and on the New England coast. In the family arms are inserted laterally into the disk and firmly fused to disk, so they do not break off easily. The disk plates are naked, often symmetrically arranged. These are mainly deep-water species, and *Ophiura* is a typical genus found in dredgings. The family Ophiodermatidae comprises brittlestars that resemble Ophiuridae, but the disk plates and jaws are covered by granules. *Pectinura* ranges warm continental shelves. *Ophioderma* is common on eastern North American coasts. In the family Ophiacanthidae the arms are inserted beneath the disk, not firmly fused to disk, so they usually break off in the dredge. Erect spines occur on the disk, and arm spines are elongate and usually numerous. They are often epizoic, clinging to deep-water and polar benthos. *Ophiacantha* is an example.

How to Recognize the Genera of Brittlestars

The foregoing notes refer to the chief families; but, unfortunately, brittlestars are difficult animals for the general zoologist or ecologist to identify. Perhaps the best way to recognize the species that occur in an area you are studying is simply to collect and preserve in alcohol examples of the species present, and then go to a publication, such as Fell, H. B., *Keys to the genera of Ophiuroidea* (1960), and so determine the genus if possible. Then consult one of the major museums as to what species have been reported for those genera from the region you are studying. If no records exist (as is often the case) then you have a ready-made topic for a research project, and you will have to use the *Zoological Record* to guide you to related literature. Ophiuroids are so numerous and so important as fish food that you should not be afraid to take up their study, for very few people know much about them at present.

20

SEA URCHINS

General biology / Cidaroida / Diadematoida / Echinothurioida / Arbacioida / Temnopleuroida / Echinoida / Cassiduloida / Clypeasteroida / Spatangoida

By the end of the Triassic period reef-building corals had become widely established as patch reefs, and a general process of reef and atoll building continued through the Mesozoic and Cenozoic eras. Conspicuous among the epifaunal occupants of the reefs were the highly symmetrical sea urchins called Cidaroida, the only order of Echinoidea that had survived the late Paleozoic extinctions. Cidaroids, together with the several orders to which they in turn gave rise, are still important constituents of the seafloor faunas of the oceans. As explained below, echinoids may be broadly divided into (a) globular forms (Fig. 20.1) having an epifaunal habit and (b) more or less bilaterally symmetrical forms (Fig. 20.2) having a predominantly infaunal habit. Both categories are associated with coral reefs, but the globular forms (or *Regularia*) live mainly on the hard coral substrate and on rock reefs or other hard substrates, whereas the others (or *Irregularia*) inhabit the soft lagoon muds, as well as the sands and silts of other parts of the seafloor. The faunas are richest in the tropics but are also quite numerous in temperate and even in polar environments.

There are about 850 living species placed in 225 genera. They are arranged in 18 orders, grouped into two subclasses. They range in size from a few millimeters across the shell to about 20 cm. They have no known excretory organs, and it is believed that the large amounts of dark or intensely colored pigment in the skin are produced by the accumulation of excretory products, rendered nontoxic by conversion into insoluble compounds. Many are dark purple or green, others are intense shades of red or orange, and some have striped spines. Deep-sea forms, however, may be white, which seems to imply a difference in the physiology.

Some tropical species, such as the diademas (Fig. 24.1) have hollow sharp spines that can break off in the skin if handled, producing painful septic wounds because they contain a toxic substance. The genus *Araeosoma* (Fig. 24.1) has ven-

JOSEPH D. GERMANO

Fig. 20.1 Tropical urchins. (a) *Prionocidaris,* Indo-West Pacific (occurred also in Caribbean prior to late Tertiary cooling). (b) *Phyllacanthus,* Indo-West Pacific (and Caribbean prior to late Tertiary cooling). (c) *Heterocentrotus* (slate-pencil urchin), Indo-West Pacific reefs. (d) *Centrostephanus,* tropical and subtropical seas. (e) *Eucidaris,* pantropical, on reefs and hard substrate.

omous spines so dangerous as to cause occasional fatalities among fishermen. These very toxic species share the character of a flexible shell that generally collapses into a pancake-shaped mass after they are dredged to the surface. Therefore, it is a good

JOSEPH D. GERMANO

Fig. 20.2 Heart urchins and sand dollars. (a) *Abatus,* with brood pouches, Antarctic seas. (b) *Breynia,* Australia. (c) *Apatopygus,* New Zealand. (d) *Fibularia,* Indo-Pacific. (e) *Rotula,* West Africa. (f) *Clypeaster,* all warm seas. (g) *Arachnoides,* Indo-West Pacific. (h) *Echinarachnius,* eastern North America.

rule not to handle such forms without thick gloves. However, the great majority of sea urchins are quite harmless to man. In many countries species are fished commercially and sold as food (the reproductive glands constituting an epicurean treat, although the Polynesians eat the entire contents of the shell, gut contents and all).

Sea urchins range all seas from sea level down to more than 4 miles. The rounded or regular urchins feed mainly on algae, often hiding by day in rock pools, covering their body with stones or shells held like umbrellas by the tube feet and emerging to feed at night or at high tide. The heart urchins, on the other hand, generally live imbedded in the mud upon which they feed. Sea urchins move about quite slowly, using their tube feet and also their movable spines as locomotor organs. They are attacked by numerous parasites, among them protozoans, nematodes, and some mollusks. There are some crabs that inhabit the rectum of certain sea urchins, and others live on the shell itself, feeding on the tube feet and spines. Other animals may shelter among the spines, especially among the long venomous spines of the diademas, which effectively protect the waifs from attack by other fishes.

EPIFAUNAL SEA URCHINS

These urchins have a globular shell (called the **test**), with the mouth located on the lower surface, surrounded by a flexible membrane called the **peristome,** and with the anus located at the center of the upper surface, surrounded by a flexible membrane called the **periproct.** The mouth houses a ring of five sharp teeth, each operated by an independent jaw—the whole apparatus constituting the so-called *Aristotle's lantern.* The following is an outline of the main features of the included orders.

The order **Cidaroida** are easily recognized (Figs. 20.3 and 20.4) because the tube feet each emerge through a single plate of the test, whereas in other regular urchins three or more tube feet emerge through one plate. Every tube foot is associated with a pair of pores on its plate.

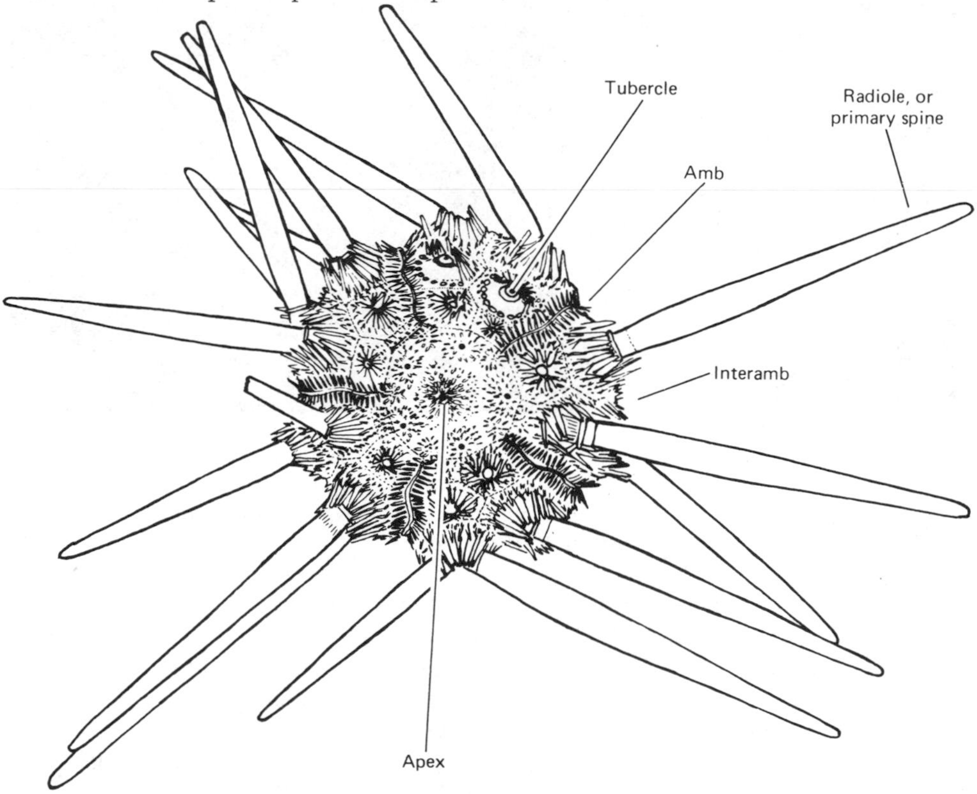

Fig. 20.3 *Cidaris abyssicola,* to show anatomical features.

JOSEPH D. GERMANO

Fig. 20.4 Details of test, regularia. (a) *Phyllacanthus*, Indo-West Pacific. (b) *Coelopleurus*, cosmopolitan, mainly in deep water. (c) *Echinometra*, pantropical, reefs. (d) *Histocidaris*, deep water, mainly in tropical and subtropical seas.

The following are the more important genera of cidaroids in the ecosystems described in later chapters. As they are illustrated in detail, the systematic characters are here limited to the briefest statements. *Histocidaris* (Fig. 20.4) is a deep-water genus in which the large tubercles that support the spines have a crinkly margin (crenulation). It is found in tropical seas. *Goniocidaris* (Fig. 27.4) is an Indo-Pacific form with horizontal grooves between the plates, and often with disk-like expansions of the tips of the larger spines. *Stereocidaris* has upper plates of shell often withhout spines (or tubercles that support the spines). It frequents the Indo-Pacific, mainly in offshore waters. Genus *Phyllacanthus* (Fig. 20.1) contains large forms living at the margins of tropical reefs of the Indo-Pacific, having long cylindrical and widely separated spines. One species, *P. imperialis*, ranges from the Red Sea to Hawaii, five others are found in Australian waters. *Prionocidaris* (Fig. 20.1) inhabits the Indo-Pacific. It has larger spines with conspicuous thorns. Both *Phyllacanthus* and *Prionocidaris* formerly inhabited the Caribbean region but became extinct there apparently when the Pleistocene coolings began. *Cidaris* (Fig. 28.1) consists of Atlantic offshore and deep-water forms, sometimes with paddle-shaped spines, as in *Cidaris blakei* of the Caribbean. *Eucidaris*, a pantropical genus having a crownlike termination on the tips of the spines, is common on coral reefs; similar species occur on the Panama and Caribbean coasts, originally indicating the former

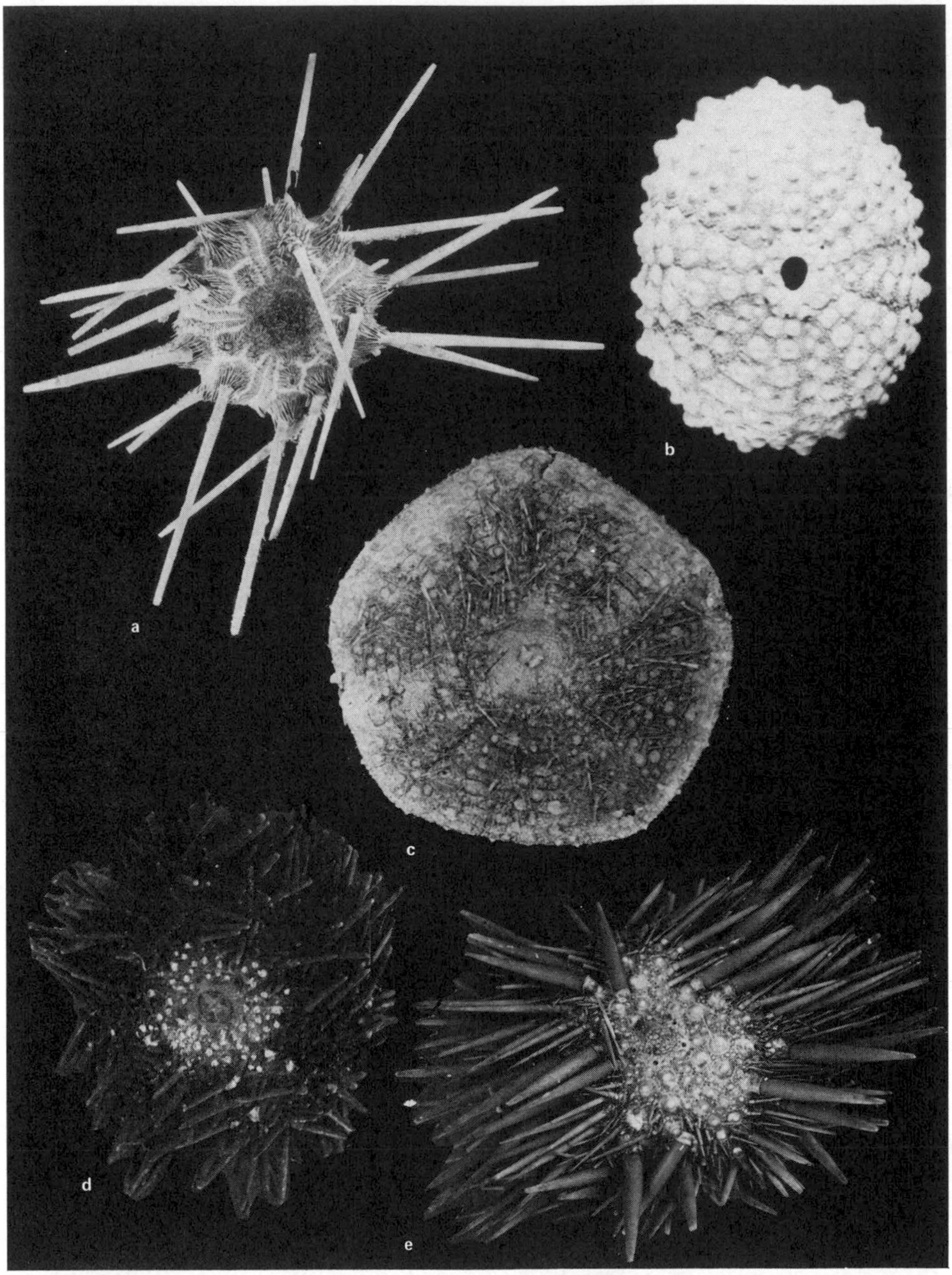

JOSEPH D. GERMANO AND GEORGE PUTNAM

Fig. 20.5 (a) *Hesperocidaris,* West-American shelf. (b) *Heterocentrotus,* Indo-West Pacific reefs. (c) *Asthenosoma,* Indo-West Pacific shelves. (d) *Tetrapygus,* Peru and Chile. (e) *Echinometra,* pantropical reefs.

existence of the Panama seaway (Figs. 10.4 and 10.5). *Hesperocidaris* (Fig. 20.5), resembling *Eucidaris* but lacking the terminal crowns, is found on west American coasts.

The order **Echinothurioida** comprises flexible-shelled sea urchins, mainly from deep water and sometimes venomous. *Asthenosoma,* having the larger spines of the lower surface with each a terminal hooflike appendage; Indo-Pacific, can inflict lethal stings from poison glands on spines of upper surface. *Phormosoma* (Fig. 23.10) occurs in deep water and has a honeycomblike appearance on its lower surface. Terminal hoofs do not occur. Members of this order have the amb plates compounded so that three or four pore pairs occur on each.

The order **Diadematoida** comprises sea urchins of tropical and subtropical seas that have long and hollow spines with a rough surface, tapering to very fine, sharp, venomous tips. These can break off in the skin of an intruder. The amb plates have three pore pairs, and the tubercles that carry the spines have a creulate margin (like *Histocidaris*). *Diadema* (Fig. 24.1) has amb tubercles in two regular series. There are no spines on the peristome. It is found in tropical and subtropical seas and in shallow-water reef dwellers. *Centrostephanus* (Fig. 20.1) is like *Diadema,* but it has spines on the peristome. *Echinothrix* is also like *Diadema,* but it has widened ambs on its upper surface and inconspicuous amb tubercles, not arranged in two series. It frequents the Indo-West Pacific and is a coral-reef dweller. *Aspidodiadema* has large spines that are curved downward. It is found in deep water, using the spines for support on soft substrate in tropical and subtropical regions.

The order **Arbacioida** comprises urchins with four plates imbedded in the periproct around the anus. The tubercles of the spines lack the perforation that occurs in the preceding orders. *Arbacia,* has tubercles in regular series. It is found in the Atlantic Ocean but is absent from Indo-West Pacific (Fig. 24.1). *Coelopleurus* (Fig. 20.4) is like *Arbacia,* but lacks spines on the upper parts, which are therefore bare of tubercles. Its range is worldwide, mainly in deep water. *Tetrapygus* (Fig. 23.10) has amb plates with four pore pairs. It is found in Peru and Chile.

The order **Temnopleuroida** comprises two major families with rather different characters, but both sharing the features that the tubercles lack perforation and the periproct lacks plates. In the family Temnopleuridae the surface of the test carries some kind of sculpture, either in the form of raised ridges, or as depressed pits, or a combination of pits and ridges. In *Temnopleurus* (Fig. 20.6) pits occur at the corners of the plates of the test. It is found in the Indo-West Pacific. In *Mespilia* the interamb toward the lower surface is naked along the midline, where transverse striations occur. It ranges the Indo-West Pacific. *Paratrema* has deep pits at the angles of the plates, and there are 5 plates imbedded in the peristome. It inhabits the Indo-West Pacific. *Salmacis* is like *Temnopleurus* but the pits are reduced to small pores. It is also found in the Indo-West Pacific. *Temnotrema* is like *Paratrema,* but there are 10 oral plates in the peristome. It is found in the Indo-West Pacific. *Pseudechinus* (Fig. 23.10) has sculpture on the test in the juvenile stage only. It has west-wind drift distribution in the Southern Hemisphere, New Zealand, Patagonia, sub-Antarctic islands, Australia, and was formerly cited as supposed evidence of continental drift (see Chapter 29). *Holopneustes* are globular thin-shelled forms with indistinct pores at the angles of plates. They are found in Australia and New Zealand.

The other family of temnopleuroids is the Toxopneustidae, characterized by having deep indentations in the margin of the shell around the peristome, the

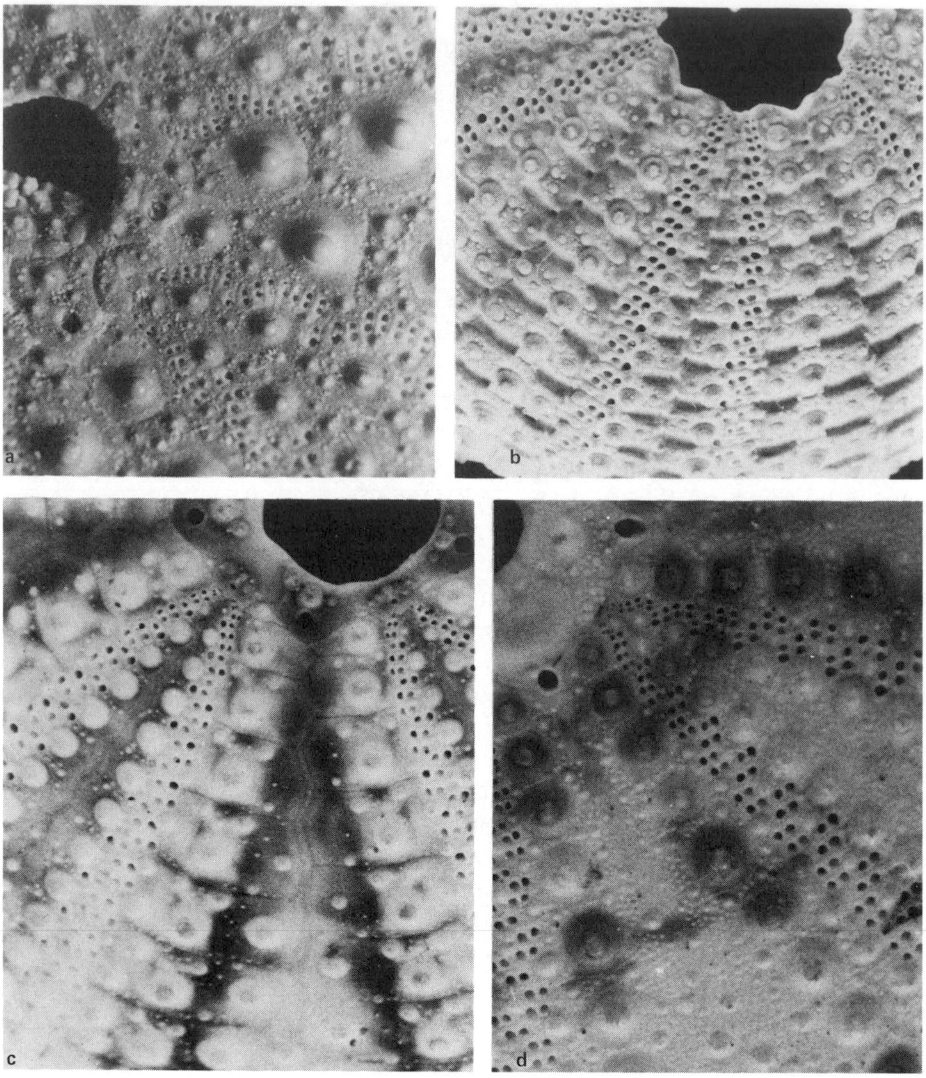

JOSEPH D. GERMANO

Fig. 20.6 (a) *Heliocidaris,* East Australia and New Zealand. (b) *Temnopleurus,* Indo-West Pacific, tropical. (c) *Lytechinus,* Caribbean. (d) *Pseudoboletia.*

so-called gill-cuts, and by lacking the sculpture on the test. Toxopneustids have toxic pedicellariae capable of stinging the human hand. Examples are *Tripneustes* (Fig. 20.7), *Lytechinus,* and *Pseudoboletia*—all found in tropical and subtropical reefs.

Members of the order **Echinoida** have tubercles lacking perforation, like the two preceding orders, but there are no gill cuts, and no sculpture occurs on the test. The genera are mainly distinguished by the number of pore pairs that occur on an amb plate and by the distribution of tubercles on the amb plates. Many of the genera are referred to in later chapters, so only a summary of the principal ones is given here. *Echinus* (Fig. 20.8) has amb plates with three pore pairs and a primary tubercle on every second amb plate. Its range is nearly worldwide, mainly deep-water except in Europe, where its species are the predominant shallow-water echi-

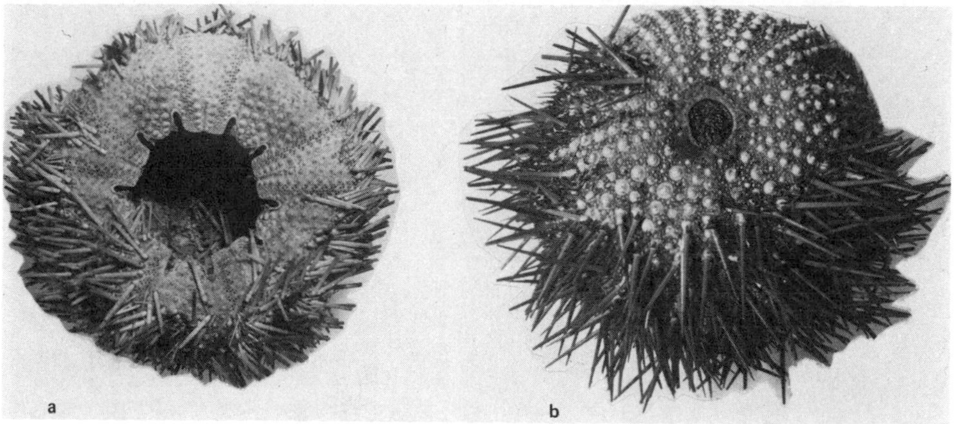

JOSEPH D. GERMANO

Fig. 20.7 (a) *Tripneustes,* showing gill cuts, pantropical. (b) *Paracentrotus,* North Atlantic.

noids. It is usually red. There seem to be no shallow-water representatives in the Americas. The genus was studied by Aristotle, and its present generic name is the same as that used by Aegean fishermen with whom Aristotle worked. Large species are edible. *Sterechinus* (Fig. 27.4) is the most common regular urchin in the Antarctic Ocean. It is distinguished by its red color and its larger spines that are developed in two sizes. *Selenechinus* occurs in the Philippines and has all the larger spines of uniform length. *Psammechinus* of the North Atlantic has a primary tubercle on every amb plate; in other respects it is similar to echinus. *Dermechinus* is a rather rare deep-water and mainly (though not exclusively) southern genus in which its height very much exceeds its breadth, some specimens resembling cucumbers placed on end. *Echinostrephus* has three or four pore pairs per amb plate, and extremely long and slender spines of the upper side; it inhabits burrows in coral reefs of the Indo-Pacific. *Echinometra* (Fig. 20.4) is one of the most common sea urchins of tropical reefs; the spines are very strong and sharp and can penetrate the foot if trodden upon. The shell is usually elliptical in the horizontal plane and the amb plates have up to 10 pore pairs per amb plate. *Pachycentrotus* is a South American genus with spines barely one quarter of the diameter of the shell and amb plates having four pore pairs. *Polyechinus* is a South African genus with four pore pairs per amb plate, and spines a little shorter than the horizontal diameter of the test. *Paracentrotus* is similar, but has five pore pairs per amb plate. It is found in the North Atlantic. *Caenocentrotrus* (Fig. 27.4) of Galapagos, Chile, and Peru is similar to *Paracentrotus*, but it almost always has a commensal crab in the rectum, causing a distortion in the growth of the test; the crab is named *Fabia chilensis. Anthocidaris* of Japan and China has seven or eight pore pairs per amb plate, the spines always longer on one side of the test, and it is the object of a major fishery. *Heliocidaris* (Fig. 27.4) of Australia and New Zealand has seven to ten pore pairs per amb plate. It is a brownish littoral reef dweller feeding on kelp. The spines are as long as the horizontal radius of the test. *Loxechinus* (Fig. 27.4) is similar, but its spines are shorter and its color is green. It ranges the Chilean coasts, where it is fished commercially. *Colobocentrotus* has an elliptical shell with spines converted to buttonlike structures. Its whole body is contoured to deflect waves (for it inhabits the outer part of Indo-Pacific reefs), and it has amb plates with eight to

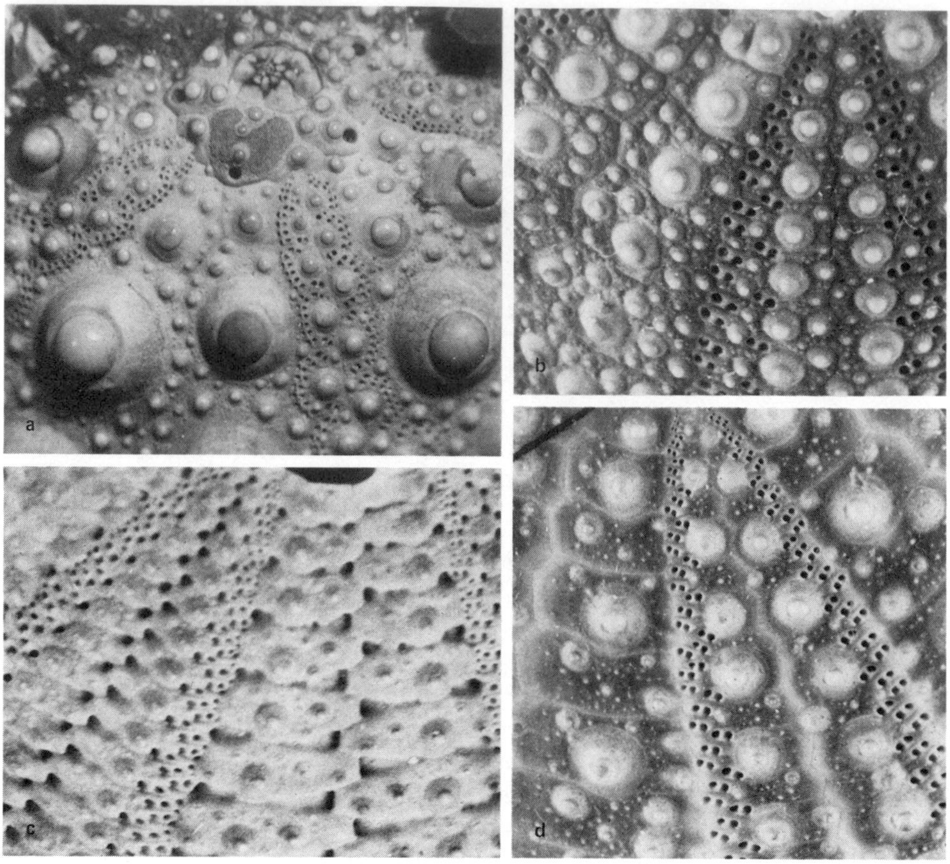

JOSEPH D. GERMANO

Fig. 20.8 (a) *Heterocentrotus,* Indo-West Pacific reefs. (b) *Psammechinus,* North Atlantic. (c) *Salmacis,* Indo-West Pacific. (d) *Echinus,* Europe in shallow water, and cosmopolitan in deep water.

twelve pairs of pores (Fig. 23.10). *Podophora* is similar, but the flattened spines are converted into a polygonal mosaic. It inhabits the Indo-West Pacific, especially Hawaii. *Evechinus* has nine pore pairs per amb plate, arranged in three vertical sets of three. It is green, and it is fished for food in Polynesia. It is found in New Zealand. *Heterocentrotus* (Fig. 20.6), the so-called slate-pencil sea urchin, has nine to sixteen pairs of pores per amb plate and massive spines. Its body is elliptical. It is found in the Indo-West Pacific. *Zenocentrotus* is like *Heterocentrotus* but its spines are smaller and those at the margin of the shell form a wave-deflecting skirt around the base. It is found in the region around the Polynesian islands.

INFAUNAL SEA URCHINS

In the Irregularia the anus (and periproct therefore) has moved into one of the interambs called the posterior interamb.

The order **Cassiduloida** comprises more or less hemispherical urchins in which the tube feet around the mouth have been enlarged to form a flower-shaped grouping (to obtain detrital food particles) called the floscelle. *Echinolampas* has a flattened lower surface, where the periproct is located. It is found in tropical and subtropical regions but is lacking from western North America. *Cassidulus*, now restricted to tropical coasts of North America (formerly pantropical), has an ovoid shell with the periproct above. *Apatopygus* (Fig. 20.2) is ovoid, concave below, vaulted above, with the periproct located in a deep groove on the upper side. It inhabits Australia and New Zealand (Fig. 20.2).

In the order **Clypeasteroidea** the shell is flattened in the horizontal plane to yield forms such as the sand dollars. The tube feet form not only a flower-shaped group (floscelle) on the lower side, but also a flowerlike set of petals on the upper side. (These are used as respiratory organs that project into the water just above the soft seabed in which the animal is imbedded for part of each day.) Sand dollars emerge from their cover by night and apparently creep about seeking detrital food material with the aid of the tube feet, they then slide obliquely under the sand again as daylight (or other unfavorable conditions) return. *Arachnoides* (Fig. 20.2) is a widely distributed Indo-Pacific sand dollar with a circular outline, the periproct at the edge above, and with one large plate occupying each interamb below. Closely related is *Fellaster* (Fig. 25.2) with several interamb plates below, occurring in New Zealand. On the west American coast a number of genera occur in which the body shape is similar to these, and one genus, *Echinarachnius* (Fig. 20.2), occurs on the east coast. It has the petals of the upper side somewhat divergent at the tips. A related form, *Dendraster* (Fig. 25.2), also having open petals, a periproct below the margin, and a system of branching grooves on the lower surface, occurs on the Californian coast. Several genera have the test strengthened by the development of vertical partitions within the shell, enclosing open holes called **lunules.** These are west coast forms, namely, *Mellita* (Fig. 25.2) with lunules in the paired ambs and in the posterior interamb; *Encope* (Fig. 25.2) in which the lunules reach the edge of the test and so form notches, except the one closed lunule in the posterior interamb; *Leodia*, resembling *Encope*, but with all the lunules closed at the edge, and the apex of the upper side is displaced posteriorly (Fig. 25.2) occurring on the east coast of North and South America; and *Melitella* with open ambulacral lunules. *Melitella* was formerly amphiamerican but is now confined to the west coast. These distribution patterns reflect the former continuity of the Caribbean and Panama seas, before the erection of the Isthmus. On the west coast of Africa occurs the genus *Rotula* (Fig. 20.2) and some related forms, all with numerous open lunules along the posterior margin of the test (Fig. 20.2).

Not all the members of the order Clypeasteroida are so flattened as the sand dollars. In *Laganum* of the Indo-Pacific, the test is flattened below but rises like a conical mound above, the periproct is below the margin, and there are internal calcareous supports for the shell. *Echinocyamus* of Europe and the Indo-Pacific has internal supports also, but the whole shell is reduced to a tiny pealike rounded contour. *Fibularia* (Fig. 20.2) of the Indo-Pacific has a similar test but lacks internal supports. In marked contrast to all the previous genera is *Clypeaster* (Fig. 20.2) with a massive test, which is usually flat or concave below. Its upper surface raised into a volcano-shaped or moundlike contour. It ranges all tropical and subtropical seas. The symmetry of its petals often very striking (so much so that in some islands the shells are traded with tourists as souvenirs). The clypeasteroids retain the biting dental apparatus of the more ancient Regularia, from which they arose in Cretaceous times. The remaining group we have now to consider, the spatangoids, have lost the teeth, and merely swallow large amounts of the substrate, or select particulate matter with the aid of the enlarged oral tube feet.

The order **Spatangoida,** in addition to the character just noted, are further characterized by a pronounced bilateral symmetry. Their mouth is on the lower side and is generally displaced anteriorly; the anus lies on a posterior periproct. These animals mainly inhabit burrows in the soft seabed and extend very elongated tube feet out on to the surrounding seafloor to search for food particles. The petaloid ambs often differ in length, and there are often tracts of ciliated spinules called **fascioles** winding like ribbons across the surface of the test, apparently producing various patterns of respiratory or feeding water currents (more research needed here). *Schizaster* has a high test with a deep anterior notch housing the middle petal; the posterior petals are the shortest (Fig. 23.11). It ranges tropical seas. In *Abatus* (Fig. 20.2) of the Antarctic continental shelf the lateral petals are sunken to serve as brooding chambers for the young (**marsupia**). *Agassizea* of the tropics lacks the anterior groove or notch of the test. *Moira* of western American coasts has deeply sunken lateral petals; otherwise it resembles *Schizaster*. *Brissus* (Fig. 23.11), with a large ovoid body like a melon, has shallow petals and no anterior notch, and the apex (i.e., point of origin of the petals) is anterior; it occurs on tropical reef lagoon muds. *Meoma* is a similar large form; it is amphiamerican, with a frontal notch. It is found in lagoons of Panama and the Caribbean and on soft seafloors—again evidence of the former continuity of the seas. *Metalia* is the Indo-Pacific equivalent of *Meoma*, having contiguous posterior petals. *Plethotaenia* has a posterior apex otherwise resembling *Meoma*, and ranges the tropical Atlantic.

Some of the spatangoids are distinguished mainly by the arrangement of fascioles. *Lovenia* (Fig. 23.11) with a striking pattern of confluent petals has a small fasciole forming a loop around the apex (**internal fasciole**). It is found in tropical soft seabeds. *Breynia* (Fig. 20.2) is a large and striking form from Australia and the west tropical Pacific, similar to *Lovenia*, but having an additional fasciole surrounding the tips of the petals (**peripetalous fasciole**) and a robust shell (Fig. 20.2). *Echinocardoum* has an internal fasciole like *Lovenia*, but otherwise is quite different in appearance, having a small delicate shell. It ranges temperate and tropical seafloors. *Brissopsis* has an ovoid shell and a conspicuous peripetalous fasciole (often tinted red) and occurs commonly as a member of deeper-water faunas in warm seas; its confluent petals give it a rather distinctive aspect, aiding recognition at sight when trawl contents are emptied on deck. *Plagiobrissus* has long narrow petals and conspicuous tubercles, occurring in tropical soft bottom faunas. *Spatangus* also has tubercles and a heart-shaped test. It is usually deep purple or purplish-red in color and has no fascioles on the upper surface. It is distributed in all seas.

21

MOLLUSKS

Characters of mollusks / Subclasses / Evolution of sea snails / Grazing herbivores / Predatory mollusks / Air breathers / Opisthobranchs / How to identify a gastropod / How to identify a bivalve / Orders of bivalves / Mollusks of tropical reefs / Trophic relationships / Antiquity / Primary heterotrophs / Secondary heterotrophs / Mollusks of the lagoon fauna / Scaphopods

Mollusks are the most abundant and most varied animals in the ocean. They occupy nearly every ecological niche available to a heterotroph, from the grazing herbivores, such as the limpets and abalones, to the largest carnivores, such as the giant squid, which is 25 m long. There are perhaps 100,000 species, arranged in a complex grouping of subclasses, orders, and families. Obviously it is not feasible to present a concise description in a single chapter that is sufficiently detailed to relate to the various ecosystems they occupy. Instead, therefore, this chapter will summarize the main evolutionary trends of those groups of mollusks that are important in the economy of the sea. For the more precise indication of the characters by which you can recognize the important families and genera refer to the chapters in the last section of this book; thus, for example, if you turn to pages 221–232 on tropical reefs, you will find there brief definitions of the most significant families of tropical mollusks. As in the case of the echinoderms and other phyla, a careful study of the numerous illustrations scattered through the book will enable you to obtain a working knowledge of the systematic arrangement of mollusks and will provide a clear means of identifying a substantial proportion of the mollusks you are most likely to encounter in the field. If possible, you should supplement this study by actual examination of specimens in the laboratory, or alternatively make a collection yourself. Specimens may be obtained by collecting your local shores, by exchanging duplicate specimens with institutions elsewhere, or by purchasing whatever is available from biological supply houses.

The Characters of Mollusks

Sea snails and clams are soft-bodied invertebrates that usually secrete an external calcareous **shell.** The shell usually shows distinctive features that are easier to observe than those of the soft parts, so nearly all classification is carried out on the basis of shell characters. The body of a mollusk comprises a central trunk, usually with a head (provided with tentacles and mouth), a **visceral hump** on the dorsal or upper side of the trunk, containing the coils of the intestine, and a **foot,** which is a creeping muscular organ on the ventral side. The coelom is reduced to several small chambers within the kidneys and gonads, the blood vascular system is well developed, and the digestive glands are associated with the gut. The position where the anus opens varies because many mollusks have undergone a secondary realignment of the visceral hump so that the anus is rotated toward the anterior end. There is a well-developed nervous system. The outer surface of the body is called the **mantle,** which contains the glands that secrete the shell.

The Subclasses

The phylum is subdivided into six classes as shown in Table 21.1. Of these six classes, the first two (the sea snails and the clams) are by far the most abundant; and in this book they are the primary groups that will concern us.

Evolution of the Sea Snails

Although the shell characters usually give a convenient and reliable indication of the affinities of any sea snail, certain details of the soft-part anatomy are of interest as throwing light on major evolutionary trends.

When a young sea snail hatches from its egg, a pair of gill tufts (**ctenidia**) lie at the posterior end of the visceral hump on either side of the anus. But after a few days the young animal begins to swing the visceral hump by its muscular effort, so that the posterior edge becomes anterior. The swings are at first temporary, reversible actions; but finally the hump remains in its new position with the anus and the ctenidia being anterior. Some sea snails are born with the reversed orientation already developed. Anatomists surmise that all sea snails that have the

Table 21.1 Classes of Mollusks

COMMON NAME	CLASS NAME	CHARACTERS
Sea snails	Gastropoda	Single shell, head, foot, and visceral hump well defined, usually epifaunal.
Clams	Bivalvia	Shell of 2 valves, head ill-defined, usually infaunal or attached to substrate.
Chitons	Amphineura	Shell of several separate valves, sluglike herbivores of shallow water.
Octopus and squid	Cephalopoda	Foot subdivided into suctorial arms, head well developed, active carnivores.
Tusk shells	Scaphopoda	Infaunal elongate detritus feeders.
Monoplacophora	Monoplacophora	Ancient limpetlike mollusks with a segmented internal structure and a single shell.

ctenidia and anus at the anterior end of the body must have evolved from ancestors originally having structures in the more usual (i.e., the posterior) position. All sea snails with the reversed structure are termed **prosobranchs**, and they are classified in a subclass of Gastropoda called the Prosobranchia. Some prosobranchs have only one ctenidium, others have none—but if the anus is anterior, their true affinity can be recognized. Most of the important marine sea snails are prosobranchs. Two other, relatively minor, subclasses may be mentioned: the Opisthobranchia, in which the visceral hump is rotated back into its original position and the shell is reduced or absent; and the Pulmonata, or land snails, in which the ctenidia are lost and instead the cavity within the mantle is modified to serve as an air-breathing lung. A few land snails have returned to the sea, to inhabit wetlands where they have free access to the air.

THE GRAZING HERBIVORES

Archaeogastropoda. From the fossil record we can recognize the oldest kinds of prosobranchs as being snails like the present-day limpets and abalones, so the name Archaeogastropoda is applied to these. They all share the common feature of having a long rasping tongue (**radula**) on which large numbers of minute teeth are arranged in transverse rows and a gill or ctenidium that is shaped like a fern frond, with filaments on either side of the gill axis (**aspidobranch ctenidium**).

The principal families of Archaeogastropoda in modern seas are the abalones (Haliotidae, Fig. 21.4), the keyhole limpets (Fissurellidae), the porcelain limpets (Acmaeidae), the pearly limpets (Patellidae, Fig. 21.1), the top shells (Trochidae, Fig. 21.4), the Turban shells (Turbinidae), and the nerites (Neritidae, Fig. 21.4). All of these are described in Chapter 26; and, in addition, references to the role of archaeogastropods in marine ecology will be found in the chapters on other regions. With the exception of the porcelain limpets, most archaeogastropods have a pearly or nacreous lining to the shell, the abalones being an outstanding example. It is believed, therefore, that the earliest sea snails probably had that feature also. Members of the group can be traced back in the fossil record for about half a billion years to Ordovician times.

A second assemblage of prosobranchs is the Mesogastropoda. All members of this group have a radula with numerous teeth in transverse rows. The structure of their ctenidium is altered so that the gill axis is attached lengthwise to the mantle, and therefore gill filaments form only along the free edge (**pectinibranch ctenidium**). For the most part these, too, are herbivores feeding on algae. Some, however, are carnivores.

The principal families included here are: the periwinkles (Littorinidae), the turritellas (Turritellidae), the ceriths (Cerithiidae, Fig. 21.9), wentletraps (Epitoniidae, Fig. 21.1), the conchs (Strombidae, Fig. 21.5), carrier shells (Xenophoridae), moon snails (Naticidae), the cowries (Cypraeidae, Fig. 21.6), the helmet shells (Cassididae, Fig. 21.5), the frogshells (Bursidae), the tritons (Cymatiidae, Fig. 21.7), and the tun shells (Tonnidae). Some of the carnivores are the moon snails, the tritons (which capture and eat starfishes), the helmet shells (which feed largely on sea urchins), and the turritellas (which hide under sand and snatch at passing organisms). The large conchs, strange to say, are relatively passive alga browsers.

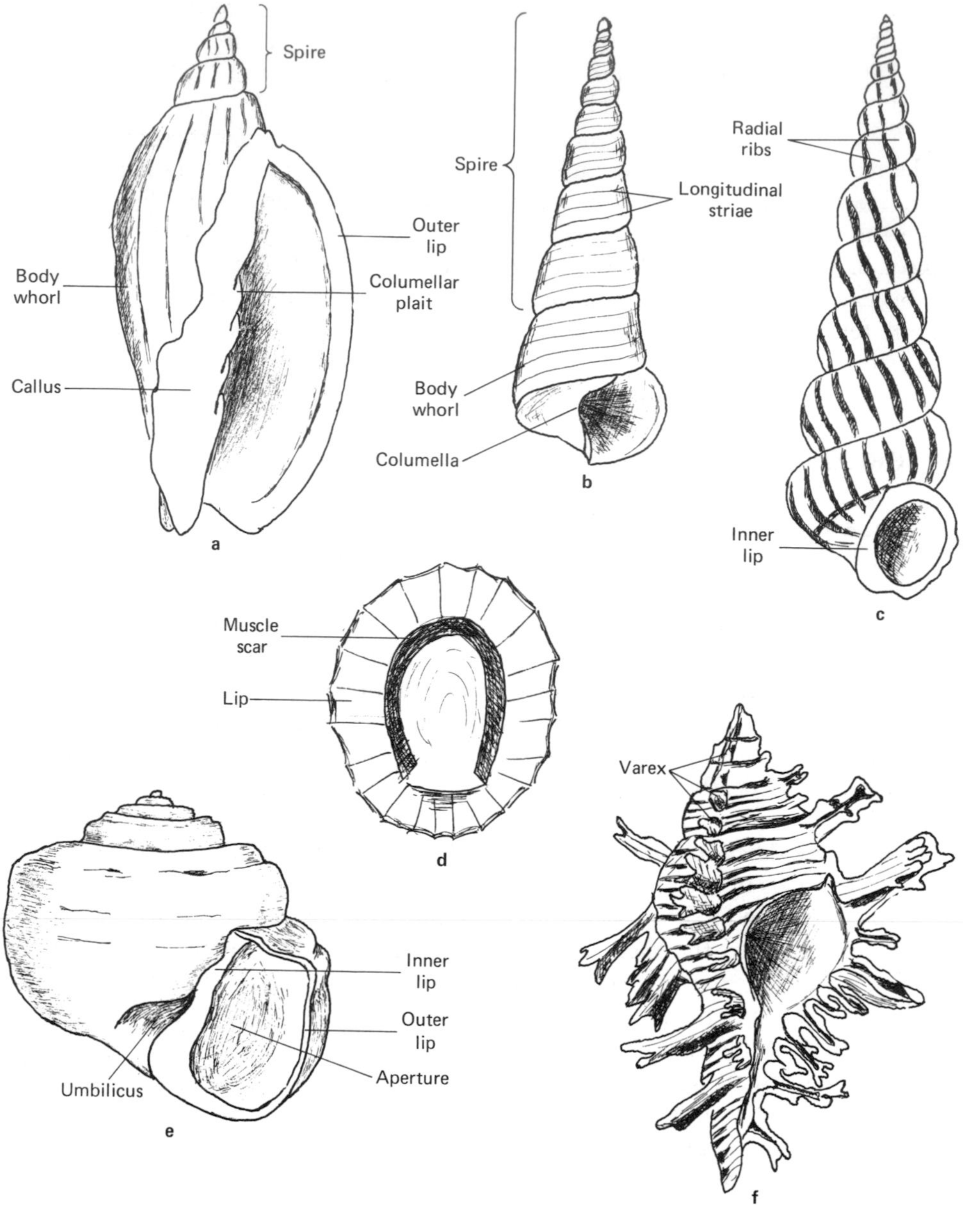

Fig. 21.1 Diagnostic characters of gastropods. (a) *Alcithoe.* (b) *Maoricolpus.* (c) *Epitonium.* (d) *Patella.* (e) *Amphibola.* (f) *Murex.*

THE PREDATORY MOLLUSKS

The third assemblage of prosobranchs are the Neogastropoda, in which the gill is of the pectinibranch type, but the radula has only three or fewer teeth in the transverse rows of the organ. These are essentially predators with well-developed sense organs for the detection of prey, a large muscular foot for leaping after the prey, and a long suctorial proboscis for sucking out the body cavity contents of the captured animals. There is usually also a well-developed **siphon,** a flexible organ extruded from the shell for maintaining the circulation of seawater through the gill chamber.

The principal families here included are: the purple whelks (Muricidae), the rock whelks (Thaididae), the coral snails (Magilidae), the northern whelks (Buccinidae), the crown conchs (Melongenidae, Fig. 21.6), the tulip shells (Fasciolariidae, Fig. 21.5), the olives (Olividae, Fig. 21.7), the miters (Mitridae, Fig. 21.7), the volutes (Volutidae, Fig. 21.7), the cone shells (Conidae, Fig. 21.8), and the augers (Terebridae).

THE AIR BREATHERS

The pulmonates, mentioned above as land snails, include a few members that have secondarily returned to the marine environment. An example is *Amphibola*, discussed further in Chapter 27.

THE OPISTHOBRANCHS

These forms, noted above as having undergone a secondary detorsion of the visceral hump so that the anus is now restored to its posterior position, include several families of importance. The names of some of the best known genera are: *Bulla* and related genera (the bubble snails), floating forms with an inflated glassy shell; *Aplysia*, a sea slug with a very reduced shell; *Limacina* and *Clione*, examples of two families that have become pelagic; the pteropods (mentioned in Chapter 13 as elements of the plankton); and *Doris*, representative of the shell-less *nudibranchs*, the largest sea slugs. Some of these are also cited in later chapters.

HOW TO IDENTIFY A GASTROPOD

In the following chapters numerous important families and genera of sea snails are mentioned as forming particular trophic components of the major marine ecosystems. Most of the genera are illustrated by photographs, so you should not find it particularly difficult to learn their names.

However, you should if possible examine their shells in the laboratory in order to see how the shell characters are used in defining the groups. Here are the more important terms employed in books on malacology (see Fig. 21.1). The part of the shell occupied by the animal is called the **body whorl.** The pointed end is the **spire.** If you hold the shell so that the spire points away from you, and the **aperture** of the body whorl is visible, then the whorl lies to your right in all normal (i.e., **dextral**) shells. The aperture is bounded by the **outer** and **inner lips.** If additional shelly material (aragonite) is deposited around the inner edge of the aperture it is called **callus.** If the callus is extensive it may obscure the **umbilicus** or the opening of the hollow interior of the **columella,** which is the name applied to the axis of the shell. In some families the columella is **plaited** like a corkscrew. The surface of the shell may have longitudinal spiral **striae** (like engraved lines), or it may have **radial ribs.** Some shells, such as those of limpets, are wide open, not spirally twisted, and these may show an internal **muscle scar**—a mark where the mantle is attached to the shell. Some shells form elaborate lips, as in *Murex*, and leave the outer lip intact when more material is added to enlarge the shell. Gastropods that have this habit produce shells on which ridges occur, parallel to the lip (representing, indeed, former lips). These ridges are called **varices** (singular **varex**). They are responsible for the singular appearance of such spiny shells as *Murex* (Fig. 21.7) and for the symmetrically ridged shell of *Harpa* (Fig. 21.5).

In learning the families you will discover that various combinations of these characters correspond to many of the best-known systematic groups. Therefore always examine any shell you may meet for the first time, and see how many of the distinctive characters you can list. This will help you to find your way through more technical books.

HOW TO IDENTIFY A BIVALVE

Figures 21.2 and 21.3 illustrate the more important shell characters. In the descriptions of the orders which follow this section, reference is made to soft-part characters; but, in general, you can rather easily discover from the shape of the shell which order a bivalve belongs to.

If both valves are similar the shell is said to be **equivalve,** if dissimilar **inequivalve.** Scallops and oysters are examples of inequivalve bivalves. The outer surface is usually covered by a brownish **periostracum** or horny layer. The outer surface may be **striated** or **ribbed.** The two valves are held together by an elastic ligament, which leaves a scar on the **hinge.** The hinge may in addition have interlocking ridges, called the **dentition.** The individual ridges (or **teeth**) may all be similar, in which case the hinge dentition is said to be **taxodont,** or there may be enlarged **cardinal teeth** and other **lateral teeth.** The two valves are attached to the soft body by **adductor muscles** that produce **scars** on the interior surface. If each valve has a single such scar the shell is said to be **monomyarian.** If there are two scars on each valve, the shell is **dimyarian.** At the hinge the shell has a projection called the **umbo;** this always points toward the anterior end of the animal (i.e., the end where the mouth is). Thus we can distinguish an **anterior adductor scar** and a **posterior adductor scar** in dimyarian shells. A slender scar often joins these two; it marks the attachment of the edge of the mantle and is called the **pallial line** or **pallial scar.** Some bivalves have the mantle folded into a posterior siphon (for conveying water away from the body when the animal is feeding by convected ciliary currents); such shells show a **pallial sinus** in the pallial line.

These anatomical terms do not give all the information that malacologists commonly include in their descriptions, but with the aid of this basic list you should now be able to find your way through the essential parts of the books you may need to consult.

An authoritative classification of Bivalvia is to be found in the multiauthored *Treatise of Invertebrate Paleontology*, R. C. Moore, Ed., Geological Soc. of America (1953–1974). However, the divisions are so complex and so many fossil forms are involved, that a much simpler arrangement is needed by the ecologist, dealing with only the living forms. The following arrangement, suggested by Kenneth Boss, may prove more useful.

Order Palaeoconcha, with the family Solemyacidae: Infaunal in soft substrates; periostracum distinct, glossy, brown; dentition reduced; foot disk-shaped, digitate. *Solemya borealis* of the New England coasts is an example.

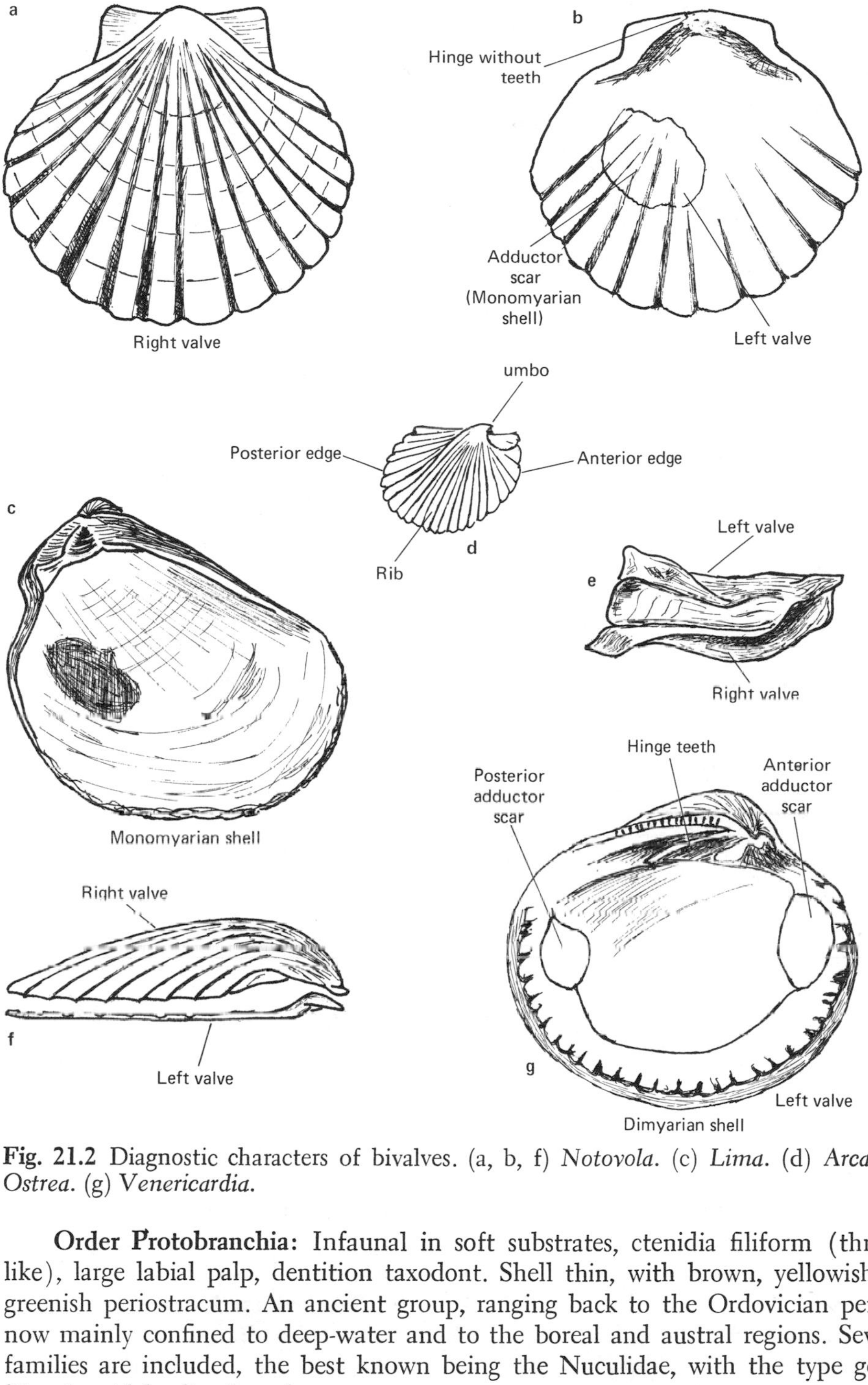

Fig. 21.2 Diagnostic characters of bivalves. (a, b, f) *Notovola.* (c) *Lima.* (d) *Arca.* (e) *Ostrea.* (g) *Venericardia.*

Order Protobranchia: Infaunal in soft substrates, ctenidia filiform (thread-like), large labial palp, dentition taxodont. Shell thin, with brown, yellowish, or greenish periostracum. An ancient group, ranging back to the Ordovician period, now mainly confined to deep-water and to the boreal and austral regions. Several families are included, the best known being the Nuculidae, with the type genus *Nucula* widely distributed.

Order Filibranchia: An assemblage of bivalve families with filiform gills, lacking complex dentition, often with a byssus for attachment to the substrate, and often with a single adductor muscle scar (monomyarian). The included families of interest are: (1) Oysters (family Ostreidae, Fig. 21.2), irregularly rounded shell, inequivalve, monomyarian; example, *Crassostrea virginica.* (2) Scallops (family

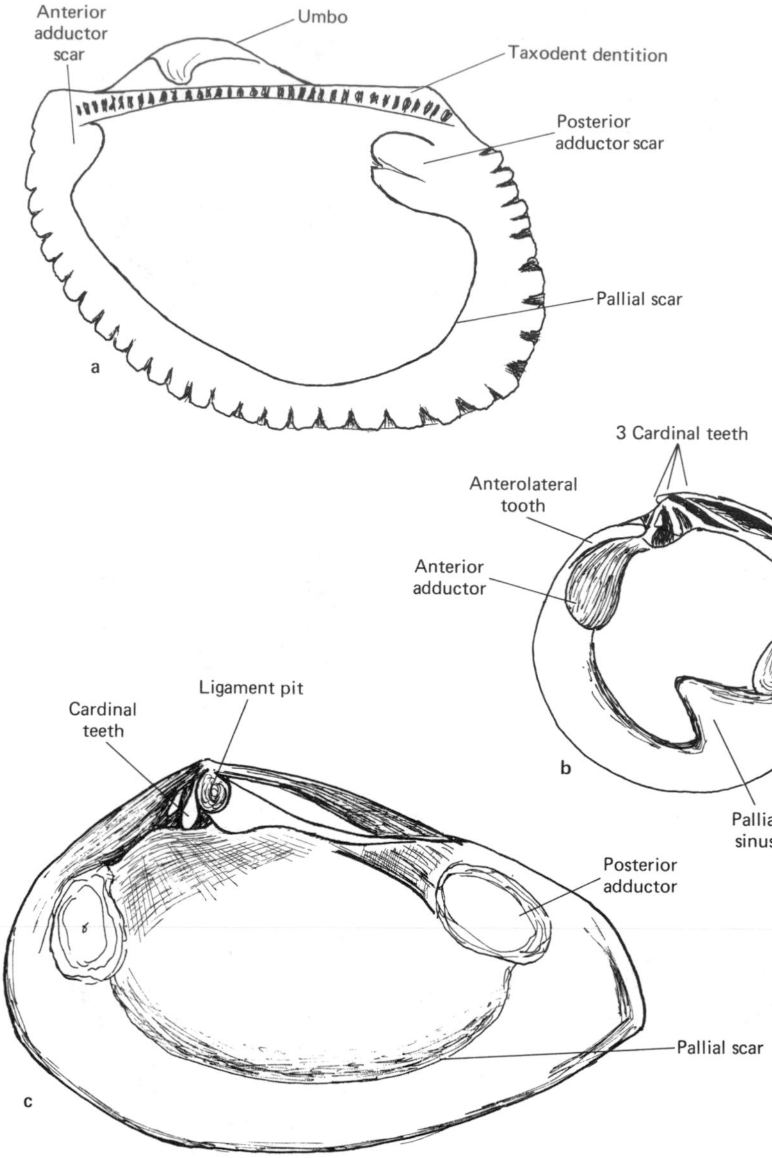

Fig. 21.3 Diagnostic characters of bivalves. (a) *Arca.* (b) *Dosinia.* (c) *Crassatellites.*

Pectinidae), fan-shaped monomyarian forms, with a byssus, found mainly in shallow water, either free-living or attached; examples, *Aequipecten, Chlamys.* (3) Mussels (family Mytilidae), colonial, intertidal and subtidal on various substrates, attached by long byssus threads, shell bluish to black, almond-shaped, monomyarian; example, *Mytilus*, cosmopolitan. (4) Horse-mussels (family Pinnidae), resembling mussels, but larger (to ca. 18 inches long), flattened, tapering to end which is embedded in soft seafloor, with byssus, crinkly shell, found in tropics and warm temperate south to New Zealand; examples, *Pinna, Atrina.* (5) Hammer oysters (family Isognomonidae), oysterlike forms, with nacreous interior, a hammer-shaped shell, tropical; example, *Malleus.* (6) Pearl oysters and wing oysters (family Pteriidae), nacreous interior, shell rounded (as in *Pinctada*) or having a winglike projection on one side of hinge (*Pteria*), tropical. (7) Ark shells (family

Arcidae), elliptical shell, often with zigzag patterns, a taxodontlike dentition, tropical, attached or free-living, with byssus; example, *Arca.* (8) Glycimerids (family Glycimeridae), like ark shells but rounded, tropical and subtropical; example, *Glycimeris.*

Order Eulamellibranchia: Bivalves with lamellibranch ctenidia. Of the numerous families may be noted: (1) Lucines (family Lucinidae), clams with rounded shells and a long thin anterior muscle scar, hinge usually dentate, infaunal, cosmopolitan but mainly tropical, eaten by fishes, a few species also eaten by man; example, *Phacoides* in the Caribbean, *Lucina* in the southeast United States, and *Codakia* in the Indo-Pacific. (2) Cockles (family Cardiidae) heart-shaped, dimyarian, shell with cardinal and lateral teeth, no pallial sinus, shallow-water infaunal; examples, *Trachycardium* in the southeast United States, and *Corculum* in the Indo-Pacific. (3) Venus clams (family Veneridae), with heavy shells, lateral and cardinal teeth, and pallial sinus, cosmopolitan; example, the quahog *Mercenaria.* (4) Sunset shells (family Tellinidae), shell elongate, rather fragile, usually with radiating sunsetlike patterns, dimyarian, strongly heterodont, deeply burrowing forms with a well-developed pallial sinus for the long siphons; examples, *Tellina,* widely distributed. (5) Surf clams (family Mactridae), large edible clams, more or less triangular or oval, hinge with a spoon-shaped depression, cosmopolitan; example, *Mactra.* (6) Jackknife clams (also called razor shells), elongate rectangular clams (family Solenidae), widely distributed; examples, *Solen, Ensis.* (7) So-called shipworm (family Teredinidae), burrowing in wood; example, *Teredo.* (8) Pholads (family Pholadidae), burrowing in soft rock or hard mud; example, *Pholas.*

THE MOLLUSKS OF TROPICAL REEFS

The chief molluscan elements of tropical reefs are the sea snails (class Gastropoda) and the clams (class Bivalvia). Other molluscan groups represented are the chitons (class Amphineura), the tusk shells (class Scaphopoda), and the octopuses and squids (class Cephalopoda).

Distribution. The distribution of mollusks on the biotope is approximately as follows: (1) **Nektonic elements** swimming over the reefs—squid. (2) **Benthic epifauna** of the reefs—most sea snails, some clams, such as *Tridacna,* some octopuses and chitons. (3) **Benthic infauna** of the sands and lagoon muds—some sea snails, such as olive shells and turret shells, most clams, and also tuck shells.

Trophic Relationships. Mollusks of the reef ecosystem may be grouped as follows: (1) **Primary heterotrophs** consist mainly of littoral sea snails of the order Archaeogastropoda and also the chitons, which are littoral elements. These mollusks have a rasping tonguelike organ called the radula, located in the floor of the buccal cavity, which is used for scraping off vegetable matter from hard substrate. (2) **Secondary heterotrophs,** the carnivorous mollusks, comprise the majority of sea snails, which are predators upon the other animals of the reef, including other mollusks. (3) **Dominant predators,** the cephalopods, occupy the highest links in the secondary heterotrophic levels of the food web. (4) **Filter feeders** are mainly the bivalves. (5) **Detrital feeders** are mainly infaunal sea snails and tusk shells.

Antiquity. Most of the genera of tropical-reef mollusks antedate the disruption of Tethys in the Miocene and all far antedate the elevation of the Panamanian isthmus. Therefore, the existing reef molluscan faunas have a decidedly pantropical aspect at and above the generic level, reflecting the former circumglobal continuity of tropical seas. As in the case of some corals, some mollusks that were formerly present in the tropical Atlantic became extinct in that ocean during the cold Pleistocene phases. The Pacific faunas, on the other hand, having a vastly greater circulating water mass, suffered much less from the Pleistocene coolings, and some genera survived there even though they disappeared in the Atlantic. The erection of the Panama land bridge prevented repopulation of the Caribbean from the Pacific, and the East Pacific barrier prevented much transfer of Indo-West Pacific stocks to the west American tropical coasts; so the Caribbean and Panamanian molluscan faunas are less varied than the Indo-West Pacific faunas. Owing to the long-continued existence of Tethys, and its unbroken connection with the Indo-Pacific, molluscan faunas present a broad similarity from the Red Sea in the west to Hawaii in the east. The richest faunas are found in the vast island archipelago of Indonesia and the adjacent reefs of northern Australia.

The following summary groups the mollusks of coral reefs under ecological categories.

Primary Heterotrophs

Most molluskan herbivores on tropical reefs are sea snails (Gastropoda) belonging to the order *Archaeogastropoda*, an assemblage ranging back in time to the early Cambrian period. The surviving genera mostly range from the Cretaceous or early Tertiary. Members of the order are large shallow-water forms, feeding on algal growth on the interface of the sea and the substrate. The radula has numerous rows of rasping teeth, and each transverse row comprises many teeth of several kinds. The most conspicuous members of this group are the abalones, the keyhole limpets, the pearly limpets, the porcelain limpets, the top shells, the turban shells, and the nerites. These seven groups correspond to seven families in the formal classification. Genera and species can usually be recognized by shell characters.

Abalones (Family Haliotidae, Fig. 21.4). The best-known tropical members are the Ass-ear abalone (*Haliotis asinina*), a small elongate species of the southwest Pacific, and the black abalone (*Haliotis carcherodi*) of west Mexico; other species, mostly small, occur in the Indonesian region. (NOTE: The large and best-known species are not tropical, but belong to the warm temperate faunas of the North Pacific, New Zealand, and the Mediterranean.) Abalones have a nacreous shell, an ear-shaped shallow spiral, a large aperture, and a row of holes for extrusion of tentacular gill filaments. They are found on rocky shores and on high reefs. *Haliotis* is the only surviving genus of this family.

Limpets. There are three surviving families of the cap-shaped sea snails called limpets. Related to the abalones are the key hole limpets (family Fissurellidae), which are cap-shaped porcellanous forms with a hole or slit in the shell to house the anus. They are found on rocky or coral reefs. *Fissurella* is a tropical example. The true limpets (family Acmaeidae) are similar, but they lack the aperture, the anus having moved down to the posterior edge of the shell; the genus *Acmaea* is an example. The pearly limpets (family Patelllidae) have a shell like that of Acmaeidae, but it is nacreous within. This family does not inhabit the Americas, but it is common elsewhere; an example is *Patella*. All groups of limpets are found in temperate as well as tropical seas.

Top Shells (Family Trochidae, Fig. 21.4). This is a large group of widely distributed archaeogastropods that secrete a low, conical, top-shaped shell, which is

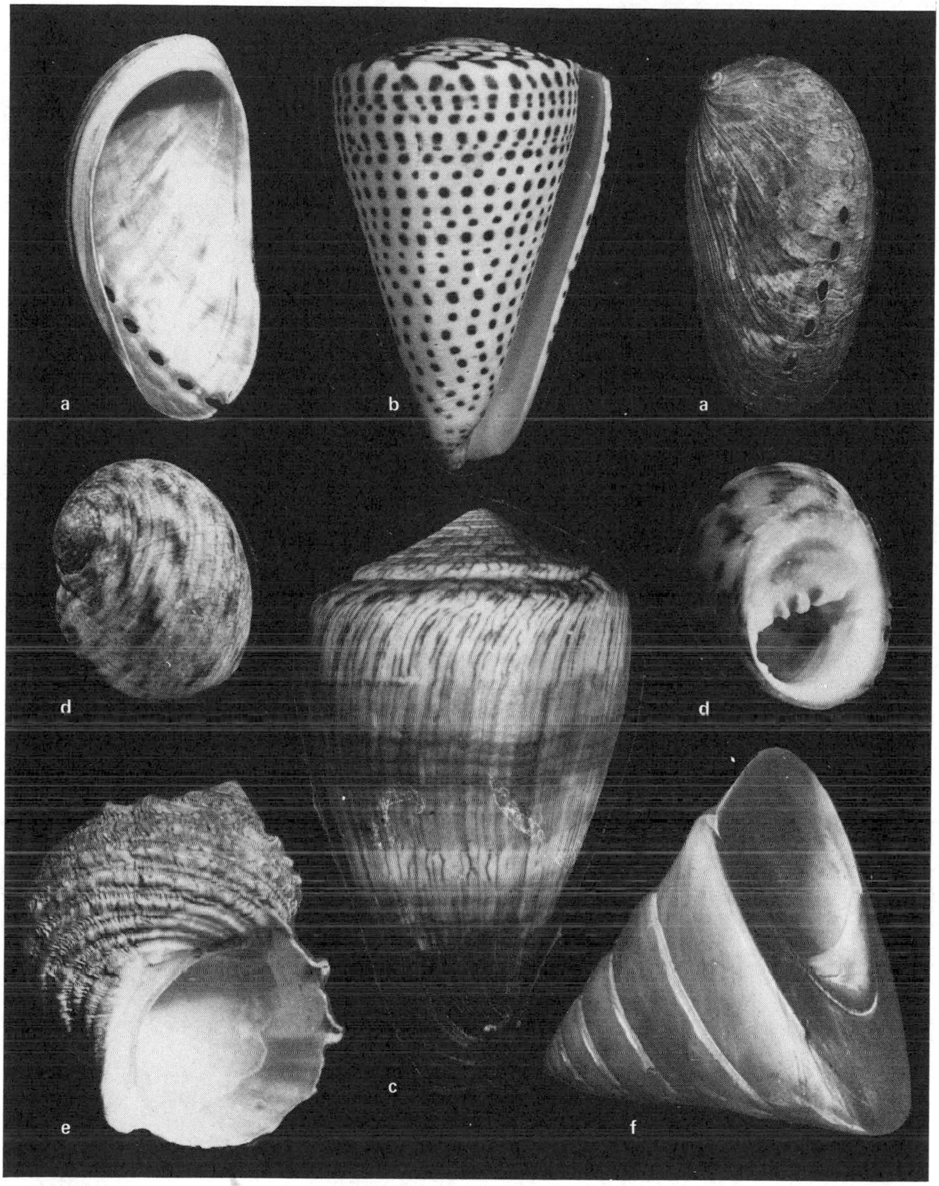

BERNARD ZIMMERMANN AND GEORGE PUTNAM

Fig. 21.4 (a) *Haliotis asinina*, Indo-West Pacific. (b, c) *Conus*, Indo-West Pacific. (d) *Nerita*, Indo-West Pacific. (e) *Turbo*, Indo-West Pacific. (f) *Trochus*, Indo-West Pacific. Note: Other species of these genera occur also in the tropical Atlantic.

nacreous within. The family is very conspicuous in the tropics, where large species occur. Some of them have been used in the pearl-button trade, where the raw shell is often referred to as "trochus." All top shells have a thin, horney operculum. The best-known tropical genera are *Trochus, Tectus,* and *Clanculus.* (Some cool-water and soft-bottom representatives of the family from the North and South Pacific coasts are mentioned in Chapter 25.)

Turban Shells (Family Turbinidae, Fig. 21.4). These resemble top shells but differ in having a thick calcareous operculum; they are commonly called cat's eyes.

Most turbans can also be distinguished from tops by having more swollen whorls, so that the spire does not present a smooth conical outline. Like most archaeogastropods, the interior of the shell is nacreous. The large tropical members are also employed in the pearl-button trade. Representative tropical genera are *Turbo, Dephinula,* and *Phasianella*. There are also cool-water representatives of the family.

Nerites (Family Neritidae; Fig. 21.4). These are snails that frequent tropical shores and have an operculum of thick calcareous material, notched to fit corresponding teeth developed on the inner margin of the aperture. An example is *Nerita*.

Secondary Heterotrophs

Nearly all other shelled tropical sea snails belong in the grouping *Caenogastropoda* (subdivided by many authors into two orders called Meso and Neogastropoda, although these details need not concern us). These have the radula modified, usually with fewer teeth, to suit it to carnivorous feeding. In addition, the margin of the mantle is commonly notched anteriorly (and the aperture similarly notched) to accommodate a suctorial siphon, which is inserted into the prey. The following are important families on hard tropical substrates.

Corkscrew Shells (Family Vermetidae). This family has a tubular and irregularly developed shell that is usually in a corkscrew spiral for the first part of its growth. These animals look like turret shells that have become untwisted (see turret shells under soft substrates, below). They are widely distributed in warm seas and often are cemented on to the hard substrate. An example is *Vermicularia* from the Floridian reefs.

Strombs (Family Strombidae). These are active, carnivorous, tropical snails with a thick, solid shell and a greatly enlarged body whorl. The aperture is long and narrow with a notch at either end; the outer lip usually thickened and expanded. In the Caribbean the strombs, such as genus *Strombus*, are called conchs (Fig. 21.5). (One such shell figured in *Lord of the Flies.*) In the Indo-Pacific the strombs include the highly characteristic spider shells and scorpion shells, genus *Lambis* (Fig. 21.6), where the outer lip is subdivided into finger-shaped straight or hooked projections.

Cowries (Family Cypraeidae). One of the most beautiful families of tropical sea snails. The shell is oval and inflated, the body whorl usually conceals the spire, and the aperture is slitlike, running the length of the shell and bordered by denticulation. The external surface of the shell is highly polished and pigmented in patterns characteristic for each species. When alive, a brightly colored mantle flap covers the shell on either side, with the flaps meeting along a dorsal midline where a line of pigment is commonly developed on the shell. Cowries have numerous species, which are much more varied and brightly colored in the Indo-West Pacific than elsewhere. All are placed in the genus *Cypraea* (Fig. 21.6).

Frog Shells (Family Bursidae). Trumpet-shaped shells, with two varices per whorl and a deep posterior notch for the siphon. Example *Bursa*.

Tritons (Family Cymatidae). Large, tropical predators with trumpet-shaped shells, often of considerable size. Shell covered by periostracum, but polished and patterned beneath. Inner lip usually wrinkled. Up to three varices per whorl, usually fewer, the varices at regular angular intervals. *Cymatium* is the Triton shell of the entire tropics.

GEORGE PUTNAM

Fig. 21.5 (a) *Fasciolaria,* Carolinas and Caribbean. (b) *Strombus,* pantropical. (c) *Harpa,* Indo-Pacific. (d) *Cassis,* pantropical. (e) *Vermicularia,* Florida.

Tun Shells (Family Tonnidae). Large, thin, subglobular shell with an enormously enlarged body whorl, mainly tropical. Example *Tonna.*

Murex Shells (Family Muricidae, Fig. 21.7). Active predators preferring trop-

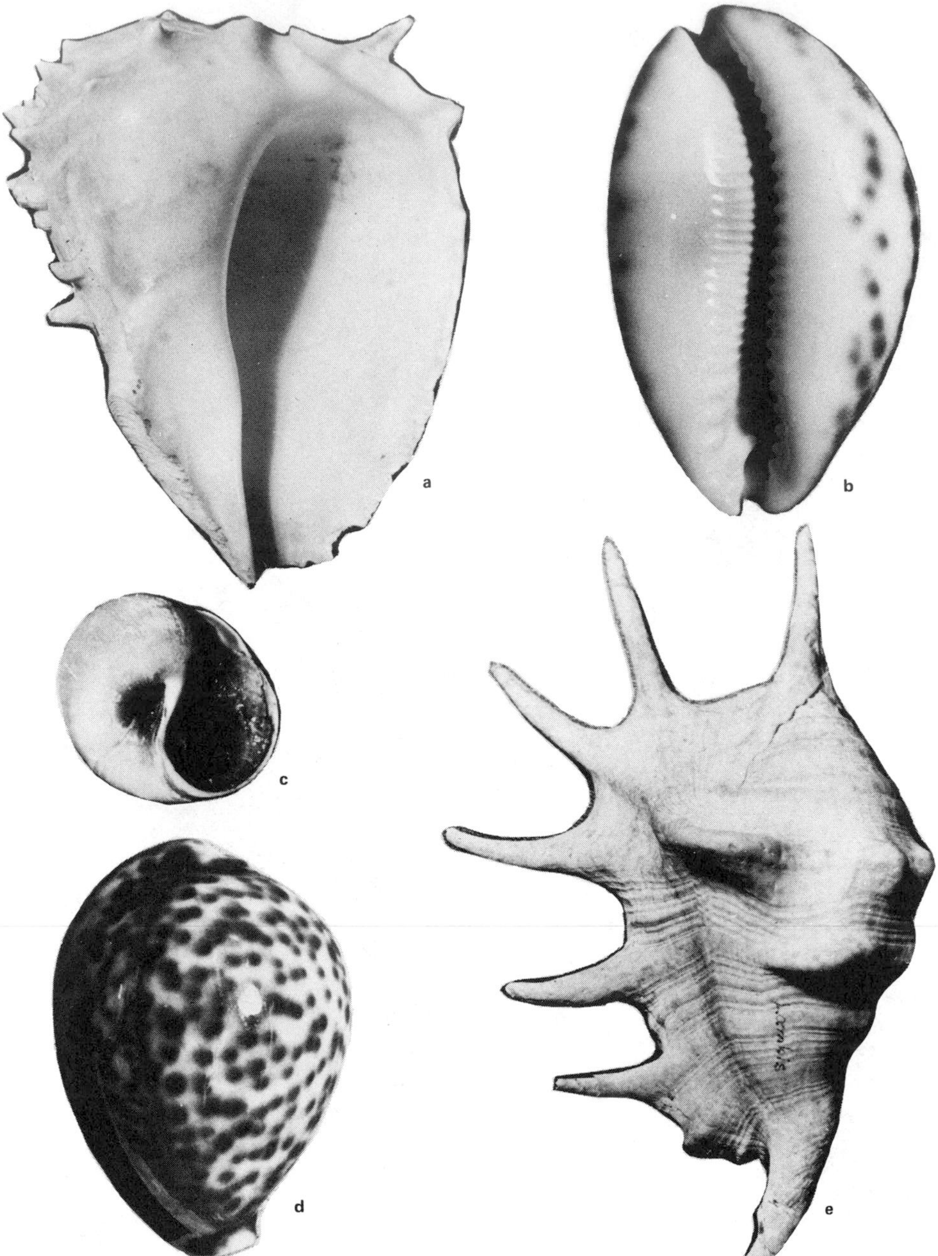

GEORGE PUTNAM

Fig. 21.6 (a) *Melongena,* Caribbean. (b, d) *Cypraea,* Indo-West Pacific reefs. (c) *Natica,* warm seas. (e) *Lambis,* Indo-Pacific.

ical hard bottom in shallow water; shell thick and usually very spiny, the spines being arranged in varices, with three or more varices per whorl. Species yield Tyrian purple (a colorless secretion of anal glands which becomes red or purple on exposure to the air). *Murex.*

Volutes (Family Volutidae, Fig. 21.7). Another family popular with naturalists on account of the beauty of the shell; mainly tropical, but with cold-water

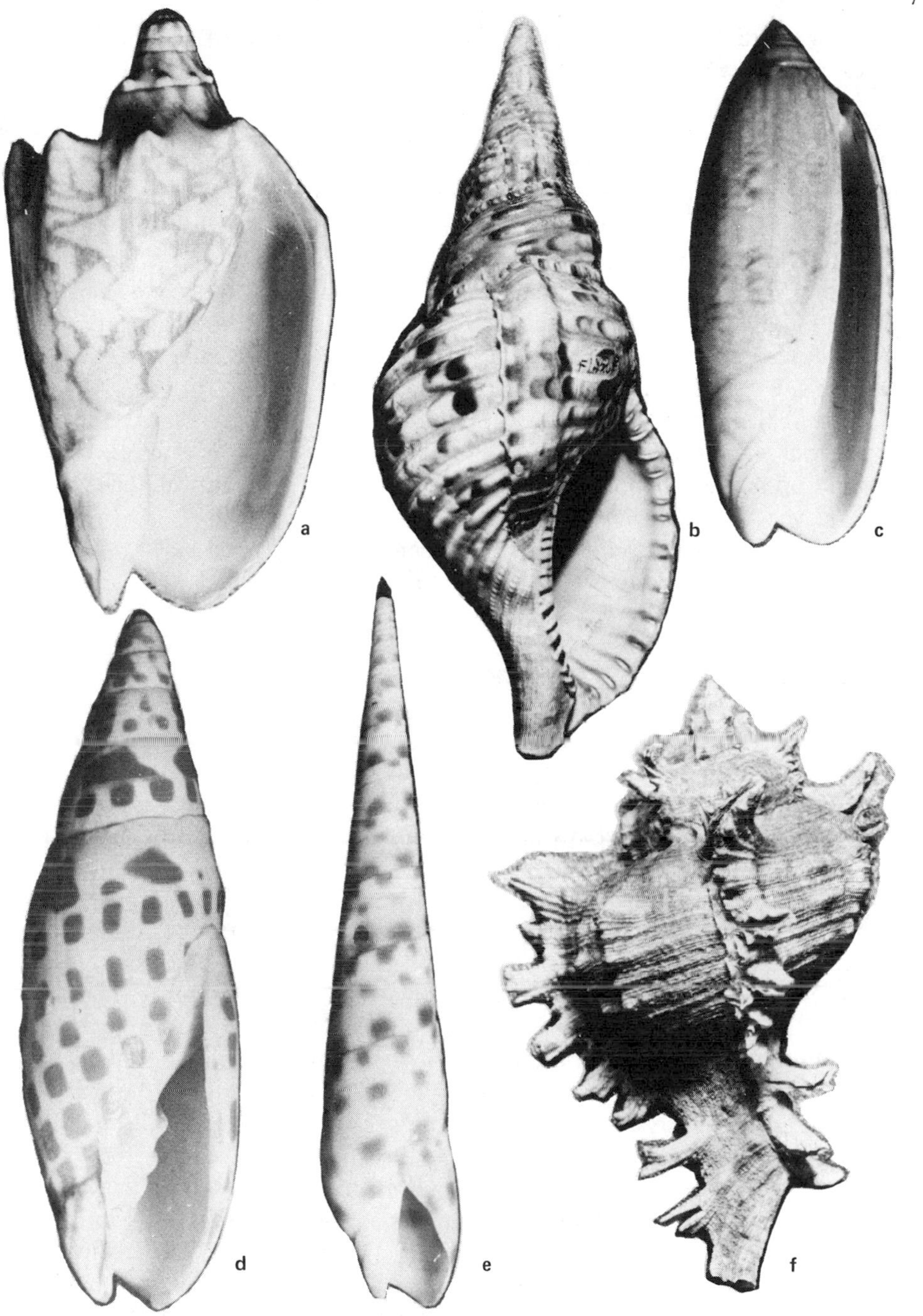

GEORGE PUTNAM

Fig. 21.7 (a) *Voluta,* Indo-West Pacific reefs. (b) *Charonia* pantropical. (c) *Oliva,* pantropical. (d) *Mitra,* Indo-West Pacific reefs. (e) *Terebra,* Indo-Pacific. (f) *Murex,* all warm seas.

species in deeper waters outside the tropics, and some warm temperate Indo-Pacific species. The volutes have the columella twisted like a corkscrew. Example *Voluta.*

Miters (Family Mitridae). Like small volutes, but the upper columellar fold is the strongest. Tropics only. Example *Mitra* (Fig. 21.7).

Auger Shells (Family Terebridae). Spire long, turret-shaped, columella plaited. Tropics only. *Terebra.*

Cone Shells (Family Conidae). Cone-shaped or fusiform (depending on whether the spire is flattened or elevated), with a very narrow, elongate aperture. Shell covered by a dark periostracum, but after removal the underlying surface is polished and brilliantly patterned in most species. A venom gland associated with the radula enables the animal to inflict a poisonous bite. Deadly species are those with orange or yellow or red contrasting pigments, or black and white or black and yellow color patterns. Only safe rule is never to pick up a cone shell in the ungloved hand, as the animal suddenly exserts the head and bites instantly. Predators of all reefs, but most conspicuous in the Indo-West Pacific. Only included living genus is *Conus,* with hundreds of species (Fig. 21.8).

Harp Shells (Family Harpidae). Spire low, body whorl enlarged and ribbed by numerous axial ridges or varices. Several Indo-West Pacific species, one Panamic, and no Atlantic representatives. "Among the most beautiful shells" (Myra Kean). *Harpa* the only genus (Fig. 21.5).

Turrids (Family Turridae). Mainly Panamic turret-shaped shells with a deep anal notch in the posterior part of the lip. *Turris.*

BIVALVES OF HARD-REEF SUBSTRATES

As stated above, most Bivalvia are filter-feeders; and their food comprises minute plankton, including diatoms as the major constituent. They can generally be regarded as *primary heterotrophs.* The various families have tendencies to inhabit particular habitats, the chief distinction being between hard-bottom (and coral-reef) habitat and soft-bottom (lagoon muds and sands). But these distinctions are not absolute; and some families may include members of either type, or which occupy both habitats. The following separation into hard-bottom and soft-bottom families is therefore only a general guide, and exceptions must be expected.

Ark Shells (Family Arcidae). Broad shells with a long hinge that carries numerous teeth (*taxodont dentition*), no siphons. Most species are anchored to hard bottom by a *byssus,* but some are free-living. Mainly tropical forms. Example *Arca* (Fig. 21.9).

Pearl and Wing Oysters (Family Peteriidae). Shell nacreous within, with an outer periostracum. Attached to rocky or coral seafloor. The wing oysters (*Pteria*) have thin, fragile shells; the pearl oysters (*Pinctada*) have a thick, black periostracum and a thick shell. Both genera characterize Indo-West Pacific tropical seas.

Hammer Oysters (Family Isognomonidae). Mainly Indo-West Pacific tropics; the shell broadened only along the hinge, the rest of the valves narrow; anchoring to rocky seafloor. *Malleus.* In mangrove sloughs the tree oyster (*Isognomon*) attaches to the lower branches.

File Clams (Family Limidae). Oval, ribbed shell, with narrow hinge. Edge of mantle produced into tentacles. Can swim by flapping tentacles and shell valves. *Lima* (Fig. 21.2).

Scallops (Family Pectinidae). Valves dissimilar, the convex one usually the left, occupying the lower side in life. Clams of this family can swim actively and have numerous eyes scattered round the edge of the mantle. Bottom substrate can vary. Example *Chlamys.*

GEORGE PUTNAM

Fig. 21.8 (a, c) *Conus*, Indo-West Pacific. (b) *Cymbium*, West Africa. (d) *Cypraecassis*, Indo-West Pacific. (e) *Cerithium*, intertidal and mangrove sloughs, tropics. (f) *Fusinus*, pantropical.

Giant Clams (Family Tridacnidae). Heavy valves with about four folded ridges and valleys on each, anchored by the hinge region to the coral reef, the free margin directed upwards, and displaying mantle tissue tinted green by symbiotic algae. Can weigh half a ton; next to giant squid, the largest mollusks. Can produce golf-ball sized pearls of nonprecious quality. Example *Tridacna* (Fig. 21.9).

MOLLUSKS OF THE LAGOON FAUNA

Gastropods

The following tropical families of Caenogastropoda are more usually found on or imbedded in soft substrates, such as the sands and silts of coral-reef lagoons and other soft-bottom tropical habitats. Some of these are detrital feeders, relying on what fragments of organic materials happen to be swept near the entrance to their burrows (e.g., Turritellidae); some are active carnivores, but prefer to live under cover of a layer of sand (e.g., Olividae); others are large wide-ranging carnivores of the epifauna which happen to occupy soft-bottom terrain (e.g., Cassididae); others inhabit eelgrass on shallow soft-bottom coasts, (e.g., Cerithiidae).

Turritella Snails (Family Turritellidae). Shell with a very long, slender spire of many whorls, with revolving striae. Operculum horny. Inhabiting muddy bottom, infaunal in warm seas. *Turritella.*

Ceriths (Family Cerithiidae). Tropical, intertidal mud snails. Shell resembling a turritellid but with a small oblique aperture and a short anterior canal. *Cerithium* (Fig. 21.8).

Moon Snails (Family Naticidae). Widely distributed carnivorous snails present in all seas, including tropics, living on sandy bottom. Shell globular, often polished and patterned with pigmentation. Foot large, often concealing the shell when the animal is active. Naticids drill into the shells of other mollusks, then insert the siphonlike proboscis into the drilled aperture to obtain soft tissues as food. *Natica*is said to drill only when sand covers it. They feed also on dead fishes. Example *Natica.*

Cameo and Helmet Shells (Family Cassididae). Active predators inhabiting sandy bottom in warm seas. Shell mostly large and inflated, heavy, thick, the aperture long, terminating in an anterior recurved canal, the outer lip usually thickened. Examples *Cassis* and *Cypraecassis* (Fig. 21.8).

Olive Shells (Family Olividae). Cylindrical or olive-shaped brightly colored and highly polished thick shell, with a large body whorl surrounding the spire and a plaited columella. Infaunal carnivores of tropical soft bottom. Olives lie just below the interface and can seize passing epifaunal animals; they can therefore be captured by baited hook and line. Example *Oliva* (Fig. 21.7).

Bivalves

These are essentially *infaunal elements,* wholly or partially buried in sand or mud, obtaining nutriment in the water currents which they generate by means of ciliated acts on the ctenidia (gills). Many worldwide families are included here, but the following are more characteristic of tropical seas.

Horse Mussels (Family Pinnidae). Occupying the floor of sheltered lagoons, with the lower half (comprising the pointed end region of shell) imbedded, the upper half in the water, and byssus threads anchoring the animal into the surrounding muds. Can occur in large populations separated each from the other by a uniform interval. Example *Pinna.*

JOSEPH D. GERMANO AND ISIDORO ZARCO

Fig. 21.9 (a) *Barnea* (Pholadidae), boring mollusk, North Atlantic. (b) *Phacoides* (Lucinidae), southeast North America. (c) *Cuspidaria* (Cuspidariidae) deep water. (d) *Nucula* (Nuculidae), deep water. (e) *Tridacna* (Tridacnidae), giant clam (to 250 kg), Indo-Pacific reefs. (f) Jackknife clam, *Ensis* (Solenidae), Atlantic. (g) *Arca* (Arcidae), worldwide. (h) *Cardium* (Cardiidae), pantropical.

Lucines (Family Lucinidae). Clams with a long, anterior muscle scar; shell often tinted within. Infaunal, tropics, and subtropics. Examples *Lucina* and *Codakia*.

Cockles (Family Cardiidae). Mainly thick-shelled, with radial ribs, infaunal. Examples *Trachycardium*, *Hemicardium*, *Corculum*, and *Cardium* (Fig. 21.9).

Venus Shells (Family Veneridae). Infaunal clams with cardinal and interlocking lateral hinge-teeth. Common throughout the world. Tropical genera include *Tapes* (lettered venus).

Sunray Shells (Family Tellinidae). Broad, fragile shells, usually polished and tinted with radiating sunraylike patterns. Example *Tellina*. Similar but duller clams from tropical muds are placed in the family Garidae.

Class Scaphopoda. The single, common genus *Dentalium* or tusk shell (family Dentaliidae) occurs as an infaunal element in soft tropical sands of the continental shelf. Food currents are maintained by cilia, and feeding utilizes plankton.

22

PRAWNS, CRABS, AND LOBSTERS

Anatomical terminology / Decapoda / Natantia / Penaeidea / Stenopodidea / Caridea / Reptantia / Macrura / Anomura / Brachyura

The Crustacea Decapoda are the most numerous marine arthropods. The order comprises some 8500 living species. Fossils are known from the Triassic onward, becoming important during the Jurassic, when lobsters and crabs first appeared; all existing groups are represented in the Tertiary. The origin of the order is unknown, but there is a close relationship with the order Euphausiacca, and the two orders are often combined in one superorder under the name Eucarida. The Euphausiacea, comprising about 90 species, differ from Decapoda in having the gills exposed, and the anterior pairs of thoracic limbs not modified as accessory jaw appendages.

TROPICAL CRUSTACEA DECAPODA

The decapod crustaceans may be grouped loosely as members of the **predatory** (i.e., secondary) **heterotrophs** in nearly all marine habitats, including therefore tropical reefs, though some members eat algae; some crabs emerge on to land at night to forage, others become wholly terrestrial in the adult stage, even ascending trees; best known of these is *Birgus,* the Indo-West-Pacific coconut crab, which has marine larval stages but frequents coconut groves to feed on nuts in the adult stage. Some crabs are scavengers, feeding on dead fishes and other carrion. In this chapter the pincer-shaped nippers are referred to as claws.

DECAPODS OF HARD TROPICAL SUBSTRATES

(a) Natantia (Shrimps)

Alphaeid Shrimps (Family Alphaeidae). Anterior pair of limbs enlarged, one of them with a snapping claw; the noise of that claw is lethal to small fishes on which these crustaceans feed. Tropical reefs, also shallow temperate shores. *Alphaeus* (Fig. 22.1).

JOSEPH D. GERMANO

Fig. 22.1 Decapoda Natantia (shrimps). (a) *Penaeus,* tropical and temperate continental shelves. (b) *Alphaeus,* tropical reefs and shallow temperate shores. (c) *Sicyonia,* pelagic in warm seas. (d) *Hippolysmata,* in seagrass along temperate shores. (e) *Palaemon,* in seagrass along temperate shores.

Family Stenopodidae. Shrimps with the third pair of limbs enlarged. Tropical shores, especially reefs. *Stenopus.*

(b) Reptantia (Lobsters, Crabs)

Spiny Lobsters (Family Palinuridae). Lobsters with no rostrum and lacking clawed limbs (except the fifth in the female). Most tropical and subtropical hard bottom. *Panulirus* (Fig. 22.2).

Prawn Killers (Family Scyllaridae). Carapace very depressed. Antennae very flat. No clawed limbs (except fifth in female). Most tropical and subtropical coasts. *Scyllarus* (Fig. 22.3).

Tropical Hermits (Family Diogenidae). Abdomen soft and coiled for holding gastropod shell. The tropical hermits of this family differ in details of the jaw structure from temperate hermits of the family Paguridae. *Petrochirus diogenes* is the great hermit of the Caribbean (Fig. 22.3).

Purse Crabs (Family Leucosiidae). Carapace hemispherical, female with abdominal brood pouch. *Persephona* (Fig. 22.4). Tropical and warm temperate coasts.

Grapsids (Family Grapsidae). Carapace subquadrate, legs flattened for running. Tropical and temperate shores, often emerging on land, also in marshes. *Sesarma.*

Fiddler and Ghost Crabs (Family Ocypodidae). Eyes on long stalks (ghosts), males with one cheliped enlarged for signaling (fiddlers). Temperate and tropical mud flats. *Uca* (Fig. 22.3).

DAVID F. MOYNAHAN

Fig. 22.2 Spiny lobster, *Panulirus,* tropical and subtropical coasts with hard bottom.

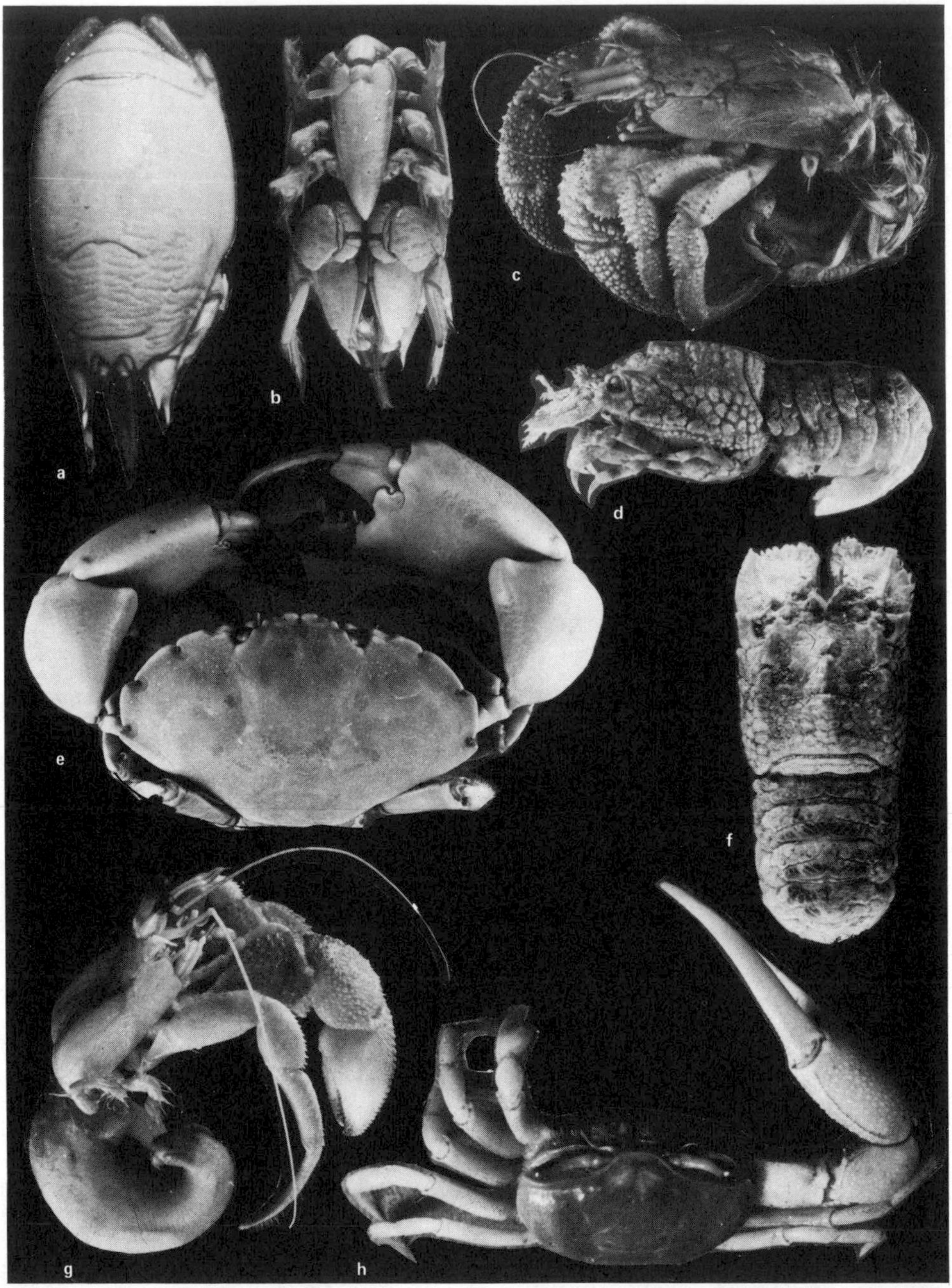

JOSEPH D. GERMANO

Fig. 22.3 (a, b) *Emerita,* mole crab, surf zone of temperate seas. (c) *Petrochirus* the great hermit crab, Caribbean. (d, f) *Scyllarus,* prawn killer of tropical and warm temperate coasts. (e) *Menippe,* temperate and tropical shores. (g) *Pagurus,* hermit crab of temperate shores. (h) *Uca,* fiddler crab, temperate and tropical mudflats.

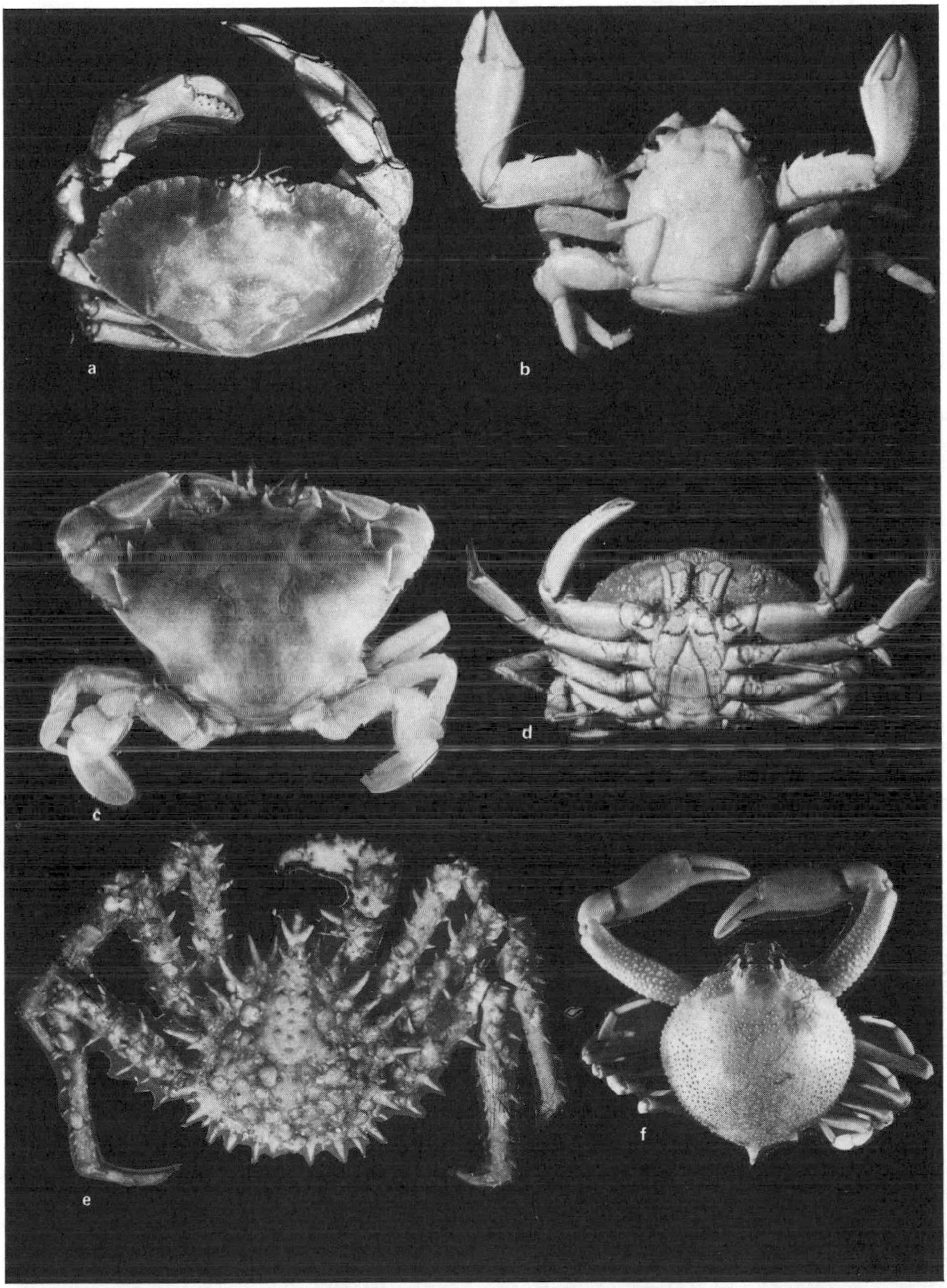

JOSEPH D. GERMANO

Fig. 22.4 (a, d) *Cancer*, edible crab, temperate coasts. (b) *Petrolisthes*, intertidal in warm seas. (c) *Ovalipes*, swimming crab of warm and temperate coasts. (e) *Lithodes*, spider crab, cold temperate shelves. (f) *Persephona*, purse crab of temperate and tropical coasts.

Maias (Family Majidae). Rostrum elongated, limbs long, spiderlike. Tropical and temperate shores. *Stenorhynchus,* the arrow crab of coral reefs.

DECAPODS OF TROPICAL SOFT SUBSTRATES

Box Crabs (Family Calappidae). Front claws extremely large, used to open gastropod shells. Tropical and warm, temperate, sandy bottoms. *Calappa* (Fig. 22.5) and *Hepatus.*

DECAPODS OF TEMPERATURE LATITUDES

Shrimps and decapods with well-developed swimming appendages on the abdomen called **pleopods.** The thoracic limbs are not used as walking legs; they may be termed **pereiopods.**

Family Hippolytidae. Shrimps with the rostrum usually well developed and serrate. Eyes exposed. Second pair of limbs with subdivided carpus. Common in grass beds along temperate shores. *Hippolyte* and *Hippolysmata* (Fig. 22.1).

Family Palaemonidae. Shrimps with a prominent, serrate rostrum. Anterior pair of legs shorter and weaker than second pair. Common in grass beds along temperate shores and also in freshwater. *Palaemonetes, Palaemon* (Fig. 22.1), and *Percilimenes.*

Family Pandalidae. Shrimps with the rostrum long, serrate. Anterior pair of limbs with minute claws. Second pair of limbs with many segments. On continental shelves in temperate seas. *Pandalus.*

Family Porcellanidae. Half-crabs, resembling true crabs, but having a well-developed tail fan. *Petrolisthes* (Fig. 22.4) is an example.

Family Portunidae. Last pair of legs broad and thin, forming swimming appendages. Often important commercially on New World tropical and temperate

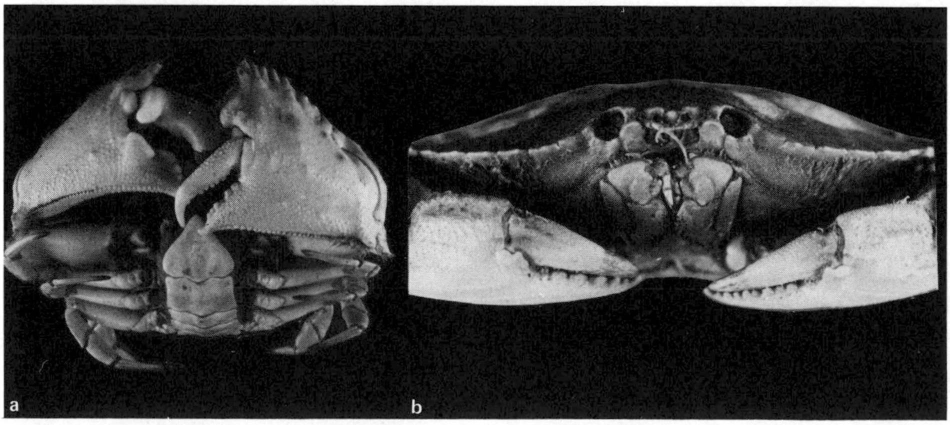

JOSEPH D. GERMANO

Fig. 22.5 (a) *Calappa,* box crab, tropical and warm temperate shelf on sandy bottom. (b) *Cancer,* cool and warm temperate coasts.

coasts. In America the family includes the blue crab, *Callinectes,* and the lady crab, *Ovalipes* (Fig. 22.4).

Family Xanthidae. Mud crabs with antennules folding obliquely. Temperate and tropical shores. *Menippe* (Fig. 22.3) is an example.

Family Crangonidae. Carapace depressed. Rostrum reduced. Along temperate shores, littoral and sublittoral. *Crangon.*

Family Nephropidae. Lobsters. Rostrum present. Three anterior pairs of limbs clawed. Continental shelf of western and eastern North Atlantic and South Africa. *Homarus.*

Family Callianassidae. Carapace with lateral longitudinal suture. One of anterior pair of limbs greatly enlarged. Pleopods broad. Mud flats along temperate coasts. *Upogebeia.*

Family Paguridae. Hermit crabs resembling *Paguristes* but differing in jaw structure. Chiefly along temperate shores. *Pagurus* (Fig. 22.3) and *Clibonarius.*

Family Lithodidae. Spider crabs. Carapace and walking legs crablike, but abdomen asymmetrical in females and posterior pair of limbs reduced, showing hermit crab relationship. Deeper parts of continental shelves in temperate seas. *Lithodes* (Fig. 22.4) and *Paralithodes.*

Family Hippidae. Mole crabs. Carapace and pereiopods adapted for burrowing, antennal flagella for filter feeding. Surf zones of temperate seas. *Emerita* (Fig. 22.3).

Family Cancridae. Edible crabs. Antennules folding longitudinally. Third maxillipeds long, overlapping mouth. Temperate coasts. *Cancer* (Figs. 22.4 and 22.5). Edible crab of many countries.

Family Pinnotheridae. Pea crabs. Females with soft, integument, and small eyes, commensal in bivalves, ascidians, and worm tubes. *Pinnotheres.* Widely distributed.

Family Penaeidae. Shrimps. Rostrum well developed. Three anterior pairs of limbs clawed. An example is *Panarus.* Common and often of commercial importance on continental shelves of all tropical and warm temperate seas.

Family Sergestidae. Shrimps with a reduced rostrum and the two posterior pairs of walking limbs are reduced and flattened. Tropical and warm temperate seas, usually pelagic. *Sicyonia* (Fig. 22.1) is an example.

23

TROPICAL SEAS AND CORAL REEFS

Productivity of tropical seas / Location and cause of cold upwelling water / Westerly trend of surface currents / Biassed isotherms / Coral reefs / The bioherm and hermatypic corals / Distribution of coral reefs / Types of coral reefs and their origins / Antiquity of coral reefs / Reef communities / The reef ecosystem / High productivity of reefs / Systematics of reef corals / Reef benthos / Life on a coral reef

Productivity of Tropical Seas

In any natural ecosystem there are usually up to five trophic levels, namely, the primary producers or autotrophs (plants), the primary heterotrophs (herbivorous animals), and two or three levels of secondary heterotrophs (carnivorous animals). Because each trophic level is dependent upon the one beneath it, and a time lapse occurs in the transfer of energy from sunlight through the successive levels of the ecosystem, measures of productivity need to be made independently at each trophic level and at various intervals of time. The standing crop at any instant varies seasonally, mainly because the primary producers receive seasonally variable inputs of solar energy. In temperate and cold seas this is illustrated by the annual spring diatom increase (SDI), which sets the rhythm of the productivity cycle.

In tropical seas the high-energy input from the sun heats the surface waters and reduces their density; on the other hand, below about 800 m there is a constant zone of higher-density cold water. Between these two strata is a well-marked temperature gradient, or **thermocline,** and a corresponding density gradient. These are permanent features, and although wind disturbances of the upper layer may produce circulation (and counter circulation below), no effective mixing takes place. Therefore, the available mineral nutrients in the upper layer are a constant quantity, and this factor sets a rigorous limit on the amount of diatom production in tropical seas. Thus, with two exceptions noted later, the primary production in tropical seas is relatively constant and at a relatively low level. Hence the generally observed fact that *productivity in all tropical open seas is low*. In tropical seas near continental or island coasts, there may be notable departures from this general rule in regions where significant upwelling of deep waters can occur and on coral reefs.

Location and Cause of Tropical Upwelling

Air over the equatorial regions of continents becomes heated, therefore rarified, and then rises. It is replaced by cooler air flowing toward the equator from the north and south. Since the replacing air is originating from a part of the earth's surface that is rotating from west to east at a lower velocity than the equator rotates, the conservation of its angular momentum requires that the new air maintain its initial rotational velocity, so on entering regions nearer the equator the replacement air appears to swing somewhat to the west. Thus, north of the equator a wind appears to blow from the northeast, and south of the equator from the southeast. During its passage toward the equator along the western side of any continent, the air–sea couple of this longshore wind imparts a similar motion to the underlying coastal water at the air–sea interface. Thus, north of the equator the surface waters flow obliquely away from the land in a southwest direction, and south of the equator they flow obliquely away from the land in a northwest direction. But the overlying waters, thus set in motion, exert a couple on the underlying waters, causing them to flow similarly but more slowly, since the energy falls off logarithmically with increasing depth.

Thus, in the tropics, water of the ocean is forced to move westward from western coastlines, the rate of flow being greatest near the surface. This water can only be replaced by upwelling from greater depths; the forces are great enough to overcome the thermocline so that mineral-rich bottom water reaches the surface. Here the added nutrients, especially nitrates and phosphates derived from bacterial reduction on the seabed, cause a corresponding bloom of the surface phytoplankton. This, of course, results in increased productivity in the surface waters of seas adjoining the western margins of continents in the tropics. The lands affected are therefore between southern California and northern Chile, and those between Morocco and southwest Africa, in whose coastal waters important fisheries exist.

CORALS

The term coral is generally in use for any member of the phylum Coelenterata that secretes a hard exoskeleton or endoskeleton. Two classes of the phylum have this property. (1) **Hydrozoa,** of which only the genera *Millepora* (Fig. 23.1) and *Stylaster* need concern us, are colonial hydrozoans in which a calcareous skeleton is laid down between and over the coenosarc. Members of this group are referred to as *hydrocorals*. (2) **Anthozoa,** of which two subclasses include members that produce hard skeletons. These are the subclass Alcyonaria, or **octocorals,** in which

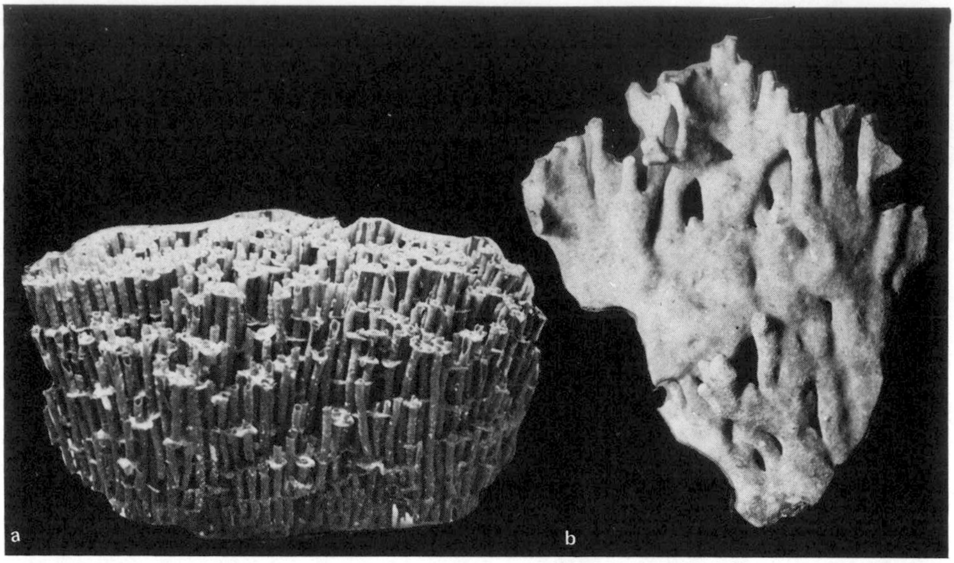

ISIDORO ZARCO

Fig. 23.1 (a) *Tubipora,* Indo-Pacific, ×0.5. (b) *Millepora,* pantropical, ×0.5.

the zooids have eight tentacles, and the skeleton takes the form of a mixture of keratin and calcareous spicules. The best-known members being the organ-pipe coral, *Tubipora* (Fig. 23.1), of the Pacific and Indian oceans, and the seafans, *Gorgonia* (Fig. 23.2), and seawhips (order Gorgonacea); and the subclass Zoantharia (Figs. 23.3 to 23.9), comprising sea anemones and **hexacorals,** in which the body has a six-part symmetry. Sea anemones are solitary and produce no skeleton, whereas the order Scleractinia comprises both solitary and colonial members, all of which secrete hard calcareous exoskeletons, constituting the vast majority of corals. The black corals or sea trees, which do not form reefs, are also classified with the Zoantharia under the name antipatharian corals.

Corals that congregate in vast communities to form reefs are referred to as **hermatypic** corals; those that do not (mainly deep-water forms) are said to be **ahermatypic.** The term bioherm is used for any organic reef, such as a coral reef. Other organisms that can, or once could, form reefs include the Precambrian and Cambrian phylum of spongelike marine animals called Archaeocyatha and the calcareous red algae called corallines (phylum Rhodophyta). Some bivalve mollusks and some extinct brachiopods are found as massed shell banks, having some analogy with reefs but lacking the massive vertical growth of typical reefs.

How to Identify Corals

John Wells (1966) has given a systematic review of the genera of stony corals (order Scleractinia) in the *Treatise on Invertebrate Paleontology.* Following is a summary of the more important diagnostic characters employed in the classification and illustrated in the photographs accompanying this chapter.

The Corallum. The cup in which an individual coral zooid rests is called the **corallite;** in the case of a solitary coral, such as *Fungia* (Fig. 23.3), the corallite comprises the entire skeleton. In colonial corals many corallites are contiguous and more or less intimately fused together to form a compound skeleton, the **corallum.**

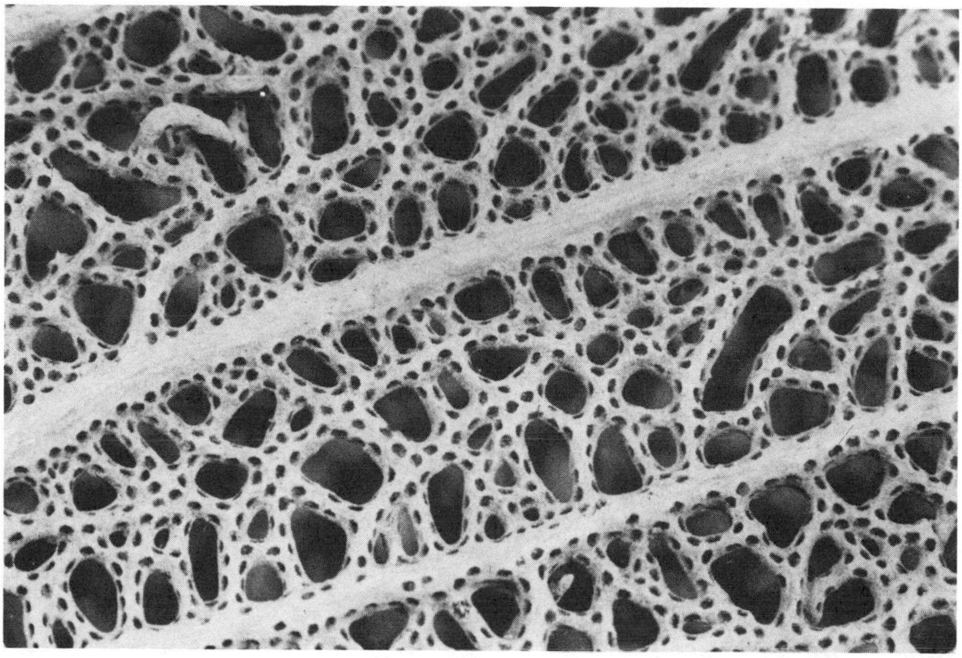

STEPHEN H. DART

Fig. 23.2 *Gorgonia,* Caribbean, ×2.

If the individual corallites remain distinct, then the corallum is said to be **phaceloid,** for example, *Euphyllia* (Fig. 23.4). The corallites may maintain their separate identity, but become fused together by the secretion of a spongy intermediate skeletal material called **coenosteum,** as in *Galaxea* (Figs. 23.4 and 23.7) and *Turbinaria* (Fig. 23.6). If the individual corallites are directly contiguous, but remain more or less cylindrical (without much distortion from the mutual compression) the corallum is said to be **plocoid,** as in *Favia* (Figs. 23.4 and 23.6). If the lateral compression through crowding forces the corallites to assume a polygonal honeycomblike arrangement, the corallum is said to be **cerioid,** as in *Favites* (Fig. 23.8) and *Goniastrea* (Figs. 23.3 and 23.6).

The gross form of the corallum varies, even with a single genus, so it is of little systematic value in identifying corals. On the other hand, this particular character is of the utmost importance in determining whether a given coral species can occupy a particular habitat. For example, a heavily pounded reef in open surf can comprise only those corals whose growth habit produces a strong skeleton. Of these, the most wave-resistant is the massive growth form, exhibited by species of *Favia, Meandrina,* and *Siderastrea* (all illustrated in Fig. 23.4); many other species in various genera are capable of growing in this way. If the corallum branches it is said to be **ramose** (for example, the well-known staghorn coral, *Acropora,* Fig. 23.4); this genus is the most abundant living coral, with some 200 species, mainly in the Indo-Pacific region, although there are Atlantic representatives also. If the corallum forms flattened plates it is said to be **foliose;** an example is *Agaricia* (Fig. 23.6). The phaceloid structure referred to above may also, of course, be regarded as a growth form of the colony, and so also the solitary habit. A particular form of massive coral habit occurs when the colony assumes a dome shape, as in *Manicina* (Fig. 23.5); such forms are described as **cupolate.**

GEORGE PUTNAM

Fig. 23.3 (a) *Mycetophyllia*, Caribbean, ×0.5. (b) *Goniastrea*, Indo-Pacific, ×0.25. (c) *Platygyra*, Indo-Pacific, ×0.25. (d) *Fungia*, Indo-Pacific, ×0.25. (e) *Psammocora*, pantropical, ×0.25. (f) *Porites*, pantropical, ×0.25.

Calyx. The cup-shaped secretion within which the individual coral usually rests in compound corals is called the **calyx**, and it comprises several parts. The wall of the cup is the **theca**, and the outermost layer of the theca is called the **epitheca.** It is in this outer layer that the growth lines laid down from day to day may sometimes be observed, as in *Manicina* (see Chapter 9). In this fragile layer of

GEORGE PUTNAM

Fig. 23.4 (a) *Euphyllia,* Indo-Pacific, ×0.5. (b) *Favia,* pantropical, ×0.5. (c) *Acropora,* pantropical, ×0.25. (d) *Siderastrea,* pantropical, ×0.25. (e) *Euphyllia,* Indo-Pacific, ×1. (f) *Meandrina,* Caribbean, ×0.25. (g) *Galaxea,* Indo-Pacific, ×0.25.

fossil corals the fundamental data on the ancient history of the moon's orbit and the earth's rate of rotation are stored. Radiating inward from the thecal wall are variable numbers of radially directed plates called **septa.** The septa form between the radially directed soft mesenteries of the coral animal; they are arranged in multiples of six. The term **monocyclic** is applied to the (very few) corals in which only six septa occur. If 12 septa can be distinguished, as in *Astreopora* (Fig. 23.7),

GEORGE PUTNAM

Fig. 23.5 (a) *Pectinia,* Indo-Pacific, ×0.5. (b) *Manicina,* Caribbean, ×0.5. (c) *Symphyllia,* Indo-Pacific, ×0.5.

the coral is said to be **dicyclic.** Higher multiples are more usual, and *Fungia* (Fig. 23.3) is an example of a polycyclic coral. In the center of the floor of the calyx there is often a projecting excresence called the **columella.** The columella may be merely a sharp projection (styliform columella) or a broad honeycombed button, as in *Turbinaria* (Fig. 23.6), or may be completely lacking, as in *Astreopora, Acropora,* and *Montipora* (all illustrated in Fig. 23.7). Sometimes the septa themselves form the theca by thickening of their substance at a given distance from the center, as in *Favia* (Fig. 23.6), or the septa may be continuous across the intervening space from one corallite to its neighbors, in which case the coral is said to be **septocostate,** as in *Siderastrea* (Fig. 23.8). In some genera transverse struts pass from one septum to its neighbor; these are called **synapticulae,** and they can be seen in *Fungia* (Fig. 23.3) and in *Siderastrea* (Fig. 23.8).

Budding. The manner in which buds of new coral zooids form from the preexisting zooids is reflected in the manner in which new corallites are formed in the corallum. If the buds branch from the outer surface of the zooid, that is, from a region outside the ring of tentacles of the zooid, then the budding is said to be **extratentacular.** In this kind of budding the smaller and younger corallites are observed to be wedged between the older ones, as in *Siderastrea* (Fig. 23.7). If, on the other hand, the new buds form within the ring of tentacles, a peculiar process called **intratentacular budding** takes place, then the result is that the preexisting theca becomes two-lobed, three-lobed, or four-lobed, according to the number of internal subdivisions produced. Such lobed outlines are seen in *Favia* (Fig. 23.6), showing that intratentacular budding occurs in that genus.

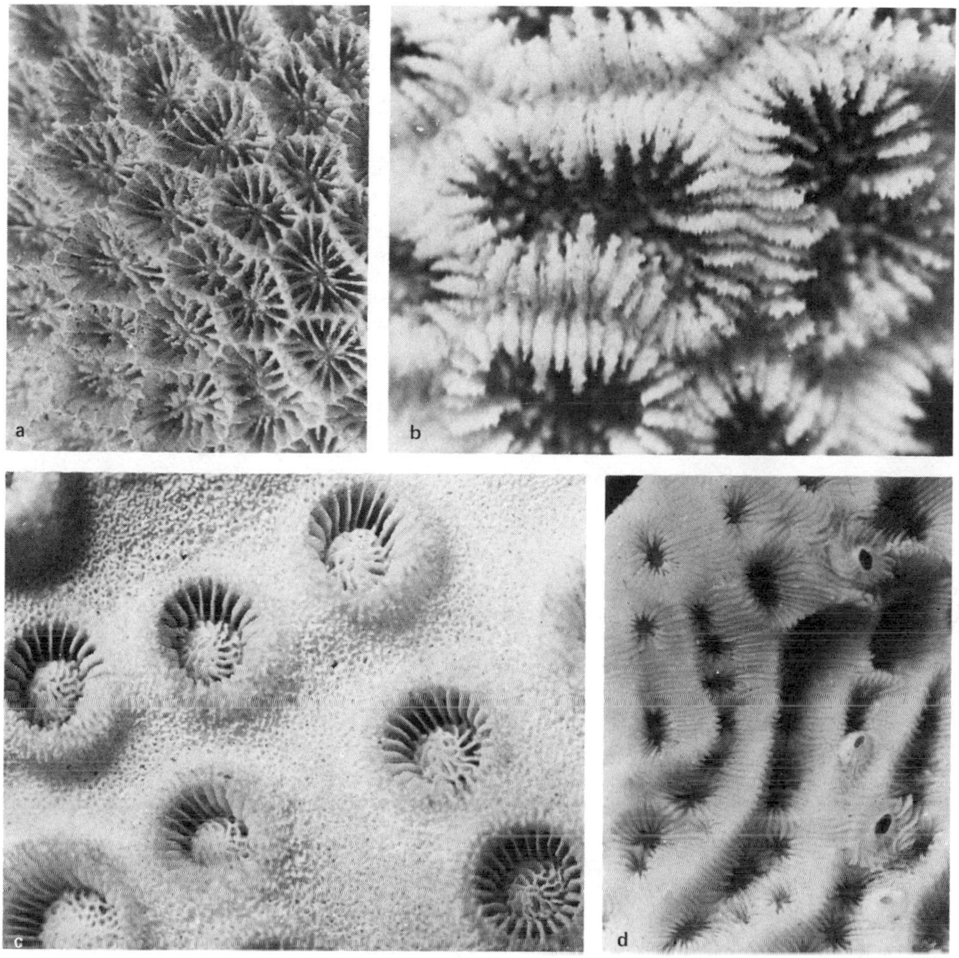

GEORGE PUTNAM

Fig. 23.6 (a) *Goniastrea*, Indo-Pacific, ×3. (b) *Favia*, pantropical, ×4. (c) *Turbinaria*, Indo-Pacific, ×2. (d) *Agaricia*, Caribbean, ×1.

Meandroid Forms. A further development of the intratentacular mode of budding leads to the strangely meandering outlines of the thecae in the so-called brain corals or meandroid forms. Simpler forms of meandroid development are exemplified by *Symphyllia* (Fig. 23.5), and more complexly intertwined patterns are seen in *Meandrina* (Fig. 23.4), *Diploria* (Fig. 23.8), and *Platygyra* (Fig. 23.9). The adoption of the meandroid form leads to the development of ridges between adjacent winding thecae, called **collines**; in other cases a depressed groove called an **ambulacrum** may form between adjacent growth centers as in *Favia*.

In the solitary corals the theca is often more or less flat, and in such cases the septa stand erect upon the flat base, as in *Fungia* (Fig. 23.3). In the phaceloid forms and some others, the thecae may develop septa that project above the top of the cup; such forms being said to be **exsert.** An example is *Euphyllia* (Fig. 23.4). Some other massive genera have exsert septa, such as *Symphyllia* (Fig. 23.9), where the edge of the septa may also be seen to be **dentate.**

The various genera are defined largely on the basis of the various possible combinations of these and similar characters. The best way to gain a practical

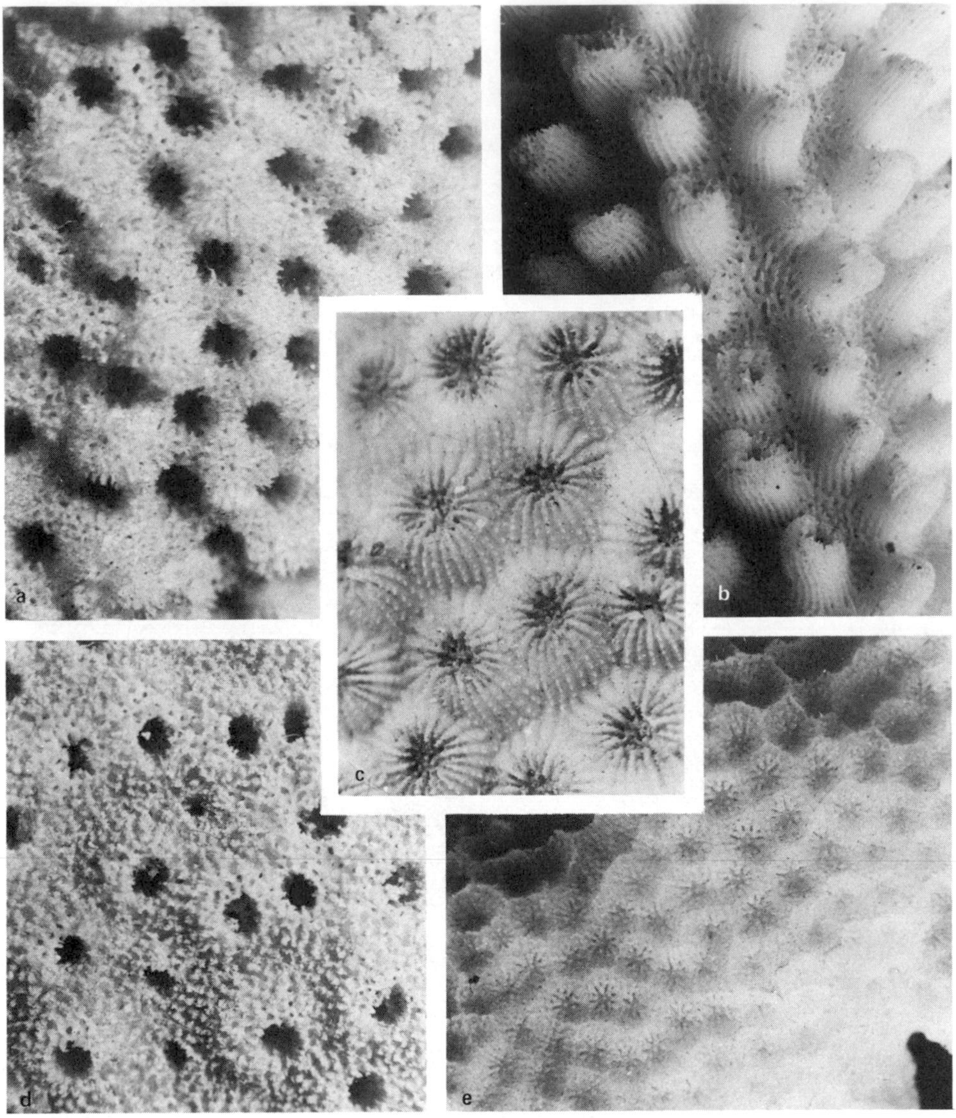

GEORGE PUTNAM

Fig. 23.7 (a) *Montipora*, Indo-Pacific, ×3. (b) *Acropora*, pantropical, ×2. (c) *Montastrea*, Caribbean, ×3. (d) *Astreopora*, Indo-Pacific, ×3. (e) *Porites*, pantropical, ×2.

acquaintance with the subject is, of course, to study actual specimens of the various coral species, in the laboratory, under a low-power microscope.

CORAL REEFS

The Bioherm. A reef is a ridge or elevated part of the seabed that approaches the sea surface. A coral reef is a reef that comprises or is capped by corals; a living coral reef has actively growing corals on its surface, the remainder of its mass made up of the accumulated skeletons of former living members. Reef corals can grow down to a depth of 50 m (below that depth the water is too cold and the sunlight

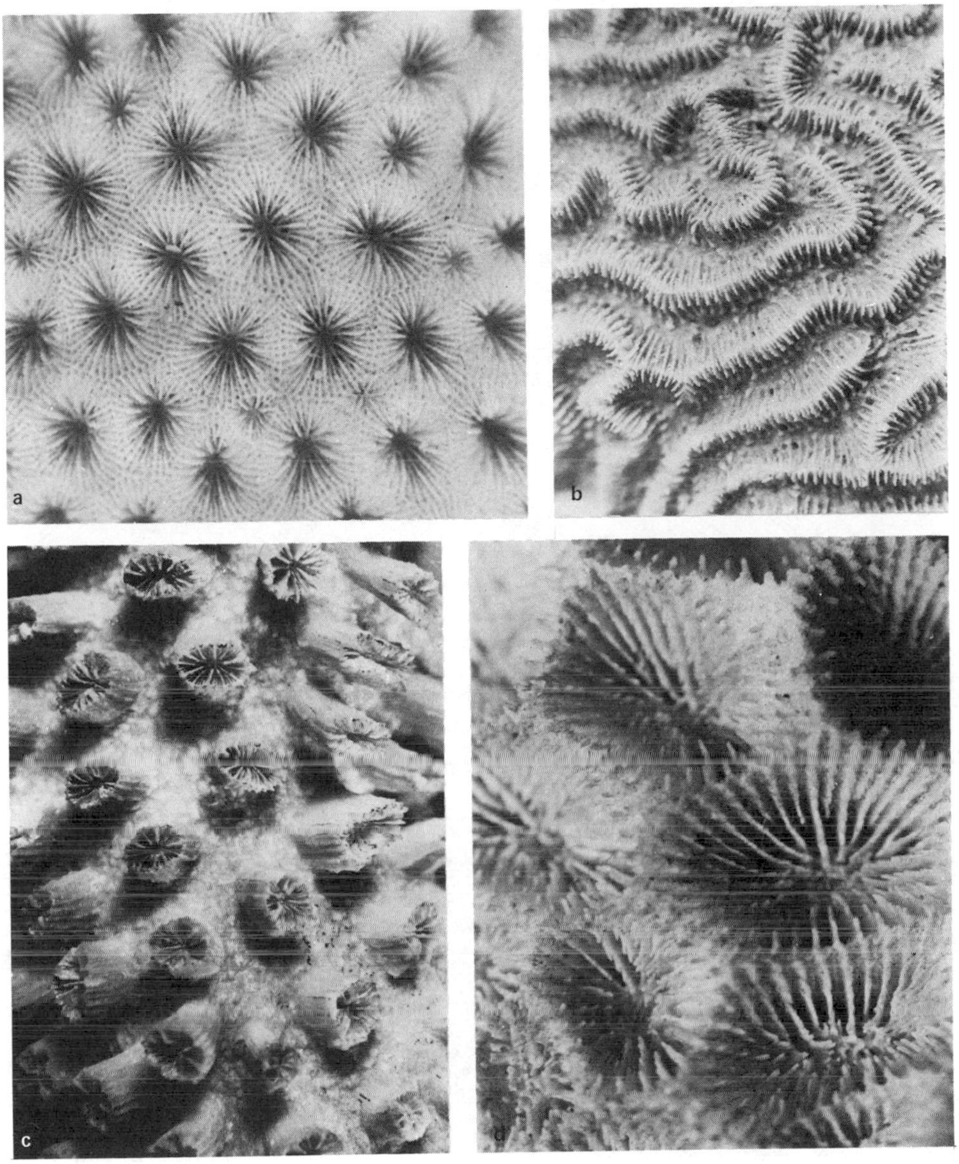

GEORGE PUTNAM

Fig. 23.8 (a) *Siderastrea,* pantropical, ×3. (b) *Diploria,* Caribbean, ×2. (c) *Galaxea,* Indo-Pacific, ×2. (d) *Favites,* Indo-Pacific, ×4.

too feeble for reef corals); but no reefs (except dead reefs) occur other than at the sea surface. This occurs because corals grow upward at a rate of not less than 1 mm per year, and sometimes as much as 100 cm per year, and, since sea level has risen 50 m over the past 12,000 years (following the end of the Wisconsin glaciation), all reefs have had adequate time to reach the surface.

Distribution of Coral Reefs. Corals require water temperatures not lower than 16°C in order to form reefs, so coral reefs are restricted to tropical and subtropical seas between about 30° north and south latitude. This belt, covering half the earth's surface, forms one of the best defined marine biogeographic regions. If the oceans were to evaporate, the belt of fossil reef coral girdling the planet would still

GEORGE PUTNAM

Fig. 23.9 (a) *Meandrina,* Caribbean, ×2. (b) *Platygyra,* Indo-Pacific, ×2. (c) *Hydnophora,* Indo-Pacific, ×4. (d) *Symphyllia,* Indo-Pacific, ×4.

remain, and even if the planet ceased to rotate, as one day it will, the position of the former axis of rotation could be determined by space visitors by studying the coral belt. By the same token, former axes and polar positions can presumably be determined by studying the distribution of reef corals fossilized from former geological periods. Evidence of this kind, discussed elsewhere, also throws doubt upon some current notions of continental drift.

Types of Coral Reefs. Three well-marked types of reef are known, namely, the *fringing reef,* the *barrier reef,* and the *atoll.*

Antiquity of Coral Reefs. Reef-forming corals belonging to extinct orders (Tabulata and Rugosa) existed in the Paleozoic era, and date back to the Ordovician period, some 500 million years ago. The Ordovician reef belt is considered (Fell, 1968) to have had the same width as at present, namely, 60°, but to have been tilted about 75° with respect to the existing belt, suggesting poles near Hawaii and off southwest Africa. Strakhof and other Russian geologists report similar evidence of polar displacement on the basis of the distribution of red desert sandstones and other climate-related rocks. At the end of the Paleozoic coral reefs suffered a severe reduction and most corals became extinct. In the mid-Triassic, about 200 million years ago, modern families of reef coral appeared, and at the end of the Triassic a reef belt is again recognizable, tilted much less than before, suggesting a north pole in east Siberia and a south pole near the Magellan Straits. Cretaceous reefs were tilted still less, and by the onset of the Eocene the reefs nearly matched those of today.

Coral-Reef Communities. Not only corals comprise the coral-reef biota. Other phyla, notably the red algae (Rhodophyta), Mollusca, and Echinodermata, have numerous and conspicuous members whose distribution corresponds to that of reef corals, and actually live on, or imbedded in, coral reefs. Since reefs are hard substrates, these associated animals and plants constitute epiflora and epifauna. However, lagoons are also physical features associated with reefs, and they have soft-bottom sediments of sand and silt. Many biota live imbedded in such soft substrate, so that a characteristic tropical soft-bottom infauna also can be recognized. Heart urchins (Fig. 23.10), sea cucumbers, bivalve mollusks, and some annelids constitute important elements of such tropical infaunas. Mangrove formations, sand, and mudflats also occur on tropical coasts with special associated fauna.

Coral-Reef Ecosystem. This has the essential elements of all ecosystems. **Energy** is supplied by solar irradiation, **material flow** is supplied by the surrounding medium—the seawater. The **autotrophs** comprise the plants, particularly diatoms (Chrysophyta), which are eaten by zooplankton, especially copepods. The **primary heterotrophs** comprise, in addition to the zooplankton just noted, benthic organisms, such as forams and the filter-feeding clams, some holothurians, and the alga-eating mollusks, such as the Archaeogastropods. The **secondary heterotrophs** comprise all the predators, including the corals themselves, which feed mainly on copepods and plankton generally, the majority of gastropods, and echinoderms generally. The **degraders** comprise, as usual, the seafloor bacteria.

High Productivity of Coral Reefs. The temperature regime demanded by reef corals favors their development on the eastern margins of tropical continents, for it is only there that shallow coastal shelves are bathed by warm currents driven by the trades and cycled to north and south as part of the major gyres. There can be no upwelling of nutrients, yet the productivity can have an approximate maximum value of 0.8 kcal m^{-2} day^{-1} or 800 gram-calories per square meter each day. This is about half the rate on temperate shelf environments, but more than ten times greater than the rate for open tropical seas. These figures relate to primary productivity.

Apparently the high productivity is caused by efficient recycling of organic and inorganic constituents within the reef ecosystem itself. Losses by emigration or accident fall into deep water being minimized by the physical nature of the reefs themselves. The great diversity of biota at all trophic levels also favors high productivity, especially when account is taken of the high proportion of symbiotic algae involved in the tissues of the coelenterates and some of the mollusks. The higher trophic levels are occupied by fishes of essentially reef-inhabiting families, rather

JOSEPH D. GERMANO

Fig. 23.10 Heart urchins of tropical lagoon floors. (a) *Lovenia,* Indo-West Pacific. (b, c) *Clypeaster,* pantropical. (d) *Schizaster,* pantropical. (e) *Cassidulus,* tropical amphiamerican. (f) *Brissus,* Indo-West Pacific.

than by casual visitors from the open sea, so little loss of biomass to other regions occurs. The bottom conditions favor the activities of local reducers, sheltering them in the innumerable protected pockets in the reef where they are not liable to be swept offshore by currents or silt slides down the continental slope. There is also perhaps some terminal input of biomass, carried westward on the equatorial currents from the regions of upwelling in the eastern parts of tropical oceans, but not continuing on indefinitely around each oceanic gyre (on account of falling temperature as the currents leave the reef zone).

Patterns of Life on a Coral Reef

The most conspicuous organisms associated with corals on the reefs themselves and in the lagoons within the reef rampart, are the brilliantly hued fishes, mollusks, and echinoderms. The two former are discussed in the tropical sections of Chapters 14 Fishes and 21 Mollusks. Notable in the case of the highly mobile and adaptable nekton are the territorial and cleaner patterns of behavior, and the various venomous fishes with associated coloration. Among the mollusks, protected as they are by the massive shell as in most tropical species, or by rapid swimming in the case of shell-less forms, there is a similar relative independence to that exhibited by the fishes. But in the echinoderms the case is different. They have responded in a much more passive, though nonetheless successful, manner to the demands of the reef environment.

Echinoderms are essentially unaggressive and relatively slow-moving animals; they are, therefore, subject to the exigencies of their environment and must be adapted to withstand the various environmental pressures because they cannot readily escape from them. These needs have governed the course of evolution, by natural selection, since organisms unsuited to a given environment must either change or quit the environment for one more suited to their needs. The total effect of the environment must always be a complex equation with terms of differing magnitude. To judge by a survey of the distribution pattern of echinoderms on coral reefs, it would seem that about seven principal variable terms are distinguishable. These are: (1) ability to withstand the destructive action of surf, (2) effects of strong sunlight, (3) intolerance of salinity change, (4) ability to withstand predator attack, (5) and (6) feeding and locomotor habits (these two are interrelated), (7) dispersal mechanisms.

There are three observed ways in which the danger from surf action is minimized. Some sea urchins spend the earlier and more delicate stages of growth in deeper water, beyond the margin of the reef, and only when the shell is fully developed do they ascend and occupy the upper regions where wave action occurs. Here they feed upon bryozoans and other encrusting organisms. Other species of urchin occur in enormous numbers on surf-pounded reefs and contrive to survive by excavating with their teeth sunken depressions in the substrate, which they permanently occupy. *Podophora* has the spines modified into a kind of armor plating, and it clings tenaciously with its suctorial tube feet to the outer reef zones. Waves are deflected by the streamline curvature of the plating (Fig. 23.11). A similar modification is seen in a fossil sea urchin from the mid-Paleozoic in North America, apparently also a reef species, for the coral-reef belt at that time traversed the eastern half of the continent. Sea urchins, and other echinoderms, which lack these modifications are obliged to occupy sheltered crevices under rocks, or in the interior of coral heads, or are restricted to the inner lagoon where wave action is minimal.

A few echinoderms are insensitive to strong sunlight, a notable example being the widespread Indo-Pacific starfish *Linckia laevigata*; its integument is densely pigmented with a richly saturated dark blue substance which apparently protects the animal from excessive irradiation, so the shallow stretches of the reefs are com-

GEORGE PUTNAM AND JOSEPH D. GERMANO

Fig. 23.11 (a) *Podophora,* Indo-West Pacific reefs. (b) *Pseudechinus,* New Zealand and Southern islands (west-wind-drift dispersion). (c) *Temnotrema,* tropical. (d) *Tetrapygus,* American seas. (e) *Phormosoma,* deep-water Pacific. (f) *Salmacis,* tropical Indo-West Pacific. (g) *Colobocentrotus,* Indo-West Pacific reefs. (h) *Phyllacanthus,* Indo-West Pacific reefs.

monly occupied in full sunlight, unlike the case in other starfishes that avoid direct sunlight. Sea cucumbers are sometimes very deeply pigmented with brown or black melanin, and these, too, seem less sensitive to light than are paler forms. In general, reef echinoderms conceal themselves from direct sunlight, and most of their activity occurs at night. Some feather stars are attracted by artificial illumination at night, a fact which suggests that a low threshold of illumination (such as moonlight) stimulates activity. Many sea urchins shelter by day by holding up shells, pieces of algae, or in regions near cities, objects such as bus tickets and crown bottle tops; these foreign objects are hold over the body by the suctorial tube feet, as if they were parasols. After sunset the objects are discarded, when feeding and locomotor activities begin. Some urchins are sensitive to the passage of a shadow over the body and direct their spines toward the source of the shadow; this is doubtless a protective response to a potential predator.

All echinoderms are intolerant to changes in salinity and avoid areas where sudden rainfall may dilute the surface water. The most sensitive are the feather stars (crinoids). These animals tend to inhabit the outer margins of the reef, where the water is less likely to vary in salt content.

The principal means of protection against predator attack are the adoption of a secretive habit and the development of sharp or toxic spines. The chief, natural predators are other echinoderms (starfishes, and some sea urchins which bite into the shell of other urchins, to feed on the soft parts within) and, more important, fishes such as toadfishes, grunts and triggerfishes. There are also some parasitic snails that attack sea urchins. Man is an important predator in the case of the sea urchin *Tripneustes ventricosus* (synonym, *Tripneustes esculentus*) the object of a protected fishery in the Caribbean. Some kinds of sea cucumbers (holothurians) are fished in connection with the trepang industry, catering to epicures in the Orient and in some western countries. The long-spined sea urchin *Diadema* sometimes lives in such dense populations that its long, hollow, brittle spines form an intermeshed plexus of considerable protective value; the more so since the pigment in the spines is also toxic, and the spines easily break off in the skin of an unwary predator. The large slate-pencil urchin *Heterocentrotus* is protected by overdeveloped massive solid spines, which contain so much mineral matter they emit a metallic tinkle when struck. The crown-of-thorns starfish *Acanthaster* inhabits reefs in the Indo-Pacific region. It is covered by large, erect spines, capable of inflicting wounds if accidentally trodden on; this animal is feared in some islands of the southwest Pacific. Starfishes (asteroids) in general are not secretive—apparently their flesh is distasteful to predators. Brittlestars (ophiuroids) are mostly secretive, and are attacked by fishes if exposed.

Feather stars prefer sheltered positions under overhanging rocks or corals. Tidal pools are avoided, probably on account of the frequent wave disturbance and the fragility of the body. When moving, their habit is to creep rather than to swim, although swimming is possible. Sea cucumbers are essentially creeping animals, but a few species can swim with an undulating motion, as in the genus *Synapta*, which swims by night though benthic by day. Some brittle stars have branched arms, such as basket starfishes, that cling to fan corals and at night fish the surrounding water by spreading their arms out like a net. A wide variety of feeding habits is found among echinoderms, with correspondingly wide variation in the structure of the feeding organs. A notable effect of some sea cucumbers is that a considerable reworking of lagoon-floor mud occurs; in Bermuda sea cucumbers are believed to turn over sediments amouting to more than 300 tons per square miles per annum.

Dispersal ability depends upon habit. Species that adhere to algae or other potentially floating objects may perform transoceanic drift voyages. If the voyager has the power of asexual reproduction, then the chances of successfully colonizing a distant reef are much increased, because it is not necessary for more than one individual to survive such a raft voyage. *Linckia guildingii* is a starfish that has spread throughout the entire coral-reef belt of the world; it has the ability to regenerate new individuals from broken arms. Perhaps its wide distribution is a consequence of accidental drift voyages by single individuals. In colder zones analogous dispersal patterns are associated with species having a viviparous habit—another character that would also favor successful transoceanic colonization because in this case a single voyage is all that is required to establish a new population, provided the founding parent is a pregnant female. These speculations are not merely fanciful, simple tests suggested by probability theory and observed cases of brood-protecting echinoderms floating on seaborne kelp have given considerable support to the idea that shallow-water echinoderms have indeed succeeded in crossing major oceans to establish new communities on distant shores. At the present time the Atlantic coral reefs are isolated from those of the Indo-Pacific, and obviously no interchange of populations can presently be occurring, unless by way of the Suez Canal (for which inference there is some evidence). However, in relatively recent geological epochs there was an open Panama seaway, and it seems very probable that forms such as *Linckia guildingii* achieved a wide tropical distribution prior to the elevation of the Panama isthmus. This, and related problems of marine biogeography, may be solved when we have accumulated sufficient information on the habits of echinoderms of the coral-reef zone.

SEA CUCUMBERS (HOLOTHUROIDEA)

The systematics of the Holothuroidea relies largely upon internal structure, requiring dissection to determine the family and, generally, an examination of the microscopic calcareous platelets in the skin to determine the genus and species. Useful keys are given by H. L. Clark in *The Echinoderm Fauna of Australia* (Carnegie Institution, Washington, D.C., Publication No. 566, 1946). This book includes the temperate and offshore species as well as tropical reef and lagoon forms. Another work by Clark, which covers the tropical forms in more detail but without keys, is his *Echinoderm Fauna of Torrest Strait* (*ibidem*, No. 214, 1921), including photographs, paintings, and a biogeographic analysis. The following notes relate to the genera that are usually most conspicuous on tropical reefs and shores.

Following characters are of systematic value: **Tentacles** are the feeding organs that form a ring around the mouth at the anterior end of the body. When a living holothurian is touched it usually withdraws the tentacles. To see their structure, either observe them in life in an undisturbed pool or aquarium, or else anaesthetize the animal slowly by placing some dissolved magnesium chloride in a dish of seawater and then placing the retracted specimen in the dish, when the tentacles will slowly be extended (after which the specimen may be preserved); use 70 percent ethyl alcohol, not formaldehyde, as the latter dissolves the skin ossicles. The tentacles may branch dendritically (order Dendrochirotidae), have only a few fingerlike branches (Apodida), or have shield-shaped terminations (Aspidochirotida). **Tube feet** may be (1) unmodified suctorial structures (pedicels), in such case most likely occurring on the two longitudinal ambulacra that are ventral (lowermost in the natural position); (2) they may be converted into conical protuberances (papillae), which are most commonly found on the three ambulacra that are, respectively, left, dorsal, and right on the upper side of the body; or (3) they may have atrophied, as in the Apodida. The **calcareous platelets** in the skin assume particular shapes in the various orders and families, such as anchors, sigmoids, tables, perforated plates, and spires. The **gonads** are usually tufted masses of tubules in the coelom; the coelom is separated into right and left sides by a vertical dorsal mesentery. Using these characters, the following genera may be noted:

Tentacles branching dendritically (order Dendrochirotida).

- 10 tentacles. Pedicels demarcate 2 ventral ambs; dorsal papillae usually replacing tube feet of upper 3 ambs. Body wall rather rigid. *Pentacta.* This genus is mainly represented in tropical Australia.

Tentacles peltate (shield-shaped at tips). Pedicels and papillae developed (order Aspidochirotida).

- Single (left-hand) gonad on left side of dorsal mesentery (family Holothuriida).
  - 5 calcareous plates guard the anus *Actinopyga*: black or red-brown species, ranging the tropics; size to 30 cm long; Bêche-de-mer.
  - No anal plates. *Holothuria.* Very numerous species, gray, brown, black, or white; length to 40 cm. Bêche-de-mer. Pantropical.
- Left and right gonads on either side of mesentery (family Stichopodidae).
  - Body-wall calcareous deposits include perforated plates or plates with 3 or 4 legs (tables). *Stichopus.* Otherwise resembling *Holothuria.* Cosmopolitan. Length usually 20–30 cm, but one species *S. variegatus* reaches 90 cm. Not used as bêche-de-mer.
  - Body-wall with only amorphous calcareous granules (otherwise resembling *Stichopus*). *Thelenota* with one or two Indonesian species (which may reach same size as *S. variegatus*). *T. ananas.* Red, used for bêche-de mer, and known as redfish.

Tentacles small, inconspicuous, digitate. Body elongate, snakelike, benthic by day, nektonic by night. Length to 1 m or more. Banded. Sluggish. *Synapta* (order Apodida).

24

NORTH ATLANTIC SHORES

Northern coasts of submergence / Isostatic glacial rebound / Southern prograded coasts / Barrier islands / Gulf Stream / North Atlantic drift / Labrador current / Environmental stress / Winter dieback / Rocky substrates / Midtidal autotrophs and primary and secondary heterotrophs / Commensalism / Lower intertidal zone autotrophs and primary and secondary heterotrophs / Purple whelks and predatory crabs / Continental shelf / Neritic autotrophs and primary and secondary heterotrophs / Sand and silt substrates / Biota of midtidal zone / Biota of lower tidal zone / Biota of continental shelf / Northern Hemisphere fishes

The general physiographic features of coasts are set out in Chapters 7 and 8, which should be read in conjunction with this section. They are addressed more particularly to readers on the eastern seaboard of North America. Many of the genera also occur on the northwest coasts of Europe, where similar ecological associations are encountered.

Some special features of the eastern coast of North America may briefly be noted. Two very distinct types of shoreline occur in the region. (1) A **northern section,** extending as far south as Cape Cod, is characterized by numerous sea sounds penetrating drowned river or glacier valleys and is bounded by long, slender, hilly peninsulas. This is a typical *coast of submergence.* (2) A **southern section,** from Cape Cod to Rio Grande do Norte, is characterized by the presence of barrier beaches in the forms of long, nearly straight, sandy islands, lying off-shore and enclosing brackish to salt lagoons. This is a coast of relative tectonic stability, but it is subject to constant erosion and progradation as the local longshore currents and onshore winds may determine.

In the far northern part of the region occur emergent coasts undergoing **isostatic postglacial rebound.** By this term is meant a readjustment of land level with respect to sea level. The region of Labrador and Greenland have only during the past few millennia become partially ice free, following the ice melt at the end of the last glacial stade. Freed from the superincumbent loads of ice that had depressed the land below its normal level, these coasts are now rising out of the sea. Thus they present the aspect of coasts of submergence (which they are), but are actually in process of emergence. Norman Brink (1974) has recently measured the rate of isostatic rebound for Greenland, and finds that it apparently rose 105 m in the first 1000 years after the ice melted. For the past 4000 years, Greenland has been rising much more slowly, at a rate that decreases exponentially, with a recovery half-life of ca. 960 years.

Because the rocky-shore habitat generated by wave action against the seaward peninsulas of the northern section presents numerous sheltered sections and sandy bays between the rocky headlands, the fauna and flora of the northern section tends to be more varied and abundant in individuals than is the case with the southern sandy coasts. Sandy coasts offer few footholds for the algae that supply much of the primary production on a strandline, and the dependent faunas are impoverished accordingly. On the other hand, the higher ambient temperatures of the southern section support a wider variety of warm-water species, culminating in the rich and varied fauna of the subtropical Floridian region. However, in this book the reef faunas are considered tropical, and the Floridian ecology is not included in this chapter.

Two ocean currents affect the region: (1) a cold southern **Labrador current,** which passes inshore along the New England and Atlantic states as far as Cape Hatteras, where it narrows and sinks beneath the surface to be overlain by (2) the warm northern **Gulf Stream,** passing inshore from the Caribbean along the coasts of Florida, Georgia, and the Carolinas, to strike offshore from Cape Hatteras, thereafter lying to the seaward side of the Labrador current, eventually to cross the North Atlantic toward Northern Europe. It is then known as the **North Atlantic drift.** The effects of these two currents are to keep the ambient temperatures of the more northern coasts low, with sea ice in winter sometimes forming on the shore as far south as Massachusetts, and to keep the southern sections relatively warm. Off North Carolina the inshore water is a narrow tongue of cold Labrador water, separating the coast from the offshore warm current. In the region of Newfoundland the warm Gulf Stream, as it becomes the North Atlantic drift, encounters the much colder Arctic water mass, which it overrides on account of the lower density of warm seawater. Here the moisture-laden air from the south, overlying the Gulf Stream, is chilled, and the moisture is yielded as fog. Fogs may also form at sea off the Massachusetts offshore islands.

The most notable feature of the climate of the northern section is the great disparity between summer and winter surface temperatures. The air temperatures (and to a lesser extent the surface sea temperatures) lead to extremes of climate, with Arctic cold in winter and tropical warmth in summer. These drastic seasonal reversals lead to mass mortality of shore animals in the winter; the summer populations are rebuilt from inshore migrant stock derived from deeper levels of the shelf. However, these deeper levels are relatively cool (ca. 5°–10°C) and support mainly a boreal fauna of relatively few species but numerous individuals, so the northeastern ecology of the United States is much less varied than that of the other coasts. The winter dieback also means that the summer populations are mainly young, and hence small, specimens. Thus the New England littoral fauna presents a decidedly impoverished and sparse aspect when compared with the Californian and Oregonian littoral fauna in the same latitudes on the west coast. Sanders considers that the northwest Atlantic ecology is that of a littoral zone under severe environmental stress.

Winter storms are occasionally very severe. The waves of the Atlantic may hurl large boulders to heights of 60 feet, lodging them in clefts or even in buildings on cliff edges. (A military structure on a Nahant cliff was on one occasion filled with boulders flung through the window in a storm.) At these times the littoral population is ground to pulp by the swirling rocks, and the countless seastars (*Asterias*) temporarily disappear, only to be replaced in equal numbers a few days later from stock sheltering in the deeper water. If animals are collected by dredging, even at such shallow depths as 15 or 20 fathoms, the samples are found to include much Arctic material.

In the following review, animals and plants are named but their systematic characters are not usually noted; these details will be found in the chapters dealing with the various phyla.

ROCKY SUBSTRATES

Most coastal areas north of New Jersey offer good examples of these habitats.

Biota of the Midtidal Zone

Autotrophs

A number of conspicuous marine algae occur in this zone, and they, together with the diatoms and other microscopic plants, comprise the base of the trophic pyramid of this part of the shelf ecosystem. Following are notable members of the community on New England rocky foreshores.

Chlorophyta or Green Algae. Most abundant in this zone are members of the family Ulvaceae or sea lettuce, forming crinkly thin green expansions reminiscent of lettuce leaves. Two genera occur: a slightly thicker form, two-cells thick, called *Ulva* (most common species, *Ulva lactuca*), and a thinner, more delicate form with only a single layer of cells, now placed in a separate genus *Monostroma*.

Phaeophyta or Brown Algae. Among the genera common on New England rocky coasts are the following intertidal forms. *Punetaria* (*P. latifolia*) has a flat straplike brown thallus and is about 5 to 20 cm long and about 1 or 2 cm wide. It usually tapers at the tip, sometimes with the thallus bifurcated. It grows attached to the shells of mussels in intertidal rock pools. Another brown alga that grows on mussel shells in the same situation is *Scytosiphon lomentaria*, forming very slender grasslike filaments up to 20 cm long, and a few millimeters wide. Best known of all the North Atlantic brown seaweeds are the fucoids (family Fucaceae), with two common genera, *Fucus* and *Ascophyllum*. The latter genus, with a New England species *A. nodosum*, forms long slender cords about 30 or 40 cm long and about 0.5 cm wide, with hollow floats developed as part of the fronds at intervals of 10 cm or so. *Fucus* is commonly represented by two sympatric species, *F. edentatus* and *F. vesiculosus*. Both species have branching, bifurcating thalli, like fronds about 15 cm or more across, the latter with conspicuous bubblelike floats developed in the thallus at intervals on the tips of some branches.

Rhodophyta or Red Seaweeds. These algae grow mainly at greater depths, but one mosslike dark purple form, *Polysiphonia lanosa,* is found in the midtidal zone where it grows as an epiphyte on *Ascophyllum.*

Primary Heterotrophs

Several species of the winkle genus *Littorina* (family Littorinidae) are here encountered (family characters, Neogastropoda resembling the tropical Neritidae, but having a horny operculum, and lacking the heavy thickening of the columella, semiamphibious herbivores of temperate and cold seas, with some species in warm temperate mangrove sloughs, climbing the stilt roots). *Littorina littorea* occurs on both sides of the Atlantic, but seems to be a recent immigrant to eastern North America; it extends south to New Jersey, and ranges northward through eastern Canada, Greenland and into northern European waters; eaten by the Viking explorers, who seem to have carried stocks of live periwinkles in their ships; however, fossil specimens of *Littorina littorea* have been found in eastern Canada dating from before the Wisconsin stade. This implies that the Vikings did not introduce the species here. Limpets (family Acmaeidae) occur, but are much more common on west American shores. (See Chapter 25, where reference is also made to the territorial behavior of marine-grazing mollusks.)

Filter feeders may be included under this head, particularly the common blue mussel, *Mytilus edulis,* that lives in great banks or sheets, attached by the spun byssus threads secreted by the otherwise functionless foot, and the intertidal barnacles *Balanus* spp. (Crustacea, Cirripedia).

Secondary Heterotrophs

Associated physically with *Mytilus* is commonly found a commensal crab, *Pinnotheres* (family Pinnotheridae, Crustacea, Decapoda). A large five-armed sea star of the family Asteriidae is characteristic as the dominant invertebrate predator of this zone in the Northern Hemisphere—*Asterias,* with *A. forbesi* and *A. borealis* on American Atlantic coasts, and *A. rubens* on European coasts.

Dominant vertebrate predators of this zone comprise mainly gulls, such as the herring gull (*Larus argentatus*), the oyster catchers (*Haematopus* spp.), and other coastal waders; also man, who, since Paleolithic times, has systematically hunted the intertidal zone for food, and still does.

Commensalism. The commensal association of *Mytilus* and *Pinnotheres* mentioned above is one of several such cases occurring in marine communities. An association exists between hermit crabs and sea anemones, but until recently no adequate explanation was forthcoming.

Naturalists have long found entertainment in the curious antics of hermit crabs —those ungainly and timid occupants of the discarded snail shells. As long ago as the second century A.D., when a Greek named Oppian wrote a book on fishing, he included a lively (and rather accurate) account of the housing problems of hermit crabs that outgrow their apartments, of the ensuing competition for a new tenement, and also of the danger of octopus attacks.

It has also been known for many years that some hermit crabs actively secure a sea anemone and transfer it to the top of the shell currently occupied by the crab. Why? Previous guesses have been that the presence of the anemones somehow makes the crab's house look more like an ordinary bit of seabed, hence less interesting to other denizens devoted to crabmeat. Writing of experiments he conducted with naively cooperative octopuses, D. M. Ross (1971) has now shown that a hermit crab with a domesticated anemone on board has an excellent—in fact about 100 percent—chance of not being eaten by an octopus (and an octopus finds hermit crabs nearly irresistible).

An octopus is singularly sensitive to the stinging cells of anemones. One octopus that incautiously took a hermit crab plus shell plus anemone into its mouth subsequently turned very pallid, and its respiration rate doubled, while ominous blisters appeared on its mantle. The trouble with that hermit seems to have been that it carelessly allowed itself to outgrow its shell, so that its abdomen was showing; this was too much for the hungry octopus to resist, in spite of the painful deterrent. Ordinary careful household management will enable a hermit crab to get rid of an octopus after it makes its first unpleasant contact with the captive anemone. At all events, that is how it works out with the hermit *Dardanus*, which specializes in the anemone *Calliactis*.

The hermit who came to a dreadful end, despite the respiratory problems of his aggressor, blisters and all, was a species of *Pagurus* and his tame anemone was a species of *Adamsia*. It could just be that a better choice of anemone might have given a different result. When hermits without anemones were shown to a hungry octopus, as many as 12 attacks were made within six hours, and none of the victims survived longer than 89 hours of confinement with the octopus. On the other hand, when hermits were similarly exhibited with *Calliactis* on the shell, the attacks were equally numerous but totally unsuccessful; all hermits remaining happily alive after periods up to 150 hours. It would seem therefore that by carrying *Calliactis* on its shell, *Dardanus* is protected.

Biota of the Lower Intertidal Zone (Below Average Low-Tide Level)

Autotrophs

Some genera of Rhodophyta are sometimes conspicuous at this level, where the bottom is only occasionally exposed at low tide. These include *Rhodymebia*, a red fan-shaped seaweed with an expanded, flat thallus that is subdivided into about half a dozen bifurcated flat lobes. It is somewhat ragged at the extremity, with the whole fan about 20 cm across, and a rather variable shape. Sometimes it is more elongated and straplike. In the lower intertidal zone, and also in rockpools, a much smaller red seaweed occurs, which is about 6 cm across the frond, is bifurcated several times, and is purple-red. This is *Chondrus crispus*, the edible sea dulse. Another red seaweed of this zone is the grasslike *Dumontia incrassata*, dark red when alive, but fading to white on preservation or when cast ashore and dried up. A mosslike feathery red seaweed is *Chondria* spp. found at a somewhat higher level, usually in rock pools. *Porphyra* closely resembles sea lettuce in shape and size, but differs conspicuously by its red-purple color.

Laminarians (Phaeophyta) may occur in this zone, but more typically belong to the deeper zone described in the next section.

Primary Heterotrophs

Sea urchins of the order Echinoida include two genera, *Echinus* and *Strongylocentrotus*, that are the most conspicuous herbivores of the zone, together with another sea urchin *Arbacia* (Fig. 24.1) of the order Arbacioida. The latter contributes an important genus *Tetrapygus* to the faunas of Chile and Peru. *Arbacia* is mainly a North American genus, although it has species in Europe too, in the warmer temperate waters. *Arbacia punctulata* ranges the eastern North American coast northward to Cape Cod. In the colder waters *Stronglyocentrotus droebachiensis* occurs on either side of the Atlantic, generally distinctive from its greenish color and polyporous amb-plates. *Echinus*, with *E. esculentus*, is conspicuous in Europe; like most species of the genus, this one is red in color. No shallow-water species of *Echinus* occur on the American side of the Atlantic. The regular sea urchins are more or less omnivorous but most of the food they encounter in the intertidal zones and subtidal seafloor comprises large algae; such smaller inverte-

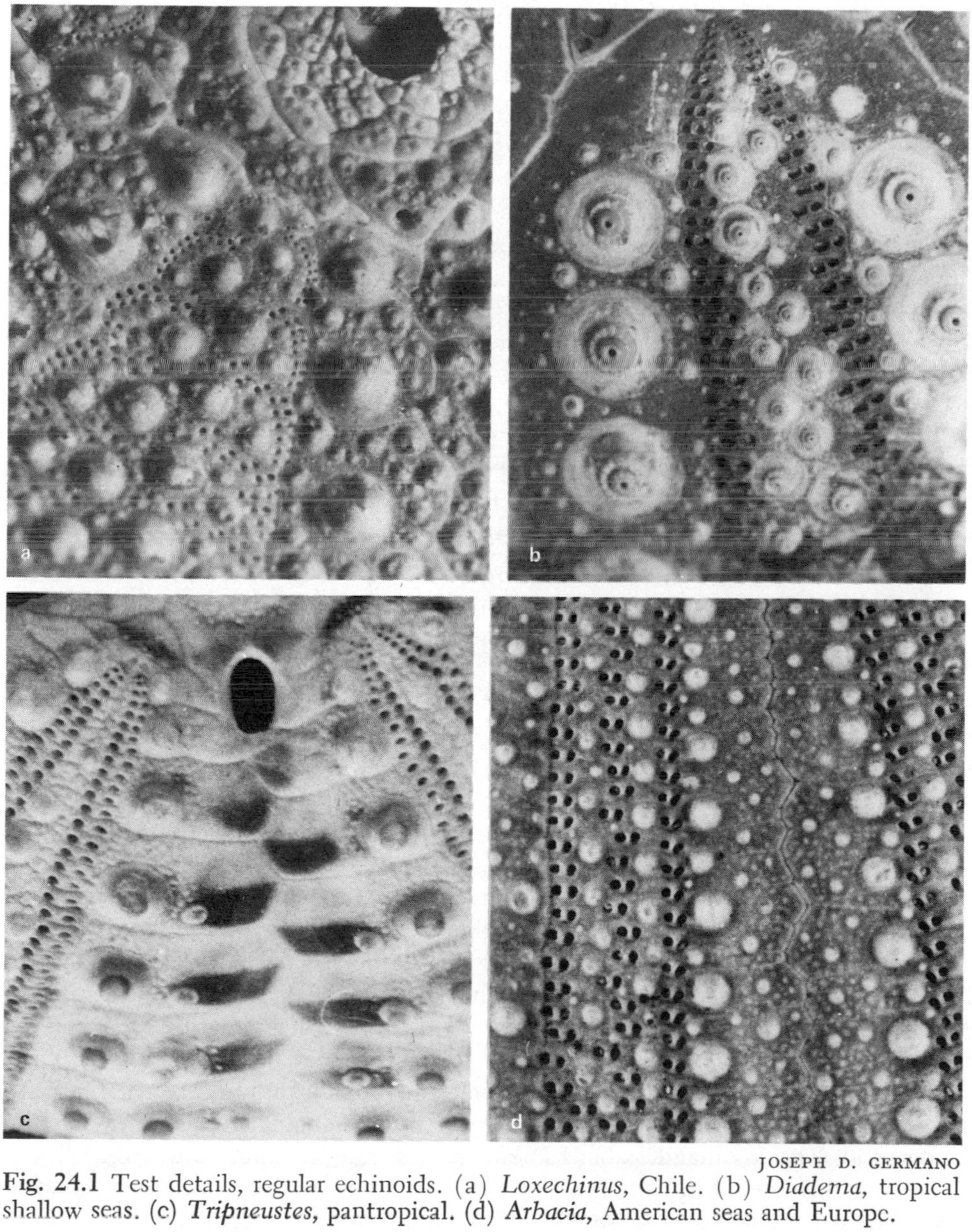

JOSEPH D. GERMANO

Fig. 24.1 Test details, regular echinoids. (a) *Loxechinus*, Chile. (b) *Diadema*, tropical shallow seas. (c) *Tripneustes*, pantropical. (d) *Arbacia*, American seas and Europe.

brates as happen to fall their way are also taken, and sea urchin stomachs contain varied assortments of worm tubes, small mollusk shell fragments, bits of paper and string and bitten fragments of the substrate, and also occasional man-derived rubbish.

Other primary heterotrophs of the zone include limpets, *Acmaea,* slipper limpets, *Crepidula,* and barnacles of the genus *Balanus.*

Filter feeders in this zone commonly include oysters, such as *Crassostrea virginica,* attached to rocks, and some holothurians, such as *Thyone* (with tube feet scattered over the surface of the body) and *Cucumaria* (with the tube feet in five long series). The holothurians capture plankton and organic detrital particles from the water by their sticky frondlike tentacles found around the mouth; these holothurians are members of the Dendrochirotida (see page 257).

Secondary Heterotrophs

The starfish *Asterias* is also conspicuous here, and another sea star *Henricia* also occurs (this is encountered on page 302 as a widely distributed member of the Echinasteridae). (See Fig. 24.2.) The brittlestar *Ophiopholis aculeata,* characterized by having rows of minute platelets around each of the main arm plates, and usually with a variegated brightly colored body, is distinctive in northern faunas; it is, like most ophiuroids, a scavenger. *Urosalpinx* is a predatory gastropod that attacks oysters and other mollusks. In order to conserve oysters from attack by *Urosalpinx,* man has in recent years become a government-sponsored predator on this gastropod. Another predatory snail, *Thais lapillus* (family Thaididae, Fig. 24.3, page 266), occurs here, but also spend a good deal of time exposed on upper tide-level rocks, apparently escaping the unwelcome attentions of a snail-hunting crab, *Carcinus maenas.* Both the snail and the crab range all northern Atlantic rocky shores, on both sides of the Atlantic. The common edible crabs of the genus *Cancer* also range this zone. The characters of the two genera can be recognized at a glance from an illustration or from specimens, but the familial characters (Portunidae, Cancridae) include structural details too specialized to note in this context. Two common species of *Cancer* on the American east coast are *C. borealis* and *C. irroratus;* the former has relatively larger chelipeds. Most of the larger invertebrate predators and the sea urchins are liable to attack by gulls. The starfish *Asterias,* however, seems to be immune from attack, presumably on account of the venom in the pedicellariae. (A dog which was seen to eat an *Asterias* later died.) Man as predator in this zone catches oysters, sea urchins, and crabs.

Thais lapillus, mentioned above, is a common intertidal sea snail on the rocky coasts and is related to the purple whelks or Muricidae, the famed purple whelks of the Phonecians. Some disrespectful commentator (was it Tacitus?) has claimed that the imperial Roman courtiers kept their distance from the emperor not so much from awe as out of a healthy regard for the *odeur de mollusque* emanating from the Tyrian robe. To judge by the shell heaps left by ancient fishermen on Mediterranean shores, the species employed in the dye manufacture were *Thais haemastoma* and *Murex trunculus.* Many other species of the family also secrete a fluid, which on exposure to the air becomes an intense purple. One of them is *Thais lapillus,* shared by North America and Europe, ranging from Norway south to Portugal. The woodland Indians knew about the reddish-purple dye that it secretes, and they used it in their decorative arts. A similar species, *Thais emarginata,* is extremely common on rocky shores of the west coast from Alaska to Mexico. The

GEORGE PUTNAM

Fig. 24.2 Sea stars of the North Atlantic. (a) *Hippasteria*. (b) *Leptasterias*. (c) *Henricia*. (d, f) *Ctenodiscus*. (e) *Asterias*. (g) *Ceramaster*. See Chapter 19 for systematics.

dye is derived from a pair of anal glands, but little seems to be known of its significance in nature. There are other curious features of purple snails, too, for example, their preference for exposed wave-washed headland, their ability to spend much of their life exposed to air at low tide, and the peculiarly heavy shell with a thickened lip around the aperture that, incidentally, can be closed by a kind of lid, called the operculum.

GEORGE PUTNAM

Fig. 24.3 Gastropods of the North Atlantic. (a) *Neptunea*. (b) *Colus*. (c) *Thais*. (d) *Buccinum*. (e) *Busycon contrarium* (note sinistral coiling). (f) *Littorina*. (g) *Lunatia*. (h) *Crepidula*. (See Chapter 21 for systematics.)

Until recently it was supposed that the heavy, thick shell was a modification permitting purple whelks to withstand the abrasive action of surf and that the operculum prevented the loss of body moisture by evaporation—both modifications

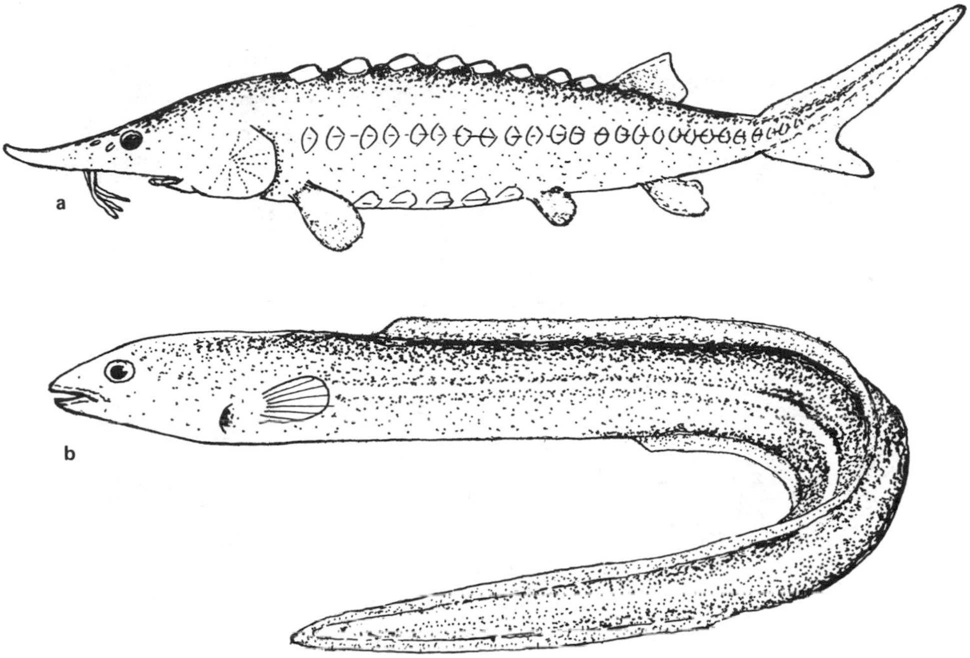

GERALD HESLINGA

Fig. 24.4 (a) Atlantic sturgeon, *Acipenser oxyrhynchus*, 1.3 m, Acipenseriformes. (b) Conger eel, *Conger oceana*, 1.3 m, Anguilliformes.

that would favor intertidal life on a rocky coast. It was further supposed that the thickened lip was a device for getting rid of excessive calcium carbonate in the body fluids. Janet S. Gibson (1970) suggested alternative explanations based on experiments she carried out on several thousand specimens of *Thais lapillus* in Wales. An initial set of tests under controlled conditions of humidity and airflow disclosed that specimens deprived of the operculum do in fact lose moisture faster than do shells which retain the structure, but the difference was surprisingly small. These experiments suggested that perhaps the operculum has some other use. Noticing that *Thais lapillus* falls prey in large numbers to the shore crab *Carcinus maenas*, the investigator determined to check whether the operculum might have some bearing in this matter. She first took 40 whelks and removed the operculum from 20 of them. All 40 were placed in fresh-running seawater; and, at the end of 25 days, all 40 of them were still alive.

Apparently the loss of the operculum is not fatal. She then took 220 whelks and removed the operculum from half. They were then placed in groups of 10 in separate tanks, half of the tanks having one crab added, and half with no crab. After 25 days, only 4 of the normal whelks had been eaten by crabs, whereas six times as many of the others had died. Evidently, therefore, operculum loss is not fatal in itself, but it is highly dangerous if a crab is in attendance. The results suggest that the chief function of the operculum is protection against predators. Gibson also found that young whelks can be killed by the crushing action of a crab's nipper, but larger specimens, in which the lip is thickened, can only be killed by a crab if it is able to force the operculum off. This suggests that the thickened lip is not so much a means of disposing of surplus calcium carbonate, as rather a means of strengthening the whole shell against crab attack.

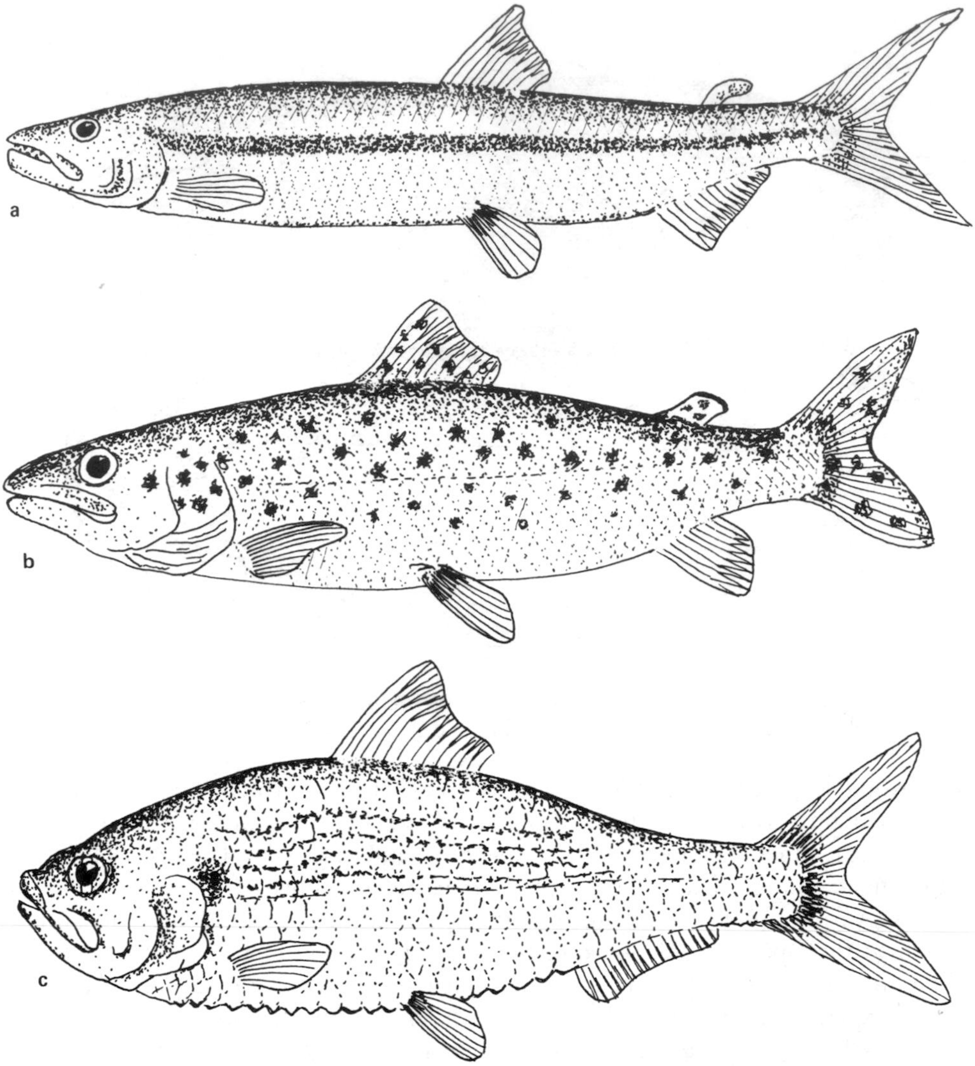

GERALD HESLINGA

Fig. 24.5 (a) Rainbow smelt, *Osmerus esperlantus*, 30 cm, Osmeridae, Clupeiformes. (b) Atlantic salmon, *Salmo salar*, Salmonidae, Clupeiformes. (c) Alewife, *Alosa pseudoharengus*, 40 mc, Clupeidae, Clupeiformes.

As for the abundance of whelks on rocky ledges above the low-tide mark, Gibson has a new explanation here too. She points out that *Carcinus maenas* moves up and down the strand twice each day, following the advancing and retreating tides. Thus, any whelk that lives near high-tide level is exposed to much less crabtime, so to speak, than a whelk living lower down the slope. Because less crabtime means more lifetime, fortune favors the whelks that clamber up the rocks and cling there when the tide goes out.

Biota of the Continental Shelf

The communities of the rocky shelf include most of the organisms mentioned in the preceding section, together with others, some of which are noted here.

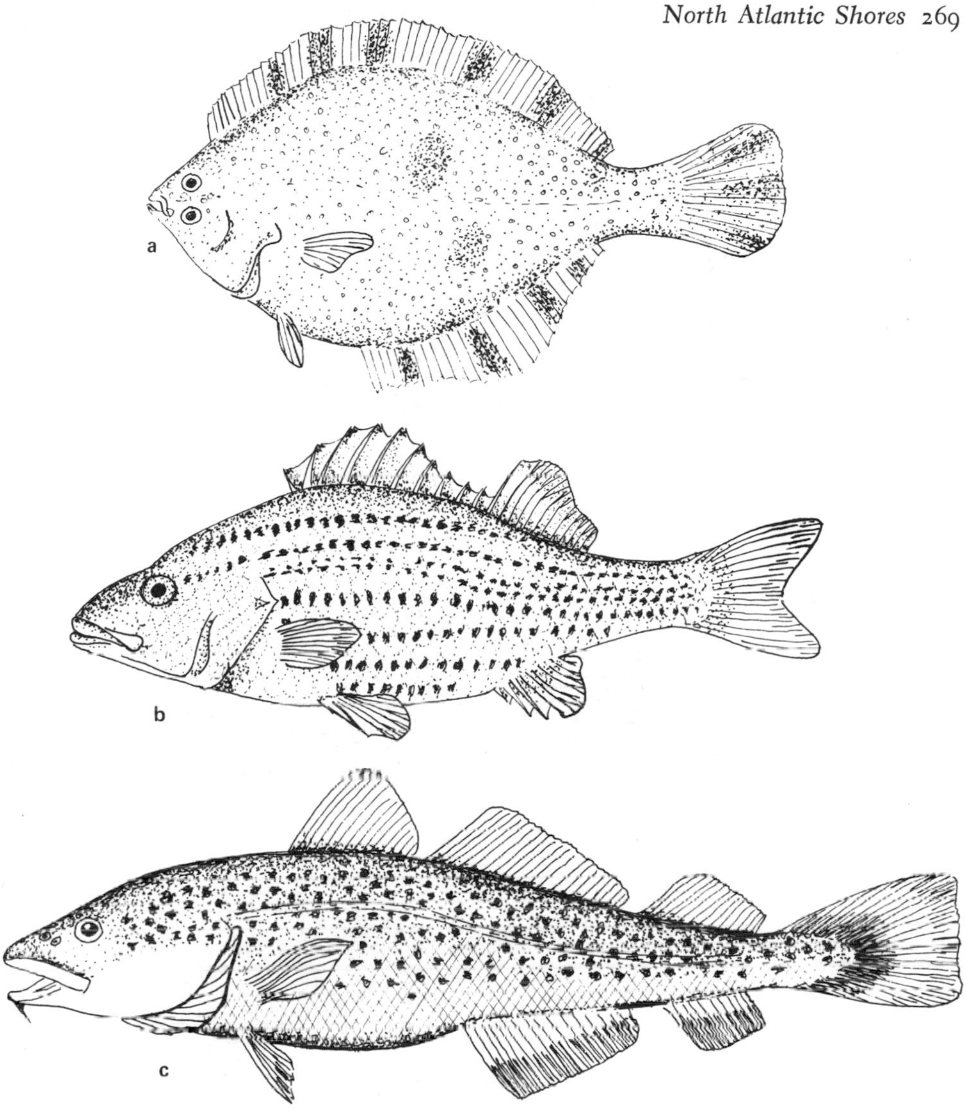

GERALD HESLINGA

Fig. 24.6 Atlantic fishes. (a) Starry flounder, *Platyichthys stellatus*, Pleuronectidae, Pleuronectiformes. (b) White perch, *Morone americana*, Moronidae, Perciformes. (c) Atlantic cod, *Gadus callarius*, Gadidae, Gadiformes.

Autotrophs

The large brown seaweeds called laminarians are conspicuous in the upper shelf region, just below the lowest low-tide zone. Chief among these in the North Atlantic are species of *Laminaria*, such as the large (1 m) brown fans of *L. digitata* and the elongate frilly undivided thalli of *L. agardhi*. The curious fronds of *Agarum cribrosum* grow in deeper water and are sometimes cast ashore by storms, with the thallus perforated by hundreds of round holes, as if peppered by grapeshot. The lower extremities of these large algae form tough holdfasts, adhering usually to a boulder. The North Atlantic species lack the numerous floats of the Pacific and Southern Hemisphere laminarians, so they do not form ocean-going rafts when they are broken from the seabed by storms.

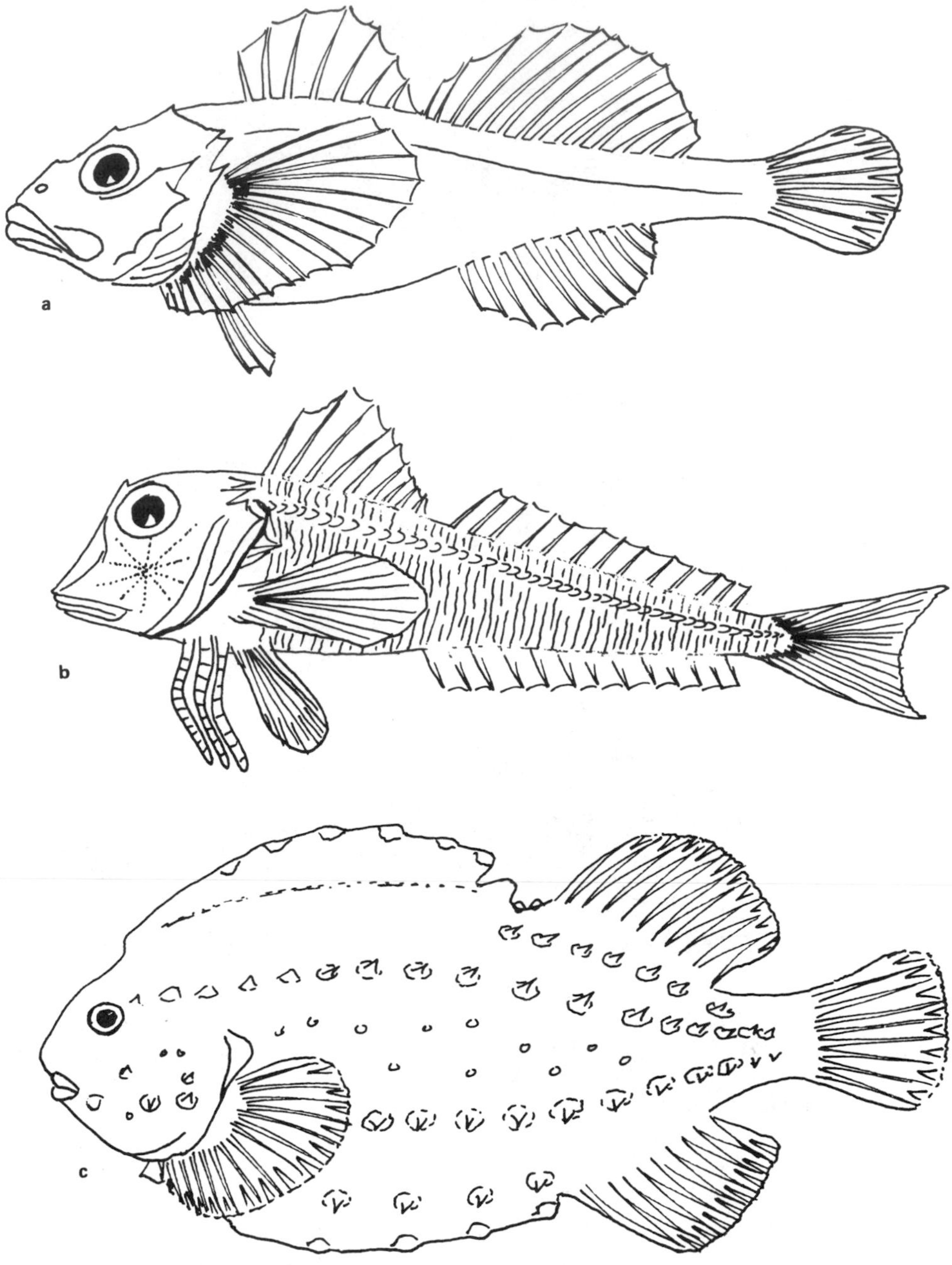

GERALD HESLINGA

Fig. 24.7 Atlantic fishes. (a) Sea scorpion or bull-rout, *Myxocephalus scorpius,* 30 cm, Cottidae, Perciformes. (b) Gurnard, *Trigloporus lastovisa,* 30 cm, Triglidae, Perciformes. (c) Lumpsucker, *Cyclopterus lumpus,* Cyclopteridae, Perciformes.

Primary Heterotrophs

A holothurian, *Psolus,* peculiar in having large calcareous plates in the body wall (an ancient character of holothurians), is found creeping or adhering to lami-

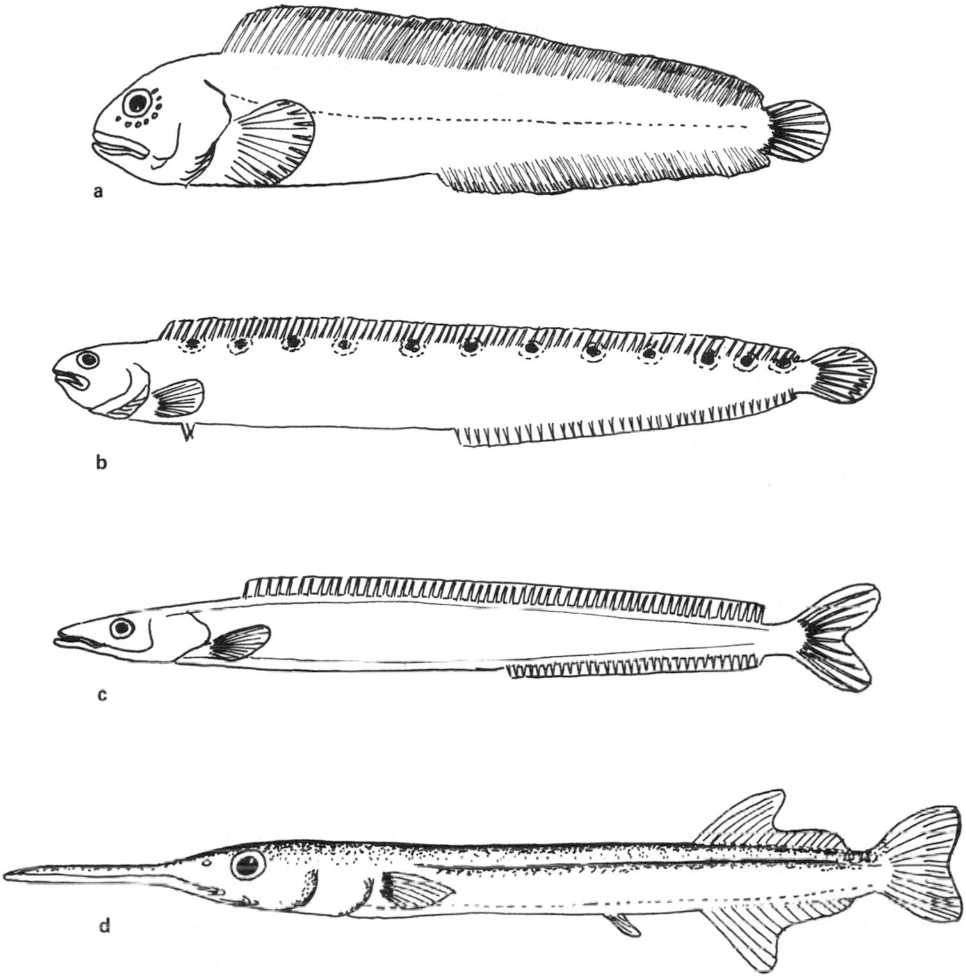

GERALD HESLINGA

Fig. 24.8 Atlantic fishes. (a) Wolffish, *Anarhichus lupus*, 1.25 m, Anarhichidae, Perciformes. (b) Butterfish, *Pholis gunnellus*, Pholidae, Perciformes. (c) Sand eel, *Ammodytes marinus*, 20 cm, Ammodytidae, Perciformes. (d) Atlantic needlefish, *Strongylura marina*, Belonidae, Beloniformes.

narians; it is a dendrochirote. Regular echinoids and other dendrochirote holothurians are conspicuous, as already noted in the previous section.

Secondary Heterotrophs

The same sea stars and crabs occur here as in the lower rocky tidal zone. In addition the lobster *Homarus* (family Nephropidae, characterized by having a pointed rostrum, and chelae on the anterior limbs) is, or was, a conspicuous predator of North Atlantic rocky seafloors. Lobsters migrate seasonally between shallow and deep-water habitats, so they may be trapped in depths beyond the edge of the shelf.

Notable gastropod predators of northern seas are the family Buccinidae, a group lacking from the southern oceans. The Buccinidae are Neogastropoda having trumpet-shaped shells, usually large with relatively few whorls, having a large aperture, usually notched anteriorly. They are active carnivores of northern seafloors.

Among the conspicuous genera of the North Atlantic are *Buccinum*, with *B. undatum* in northern European and northern American waters, also in the Arctic Ocean; *Neptunea*, with *N. decemcostata*, having the same distribution, and *Colus*, with *C. stimpsoni*, in American waters. These three genera are represented in the offshore communities in the latitude of Boston, but become progressively more shallow water in distribution as one travels northward, till in the region of northern Maine the species of the genera occur as littoral elements. These are evidently stenothermal forms, restricted to cold Arctic water and rising into the littoral zone in far northern waters where the temperatures are low enough.

SAND AND SILT SUBSTRATES

Substrates of a sandy or muddy nature commonly occur south of Boston and along much of the eastern seaboard and south of Sandy Hook. In general the fauna and flora tend to be less varied and less abundant than on hard substrates.

Biota of the Midtidal Zone

Autotrophs

A dark, bluish-green scum may be found covering tidal mud flats or adhering to occasional boulders or pilings—this is generally due to the presence of blue-green algae (*phylum Cyanophyta*). These have no specialized sex cells and reproduction is strictly by vegetative cell division; the nuclei are not differentiated, and genetic materials are scattered through the cells.

Chlorophyta. A much-branded green alga called *Codium* may occur, with cylindrical lobes of the thallus, soft and velvety to touch. Filamentous green algae, such as *Bryopsis*, may occur on soft substrates (as also on hard ones) and a creeping green, *Caulerpa*, with erect strap-shaped thallus lobes may occur.

The other algal groups are generally inconspicuous or absent. Some flowering plants, such as eelgrass *Zostera marina*, grow on muddy bottom, including estuaries and brackish-water, notably in the Baltic Sea. Beds of eelgrass continue into the subtropical and tropical region. They often form a shelter for young or adult stages of fishes (to be studied in a separate chapter). As everywhere, diatoms make up a considerable part of the biomass, especially during the spring.

Primary Heterotrophs

In keeping with the reduction in the biomass of plants, there is a general reduction in the plant-eating species of animals. A few scavenging caenogastropods eat vegetable materials, among other matter—notably the small ceriths (see page 229), such as *Cerithiopsis* and *Seila*, which live on eelgrass and other soft-bottom vegetation from Massachusetts southward. A periwinkle (family Littorinidae, page 266), *Littorina irrorata*, is found on vegetation between tides, in New England and the southern states (this latter species also occurs in hard-bottom communities, as on the New Hampshire coast at Hampton Beach). Much more important in the primary heterotroph communities of soft-bottom habitats are the various filter-feeding bivalves, which live mostly buried in mud or sand, except for their protruding siphon when the tide covers the bottom. One family, however, the Mytilidae (page 220), has the ability to live on both hard and soft bottom, and does not bury itself in the mud in soft-bottom habitats. Instead, the spun byssus threads form a tangle anchor in the mud, holding the shell with others in great sheets or masses on the surface. The other intertidal families most commonly encountered are infaunal. The following key may assist in identifying them:

A spoon-shaped tooth on hinge of left valve	Myacidae (soft-shell clams)
Not so. Shell valves elongated, gaping at each end	Solenidae (razor clams)
Not so. Highly colored, polished valves, rounded anteriorly, narrower behind	Tellinidae (tellinas)
Not so. Dull colored, weak almost toothless hinge, the animal lives imbedded in clay substrate	Petricolidae (borers)
Equivalve with 2 cardinal plus lateral teeth	Mactridae (surf clams)

On the American Atlantic coasts following are some conspicuous intertidal representatives of these families:

Mytilidae. The most common genera are *Mytilus* and *Modiolus*. *Modiolus* tends to attach its byssus to pebbles in the sand and sometimes to bury itself. Both genera range these habitats in the temperate oceans, and *Modiolus* also occurs in the Arctic.

Tellinidae. *Macoma* occurs intertidally from Arctic to Georgia, and *Tellina* is found more often at or below low tide, in the same range. They are the most colorful bivalves on North Atlantic temperate and cold coasts.

Petricolidae. As indicated in the key, these are infaunal excavators of semi-compacted soft materials, such as clay. The only conspicuous member of West Atlantic communities is *Petricola pholadiformis*, a dull-colored bivalve, about 5 cm long, that is found often in peaty substrates, as on the Hampton Beach shoreline at midtide. Note however that another family, the Pholadidae or Piddocks, have similar habits and may be locally conspicuous. Included genera are *Barnea* and *Zirfaea*. These bivalves have as the family character radial ridges making toothlike margins to the anterior margins of the valves, and they gape at either end.

Solenidae. Razor clams are recognized by their elongate body. Several genera, such as *Ensis*, up to 15 cm long and six times as long as wide; *Siliqua* and *Solen*, both much smaller and relatively less elongate; occur on sandy bottom, the dead shells often washed ashore.

Myacidae. The soft-shell clams, *Mya arenaria* and *M. truncata*, are readily recognized by the spatulalike tooth mentioned in the key. *Mya truncata*, which is the smaller species, occurs in the Arctic (Greenland coasts); both species range the eastern coasts of the United States. As an important seafood for man, these species are often sold as steamers.

Mactridae. The surf clams, with the main genus in these habitats *Spisula*, two common species. Occurring under sand in surf-racked conditions in the lower part of the intertidal zone. They also extend to soft bottoms offshore from the Arctic to southern Atlantic east American states. Often piled ashore in millions after rough inshore storms. The meat is eaten as chowder, and the large valves of *Spisula solidissima* are well known as ashtrays.

Secondary Heterotrophs

The clams are able to exist in the intertidal region because of their infaunal habit, which enables them to retreat beneath the surface when the tide goes out. However, most predators cannot behave in this way because they roam freely. Thus relatively few marine predators come up on to the intertidal sand and mud flats.

Among the few gastropods falling under this head are the Nassa snails, genus *Nassarius*, usually classified among the thaids. They are found in eelgrass on mudflats, sometimes in great numbers. They seem mainly to be scavengers.

Another scavenger, also feeding on live invertebrates in the substrate, is the horseshoe crab, *Limulus*; however, this animal does not normally frequent the intertidal region, save occasionally in the breeding season in early spring, when the animals come up from deeper water to deposit their sexual products in shallow water.

The crabs *Uca* and *Sesarma* are very mobile, and so they commonly enter the intertidal region at times when the beach is deserted in order to scavenge, especially under cover of darkness. These crabs normally inhabit salt marshes. Both genera have already been noted on pages 235–236 under tropical habitats.

Chief predators of the region are therefore the terrestrial vertebrates, including man especially, and the shore birds. However, many of the organisms of the soft-substrate intertidal region also range into deeper water, below the low-tide mark, and here they come under attack by a wider range of predators.

BIOTA OF THE LOWER TIDAL ZONE (BELOW AVERAGE LOW-TIDE LEVEL)

As indicated above, the biota include most of those already noted. To them may be added such free-roving clams as the Pectinidae here represented by genera such as *Aequipecten*. Several starfishes, notably species of *Astropecten*, may occur of soft bottom (see page 187). Some of the predatory gastropods enter the zone when the tides are at the neap, notably the large naticids (page 266), such as *Polinices* and the carnivorous moon snail. On eelgrass the bubble snail *Haminoea* may occur. In general, these animals are opportunist invaders from the shelf, which quit the region when spring low tides render the zone uninhabitable for them. Some nektonic elements similarly enter the region to graze on bottom invertebrates when tides are suitable, but quit it at other times.

Biota of the Continental Shelf Soft Bottom

Autotrophs

These are commonly restricted to the microscopic forms, unless there are scattered buried pebbles or boulders in the soft substrate. In the latter case quite extensive algal growth may develop, including even brown laminarians, but if such scattered hard substrate is missing, then little algal colonization can take place. Where substrate includes boulders or cobble partly imbedded in mud then encrusting red algae often cover the exposed portion of the boulders, and sessile invertebrates (sponges, hydroids, anemones) are likely to be found.

Primary Heterotrophs

On soft bottoms sea urchins, such as the sand dollars (with *Echinarachnius parma* on the American coasts), occur as alternately epifaunal and infaunal elements, depending upon the time of day; heart urchins of various families appear as infaunal elements, imbedded in mud or sand and extending elongated tube feet from the entrance to the burrow to obtain particulate organic debris, small organisms, diatoms, forams, etc., as food. The large hard-shell clams of the family Veneridaeoccur are found on the upper part of the shelf; these are distinguished by the thick, massive shell, the elaborately interlocking hinge teeth, the powerful hinge and counteracting adductor muscles (leaving very strong adductor muscle impressions on the interior of the shell), and by the large size and often bright purple pigmentation in the shell interior. Most notable among these are the quahog (*Mercenaria mercenaria*) of West Atlantic coasts, the object of the hard-shell clam fishery, and also used formerly as the raw material from which the American Indians cut wampum beads. During early currency shortages in the infant colony of Massachusetts, the colonists used wampum.

Secondary Heterotrophs

Because most predators are roving animals, and adapted therefore to encounter a variety of substrates, there is no great difference between the predators of the various substrates. The dominant carnivorous mollusks on the northern shelves tend to be buccinids, *Colus*, *Buccimum*, and *Neptunea* that favor both hard and soft substrates. The Asteroidea are more selective because forms such as the Astropectinidae and Luidiidae with nonsuctorial tube feet have an advantage on soft bottom; hence these two families tend to supply the genera of main carnivorous starfishes on such soft substrate. Many ophiuroidea favor soft bottom, in which the disk may be buried while the arms alone seek food material; others again creep over the surface of the bottom and rely upon their hard skeletal plates to discourage predator attack. As noted in the Southern Hemisphere (page 299), various burrowing anemones occur, and some burrowing worms are also predators.

NORTHERN HEMISPHERE FISHES

The fish faunas of the North Atlantic, and those of the North Pacific, present many similarities with those of the southern parts of the Southern Hemisphere. If we compare them with the summary of important southern families given on page 311 we find, for instance, that the following generalizations can be made:

1. Much the same family representation occurs for the various orders, although the included genera may differ considerably, and the representatives fulfill much the same roles in the ecology (mainly, of course, as predators).
2. Some important Southern Hemisphere families tend to be lacking, or only poorly represented. Notable absentees are: Galaxiidae (replaced by Salmonidae in the northern seas), Gempyllidae, Platycephalidae, Cheilodactulidae, Arripididae, and Girellidae. Some families, such as Dussumieriidae (round herrings) are well represented in the North Pacific but appear only feebly so in the North Atlantic.

3. Some conspicuous Northern Hemisphere families contribute to the fish fauna. Notable among the northern families are: the abundance of herrings; the abundance and varied nature of codfishes; presence of peculiarly northern forms such as the eellike families of wrymouths, rock eels, sand launces; the more common occurrence of eel pouts (elsewhere mainly deep-water forms); and the greater variety of the armored demersals, with several families resembling gurnards or sea robins). A peculiar fact is the presence of eels on the Atlantic coasts and their absence from North American Pacific coasts (where lampreys are called eels).

Table 24.1 is a key to the chief families of North Atlantic neritic fishes. The key is artificial (i.e., for ease of quick determination it uses superficial characters such as body shape).

Table 24.1 Key Characters of Some Important Teleost Families of the North Atlantic

Body elongate, flexible, eellike; jaws and pectoral fins present	
No pelvic fins	
Tail fin distinctly separated from dorsal and anal fins	
Body ribbonlike, head not distinctly enlarged (rock eels)	Pholidae
Head conspicuously wider than body (wrymouths)	Cryptacanthodidae
Tail fin, dorsal and anal fins continuous (conger eels)	Congridae
Pelvic fins present	
Tail fin forked (sand launces)	Ammodytidae
Tail fin tapering, not forked (eel pouts)	Zoarcidae
Body fishlike, flexible	
No fin spines	
Pelvic fins abdominal	
Jaws not forming a beak	
2 dorsal fins, the second small, without rays (adipose fin)	
First dorsal fin arises above pelvic fins (smelts)	Osmeridae
First dorsal fin arises anterior to pelvics (salmon)	Salmonidae
1 dorsal fin	
Belly with keel along ventral midline (herrings)	Clupeidae
No ventral keel (anchovies)	Engraulidae
Jaws extended to form a beak	
Both upper and lower jaws form beak (needlefishes)	Belonidae
Only lower jaw forms beak (halfbeaks)	Hemirhamphidae
Pelvic fins jugular, barbels usually present (codfishes)	Gadidae
Fin spines present	
Both eyes on one side of head, body flattened	
Eyes on right side of head (right-eye flounders)	Pleuronectidae
Eyes on left side of head (left-eye flounders)	Bothidae
One eye on either side of head	
Dorsal fin divided into 2 or more separated parts	
Dorsal fin rays form threadlike extensions (dories)	Zeidae
Pectoral fins each with 3 free spines (sea robins)	Triglidae
Not so, but head armored with spines	
Teeth large, skin prickly (sea ravens)	Hemitripteridae
Not so. Body with light and dark spots (sculpins)	Cottidae
Head not armored with spines, pectoral fin normally formed	
Soft dorsal fin followed by finlets or fringe	
No preanal free spines (mackerel, tunnies)	Scombridae
2 preanal free spines (jacks)	Carangidae
No pelvic fins (butterfishes and harvestfishes)	Stromateidae

Table 24.1 *(Continued)*

Soft dorsal fin followed by spinous fin (sea trout)	Otolithidae
Dorsal fin continuous, not divided into separate fins	
Body widest anteriorly, tapering toward tail	
First part of dorsal fin over head, as lure (anglers)	Lophiidae
Not so, dorsal fin forms long continuous band	
Dorsal fin rays rigid to their tips (blennies)	Blenniidae
Dorsal fin rays soft at tips (wolf fishes)	Anarhichidae
Body of usual fish shape, not tapering continuously behind head	
Caudal fin homocercal	
11–12 caudal rays, scales cycloid (wrasses)	Labridae
15 or more caudal rays, scales ctenoid	
Maxillary not sliding under cheek (groupers)	Serranidae
Similar, with 1–2 opercular spines, 3 anal spines	Moronidae
Maxillary sliding under cheek when mouth is closed	
Snout convex, palate toothless (porgies)	Sparidae
Snout flattened, palate toothed (snapper)	Lutjanidae
Body rigid, enclosed in bony plates, without ventral fins	
Mouth tubular (pipefishes and sea horses)	Syngnathidae
Not so. Biting mouth, body fish-shaped (cowfishes, trunkfishes)	Ostracionotidae
Body soft-skinned, swollen, dorsal fin posterior in position	
No pelvic fins (puffer fishes)	Tetraodontidae
Pelvic fins converted into a sucker (lump suckers)	Cyclopteridae

25

NORTH PACIFIC SHORES

Trend of coasts / Latitude and temperature the principal variants / Offshore cold current / Upwelling / Effect of alternating headlands and bays / Faunal provinces / Rocky shore biota / Autotrophs / Primary heterotrophs / Territorial behavior / Kelp–abalone–urchin–otter food chain / Mollusks / Echinoderms / Crustaceans / Soft substrates / Sand dollars and bivalves / Fishes

The physiography of coasts, as set out in Chapters 7 and 8, is to be read in conjunction with this section, which is addressed particularly to readers living in the western seaboard states of North America. Many of the genera noted occur also on the shore of China, Japan, and eastern Siberia.

Some special features of the west coast of North America may briefly be noted. As John Garth (1973) has pointed out the west coast of North America forms part of one continuous coastline trending northwest–southwest from Alaska to the Straits of Magellan—the longest uninterrupted coastline in the world. It is isolated by deep seas from the east coast of Asia, except in the far north where the Aleutian Island arc constitutes a land bridge open to settlement by shallow-water biota, and available therefore as a migration path in either direction.

Along the west American coast the principal variant is latitude, which may be correlated with temperature. This is an ideal setting, therefore in which to study the effects of temperature gradients on the distribution of organisms. The same comment might be made on the interaction of temperature gradient and speciation gradient. Bipolarity (Chapter 11) is exhibited, similar forms occurring on the west coast of North America compared to those occurring at corresponding latitudes on the west coast of South America. This also means that when the southern Pacific biota are considered as such (Chapter 27), the South American faunas are sometimes quite markedly distinguished from those of South Africa, Australia, and New Zealand, for the three last-named lack the American elements that Chile and Peru share with Mexico, California, and British Columbia.

A further similarity between the western North American coasts and those of western South America lies in the presence of offshore equatorward-flowing cold currents (for the California current matching the Humboldt current, see Chapter 10).

Upwelling. Cold intermediate water (Chapter 10) upwells off southern California and Baja California (again matching an upwelling off southern Peru and Chile). The predominantly westerly winds, laden with Pacific moisture suffer fog condensation as they pass over these cold areas, lead to coastal fogs and consequent inland deserts.

Marine Provinces. Naturalists recognize a boreal faunal region including the Alaskan coasts, a north temperate region extending south to about 34°N, and a tropical region extending south to include the Bay of Panama and the northern coasts of western South America. As noted elsewhere (Chapter 29), the Panamanian fauna shares matching species with the Caribbean fauna, which is one of the major reasons for believing that a Panama seaway formerly extended through the Isthmus. The continuity of the coast tends to reduce hiatus speciation and to replace it by temperature-related speciation gradients. In other words, the genetic differences tend to change more as a smooth series of minute intergradations rather than as a step function.

An alternation of rocky headlands with shallow bays has particular effects upon populations if, as on the Californian coast, a longshore cold current is present. Cold-water biota occur on the headlands; and warm-water biota are found in the embayments (where insolation raises the water temperature), thereby producing an alternation of faunules. According to Garth this condition dates from the end of the last interglacial period, warmer currents prior to that time having lessened the effect of the coastal profiles.

ROCKY SUBSTRATES

Autotrophs

The most conspicuous difference of North Pacific shelf floras from those of the North Atlantic rest in the much greater development of the laminarian algae on Pacific coasts, especially on those with rocky reefs that provide a firm attachment for the holdfast. The large bladder-kelp *Macrocystis* dominates the flora. Its species include *M. pyrifera,* an alga that has a circumpolar distribution in the southern oceans. There can be little doubt that the same genus originated in the southern oceans and was dispersed across the tropics during glacial stades of the Pleistocene, when equatorial waters cooled by some 6°C to a level tolerated by the alga.

Other large laminarians include *Nereocystis,* in which a single, large pneumatocyst carries about six or eight straplike divisions of the thallus and is itself carried at the end of a long flexible stem with a basal holdfast; and Postelsia, a palm-shaped kelp with a cluster of straplike segments developed at the tip of a robust trunklike stem. There is also the same development of smaller Phaeophyta, with Chlorophyta and Rhodophyta, as elsewhere in northern seas.

Primary Heterotrophs

With so great a development of autotrophs on the reefs, there is a corresponding abundance of herbivorous animals. The two most conspicuous groups of these are (a) the alga-eating regular sea urchins and (b) the abalones, which here reach their maximum size.

The abalones (family Haliotidae, page 222) include such massive species as *Haliotis rufescens*, the red abalone of California, up to 30 cm long and 20 cm broad; *H. fulgens*, the green abalone, up to 20 cm long. These are the two largest known species. Other North Pacific abalones include *H. kamtschatkana* of Japan and Siberia as well as western North America. Like all Archaeogastropoda, the inner nacreous layer of the shell is well developed, although the colors of the northern species do not equal the opaline and sapphirelike quality of *H. iris* (see page 306). The Californian abalones are subject to legal conservation measures, following the drastic depletion of stocks last century; but the Mexican beds are fished for the oriental market and are headed for rapid extinction.

The sea urchins that comprise the other main herbivore assemblage feed upon the kelp beds of the North Pacific. They nearly all belong to the genus *Strongylocentrotus*, with about 12 North Pacific species, about half a dozen on the Asian side, and the rest on the Alaskan, Canadian, and American coasts. The genus has its major development in the North Pacific, and it is evident that the two species of the North Atlantic are relatively late emigrants from the North Pacific, by way of the Arctic Ocean, around whose coasts the emigrant species also occur. The Californian species include some very large and spectacular long-spined sea urchins, notably *S. franciscanus* and *S. purpuratus*; these occasionally yielding specimens some 30 cm across the spine span, which are among the largest regular urchins in the world. Indeed, this predilection to large size characterizes many of the Californian marine invertebrates, probably due to the stability of environmental temperatures throughout the year. No winter kill or dieback occurs, in contrast to the violent climatic extremes of the New England coasts, where the winter kill is so severe that stocks must be renewed each spring by migration from deeper water.

An interesting food chain exists, connecting the kelp, the sea urchins, the abalones, and the sea otter. During the recent period of extermination of sea otter, the sea urchins increased in number (having no natural predator any longer) and attacked the holdfasts of the kelp, which drifted out to sea, thereby reducing the abalone population. The return of the sea otter to Monterey has meant that the sea urchin population is now being controlled, the kelp beds are regenerating (so far as oil spills permit), and the abalone stocks are increasing.

Other primary heterotrophs of the zone match those of the North Atlantic, and the notes on page 272 apply here, except that the common North Pacific littorinids are *Littorina scutulata* and *L. planaxis*, and the filter-feeding mytilids are *Mytilus californianus* (ranging from Aleutians to Socorro Island, and Mexico). In addition species of the turbinid *Tegula* are extremely numerous, and chitons are conspicuous on west American coasts. The largest and most spectacular chiton is *Amicula stelleri* (formerly placed in the genus *Cryptochiton*), reaching 30 cm in length, which has valves hidden in a leathery reddish-brown skin. It ranges Japan, Alaska, and south to southern California. Chitons of the genus *Mopalia* reach a length of about 5 cm and usually have bristles or hairlike structures on the girdle that surrounds the central eight valves of the shell. Another large and striking member of the herbivorous mollusks is the great keyhole limpet, *Megathura crenulata*, up to 10 cm long; and there are other species. For general characteristics of these Fissurellidae see page 222; and for a detailed account see R. Tucker Abbott, *American Seashells* (1954), which also discusses other families that are noted here briefly.

Of the true limpets, family Acmaeidae (page 222), the most notable west American genus is *Lottia*, with *L. gigantea* the owl limpet, to 10 cm long, ranging the rocky foreshores of most of California. Several species af *Acmaea* range from Alaska to Baja California.

Territorial Behavior. Limpets exemplify grazing organisms that take possession of a territory and defend it. Noticing that apparently each owl limpet on the Eagle Canyon coast occurs near the center of a foot-square area of yellow-green algal film on the rocks of the intertidal zone, John Stimson of the University of British Columbia, decided to investigate why no other animals seemed to be present on these areas. His conclusion, in short, is that each owl limpet grazes its own territory and drives off intruders that might otherwise compete for the alga harvest. One of his first experiments was to stain the alga film with a harmless blue dye. Then, by keeping a watch on each limpet occupier, he found that the tooth marks made by scraping actions of any feeding limpet never extended beyond the confines of the bare territory it occupies. Evidently each limpet harvests its own field, but not beyond it. Stimson next introduced intruder limpets of a different species and discovered that, whenever *Lottia* encountered an intruder, it paused briefly and then set about shoving it off the property by sheer bulldozery (a well-grown *Lottia* is large enough to deal effectively with most fellow denizens of its habitat). Similarly, short shift was dispensed to other intruders, such as snails, anemones, and barnacles. A *Lottia* landowner is equally inhospitable to an intruder of its own species, too. In 1965 Galbraith had already noted that might is right, the larger *Lottia* pushing away the younger or smaller ones. When Stimson placed inanimate foreign objects on a territory, such as steel pegs, pencil erasers, and also his own finger, they were ignored by the owner. These and some other experiments lead Stimson to infer that a relatively large organism such as *Lottia* can only survive in its rather impoverished habitat if it defends its grazing area from all other grazers and from organisms, such as anemones and mussels, that would occupy valuable grazing space or render its employment ineffective. Predatory snails are also discouraged by pinching their feet and by similar rebuffs classified by Stimson as a defense mechanism.

Of the top shells (family Trochidae, page 223), the genus *Calliostoma* is notable (this genus also occurs in Southern Hemisphere communities, see page 222). The most common species is *C. ligatum*, the ribbed top shell ranging littorally from Alaska to San Diego. Abbott (1954) includes *Tegula* here, though its appearance is more that of a member of the Turbinidae: *T. funebralis* is the commonest species, see also page 282.

Several species of the snail *Lacuna represent the Neritidae* (page 223) on the west American coasts.

Secondary Heterotrophs

As might be expected from the exuberance of the primary heterotroph elements, the carnivorous predators of the North Pacific rocky coasts are similarly conspicuous and varied and also very abundant. The fishes and sea otter have already been mentioned (pages 151 and 182); following are notes on the invertebrate predators.

Of the Buccinidae (which are also represented in the North Pacific), *Bucci-*

num undatum does not occur on the west American coast, but other species of the same genus range from the Arctic to Washington State, for example, *Buccinum plectrum*. Similarly *Colus spitzbergensis*, *Neptunea lyrata*, and *N. pribiloffensis* are western American species of genera already noted in the Atlantic from other species. The North Pacific also has its restricted buccinids, for example, *Kelletia*, with *K. kelleti*, ranges California and West Mexico but not the North Atlantic. *Busycon* and *Melongena*, on the other hand, are totally absent from the west American fauna, although important in the east. Although *Busycon* is lacking a large left-handed buccinid, *Volutopsis harpa*, occurs in Alaska.

Of the family Cymatiidae (page 224) the genus *Argobuccinum* occurs from the Bering Sea to California; species of that genus range southward to New Zealand. The Naticidae (page 215) include *Polinices reclusianus* among the west coast carnivores matching *P. heros* of the east coast.

Asteroidae. The family Asterinidae (page 190) is represented by a fine species, *Patiria miniata*, easily recognizable by its pentagonal form with somewhat emarginate sides (Fig. 25.1). The Echinasteridae (page 191) include *Henricia leviuscula*, and the Asteriidae (page 192) includes the large many-armed sunflower starfish, *Pycnopodia helianthoides*. Also in that family are the five-armed *Pisaster ochraceous* and *P. gigantea*, which are covered by granulelike spines. All these are active predators, with a particular predilection for swallowing *Tegula* whole. Other genera occur in deeper water, particularly Astropectinidae and Goniasteridae, such as *Mediaster* (page 283). Further south in the subtropical waters *Linckia* and *Oreaster* appear, matching the species of the same genera that occur in the southeastern United States (see page 190).

Long-Range Detection of Sea Stars by Sea Snails. West coast starfish gazers familiar with *Pycnopodia helianthoides*, the sunflower starfish, may be tempted to press its central disk with the index finger, often discovering as a result that mysterious hard lumps, about an inch across, exist within its body. Further investigation shows that the lumps are caused by the presence, within the stomach of the starfish, of several dead sea snails, which are identifiable as the appropriately named *Tegula funebralis*. It should be added that the sea snail probably acquired its name from the fact that the shell is coated by a blackish horny layer, rather untidily worn out in places, to disclose the underlying nacre. *Tegula* lives in vast numbers between tide marks, so it seems to have survived the last few million years quite successfully, despite the fact that it is apparently gourmet fare for sea stars. Roger Szal at Stanford discovered that *Tegula*, and some related sea snails, have a special sense organ that seems to act as a monitoring device, warning of the presence of starfishes. It is a pocketlike infolding of part of the gill, and experiments suggest that it responds to the presence in the seawater of chemical substances secreted by starfishes. The particular starfish that was tested in Szal's experiments was the five-armed species, *Pisaster ochraceus*, but one might hazard the guess that a similar result might have been obtained for *Pycnopodia*. Much the same kind of sense organ was found afterward to be present in other sea snails, including various kinds of keyhole limpet and abalone, all of them forms whose ancestry ranges far back into the Paleozoic. Starfishes also originated in the Paleozoic, and they seem to have changed very little over the past 500 million years. So the interactions noted by Szal may also be very ancient behavior patterns.

The sense organ, which Szal calls a **bursicle**, is a pocket on each gill flap in the path of incoming respiratory currents of seawater; and each is lined by cells of

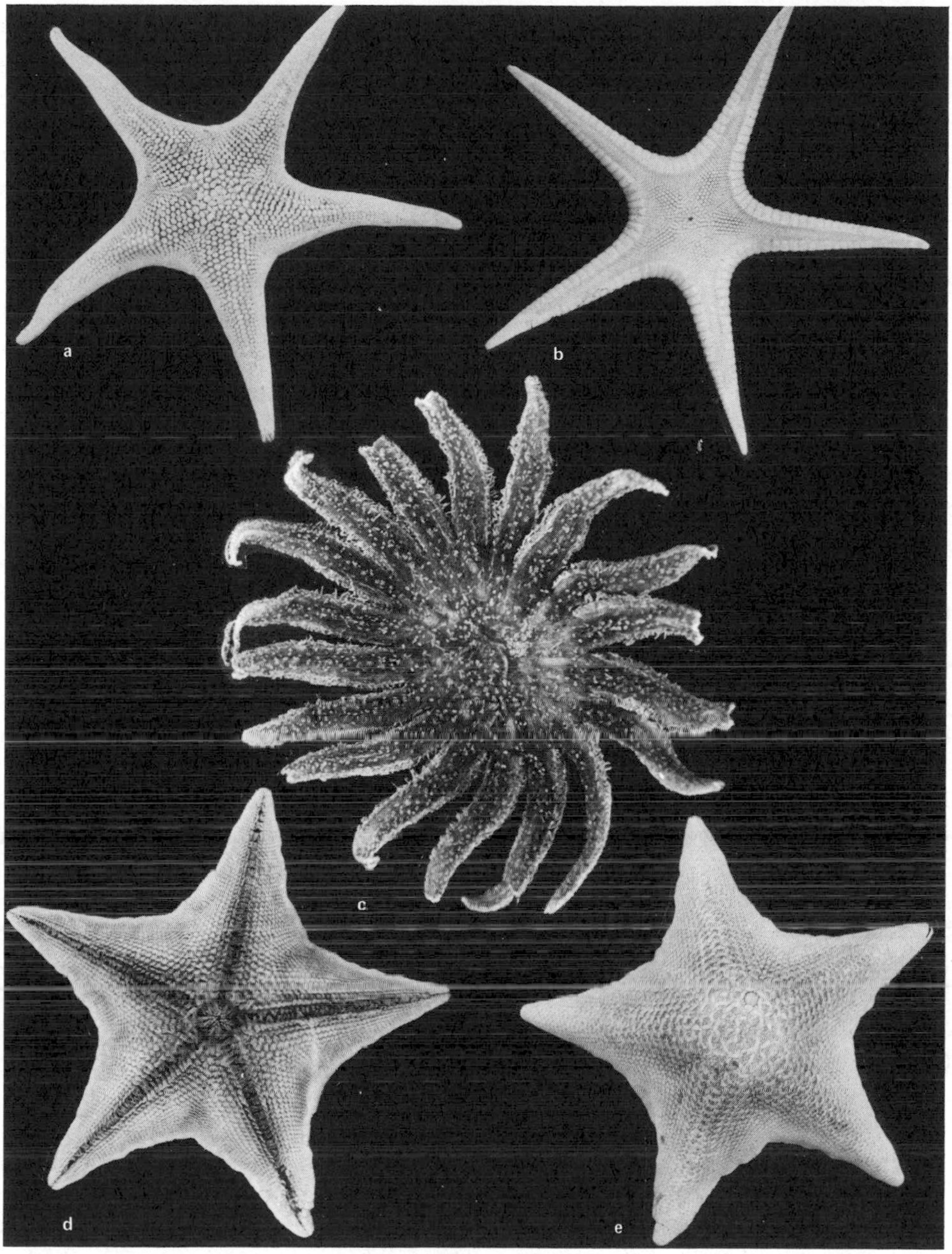

GEORGE PUTNAM

Fig. 25.1 Californian sea stars. (a) *Mediaster* (genus also ranges warm seas). (b) *Pseudarchaster* (genus also ranges warm seas). (c) *Pycnopodia helianthoides*. (d) *Patiria miniata*.

several types. One type traps incoming particles and ejects them; another type of cell sends extensions into a band of nerve fibers adjacent to the bursicle. It is the latter that Szal supposes to be the detectors, reporting the presence of chemical substances exuded by starfishes in the vicinity to the sensory nervous system. When water carrying the scent of the predator *Pisaster* is allowed to flow into an aquarium containing *Tegula funebralis*, the snails respond by rapidly crawling out of the water.

Crustacea Decapoda. The North Pacific lobsters belong to the family Palinuridae (page 235), lacking the large chelipeds of the Atlantic *Homarus* and having conspicuously enlarged antennae. The common west American species is *Panulirus interruptus*, and related to it are several Japanese species of the same genus, such as *P. japonicus, P. ornatus, and P. fasciatus.* The Asian lobsters also include a smooth-carapaced form, *Linaparus trigonus* (note that genera of this family have anagrammatic names formed from *Palinurus*, the Langouste of European waters). Prawn killers (family Scyllaridae, page 235) are represented by the genera Scyllaridaes (body rectangular in outline), *Thenus* (body widest anteriorly, gently tapering behind), and by two other genera that have a strongly serrated margin to the plates. Of these, one, *Ibaccus*, is about as broad as it is wide, and the other, *Parribaccus*, is twice as long as wide.

The crab fauna of west American coasts is extensive. The Grapsidae, shore crabs, are the dominant family. The most conspicuous genus on rocky substrates is *Pachygrapsus*, and on gravel (and mud substrates) is *Hemigrapsus*. In southern California and Central American Pacific coasts the tropical genera *Grapsus* and *Geograpsus* live on rocky substrates.

On the cold northern shores and shelves the stone crab or giant spider crab *paralithodes* occurs, with *P. camtschatica* the object of a destructive fishery. In the true spider crab family, Majidae, *Oregonia*, *Hyas*, and *Chionoecetes* range the northern American coasts as well as the corresponding coasts of Asia. On the *Macrocystis* beds of southern California *Taliepus* occurs, with species also on rocky kelp-bearing coasts of Chile. Some genera are amphiamerican with similar species on both the Pacific and Atlantic coasts, such as *Mithrax* and the tropical *Acanthonyx* frequenting floating sargasso weed. *Pugettia* ranges the cold coasts of northwest America; similar species are found in Japan. The related family of spider crabs, Partenopidae, is represented in southern California by *Heterocrypta*.

The Cancridae are represented by about nine species, including the so-called edible crab, *Cancer magister*, now the object of fishery off San Francisco. The Xanthidae are found as far north as Monterey, with *Cycloxanthops* and *Paraxanthias* extending south to tropical America. Species commensal with the corals *Poecillopora* belong to the genera *Domecia* and *Trapezia*. In the western tropical North Pacific *Tetralia* occurs on *Acropora*, but neither the host nor the crab range the American coasts.

Other carnivorous invertebrates that are conspicuous on west American coasts are the shell-less sea snails, or Nudibranchs, which are upper trophiclevel forms, feeding mainly on sea anemones (themselves also secondary heterotrophs). Some nudibranchs are brilliantly colored, and some arm themselves with the nematocysts that they acquired from their anemone victims. The sea anemones are also highly differentiated predators on west American coasts. In general they resemble ecologically the southern Pacific anemones (page 299).

SOFT SUBSTRATES

These yield communities generally resembling those of the North Atlantic but more varied in content.

The sand dollars are best known on the temperate west coast of North America from species of *Dendraster* (Fig. 25.2), a genus that lacks lunules and has the

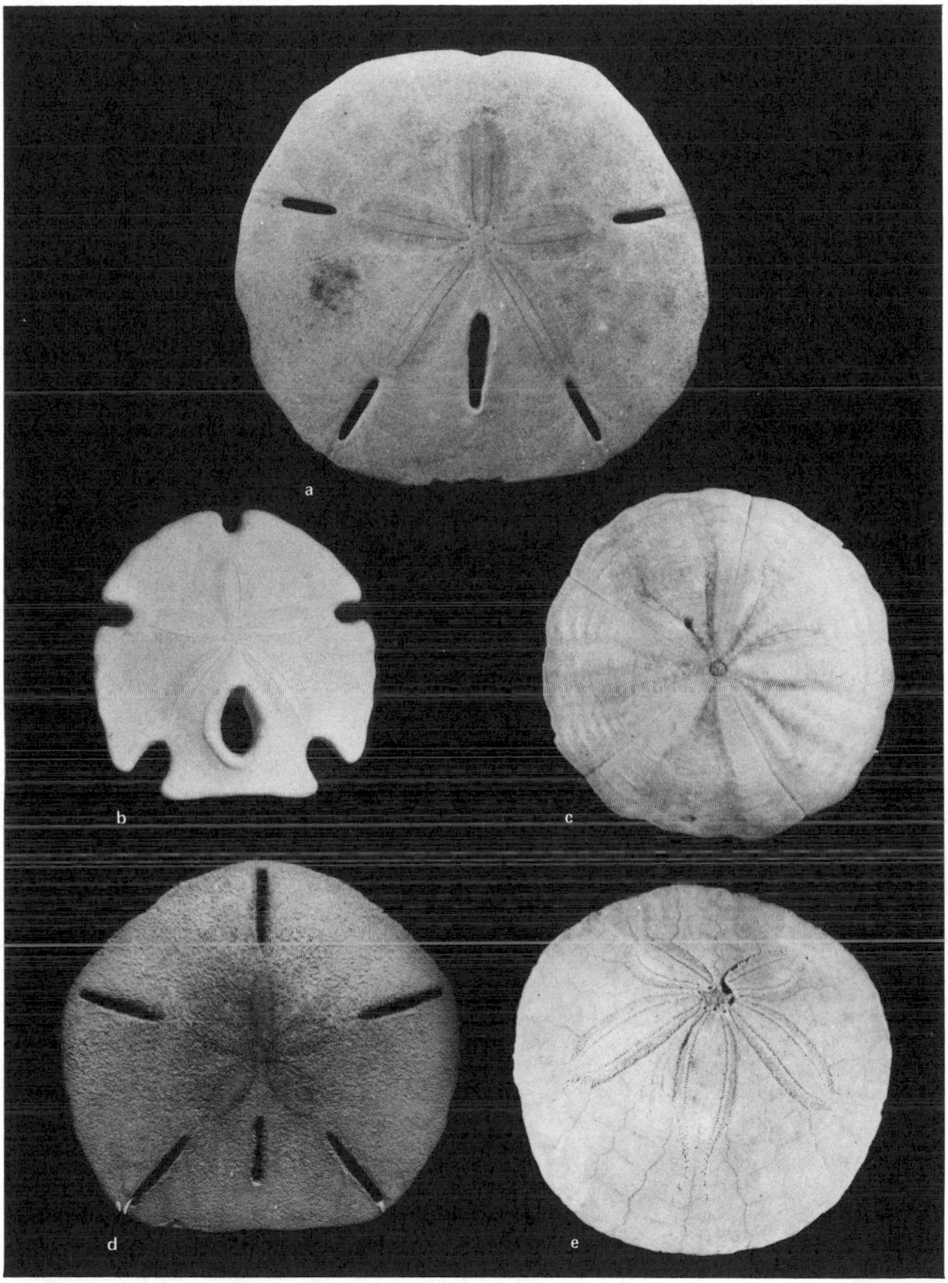

JOSEPH D. GERMANO

Fig. 25.2 Pacific sand dollars (see Fig. 20.2 for Atlantic sand dollars). (a) *Mellita.* (b) *Encope.* (c) *Fellaster..* (d) *Leodia.* (e) *Dendraster.* (a, b, d, e) are all west American; (c) is from New Zealand.

anus placed on the lower side, where the food grooves are conspicuously branched to produce dendritic patterns. The best-known species is *D. excentricus.* Like other sand dollars, they live as epifauna and infauna on sandy substrates. Further south on the west coast occur the tropical sand dollars already noted on page 211.

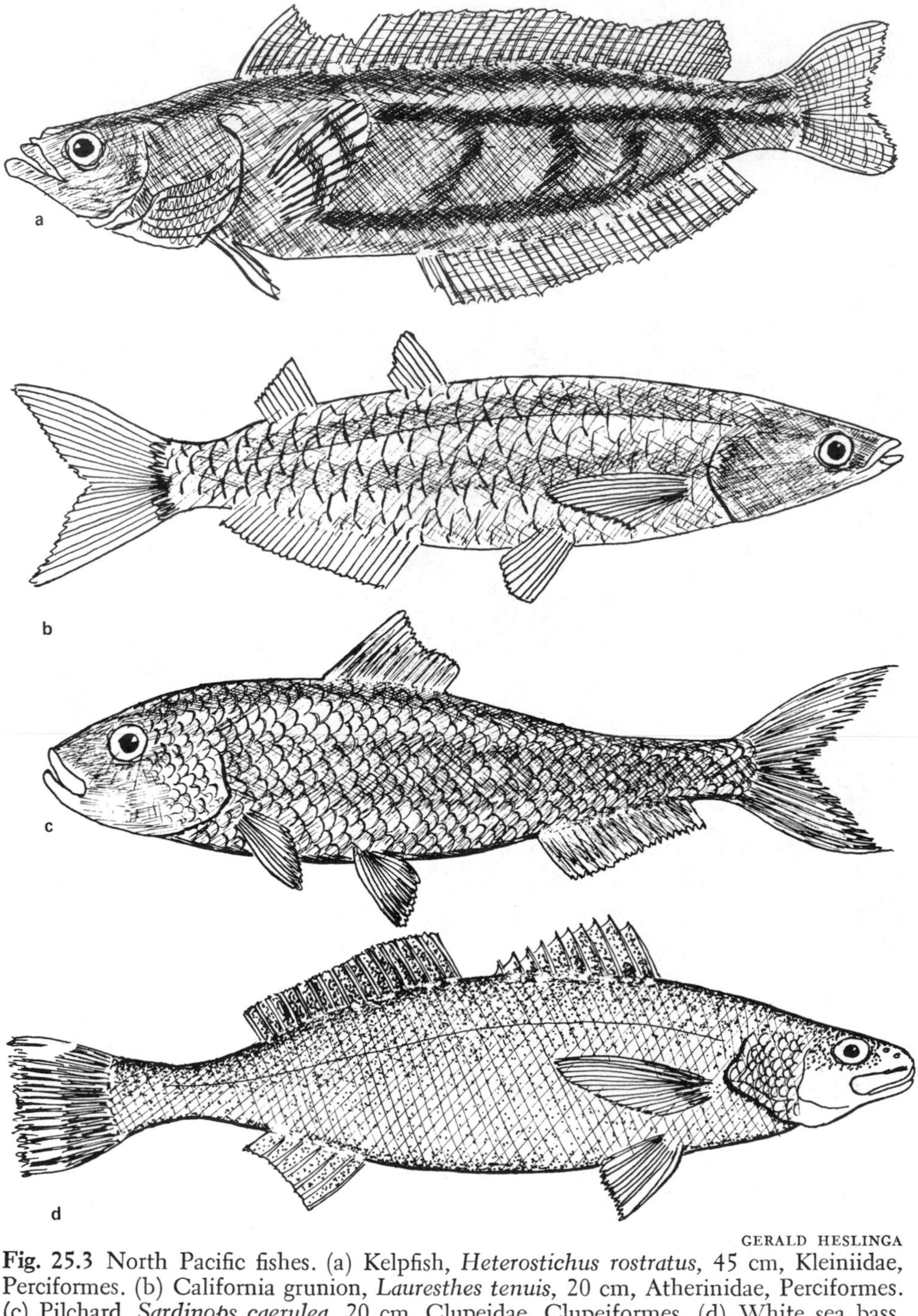

GERALD HESLINGA

Fig. 25.3 North Pacific fishes. (a) Kelpfish, *Heterostichus rostratus*, 45 cm, Kleiniidae, Perciformes. (b) California grunion, *Lauresthes tenuis*, 20 cm, Atherinidae, Perciformes. (c) Pilchard, *Sardinops caerulea*, 20 cm, Clupeidae, Clupeiformes. (d) White sea bass, *Cynoscion nobilis*, 1.5 m, Otolithidae, Perciformes.

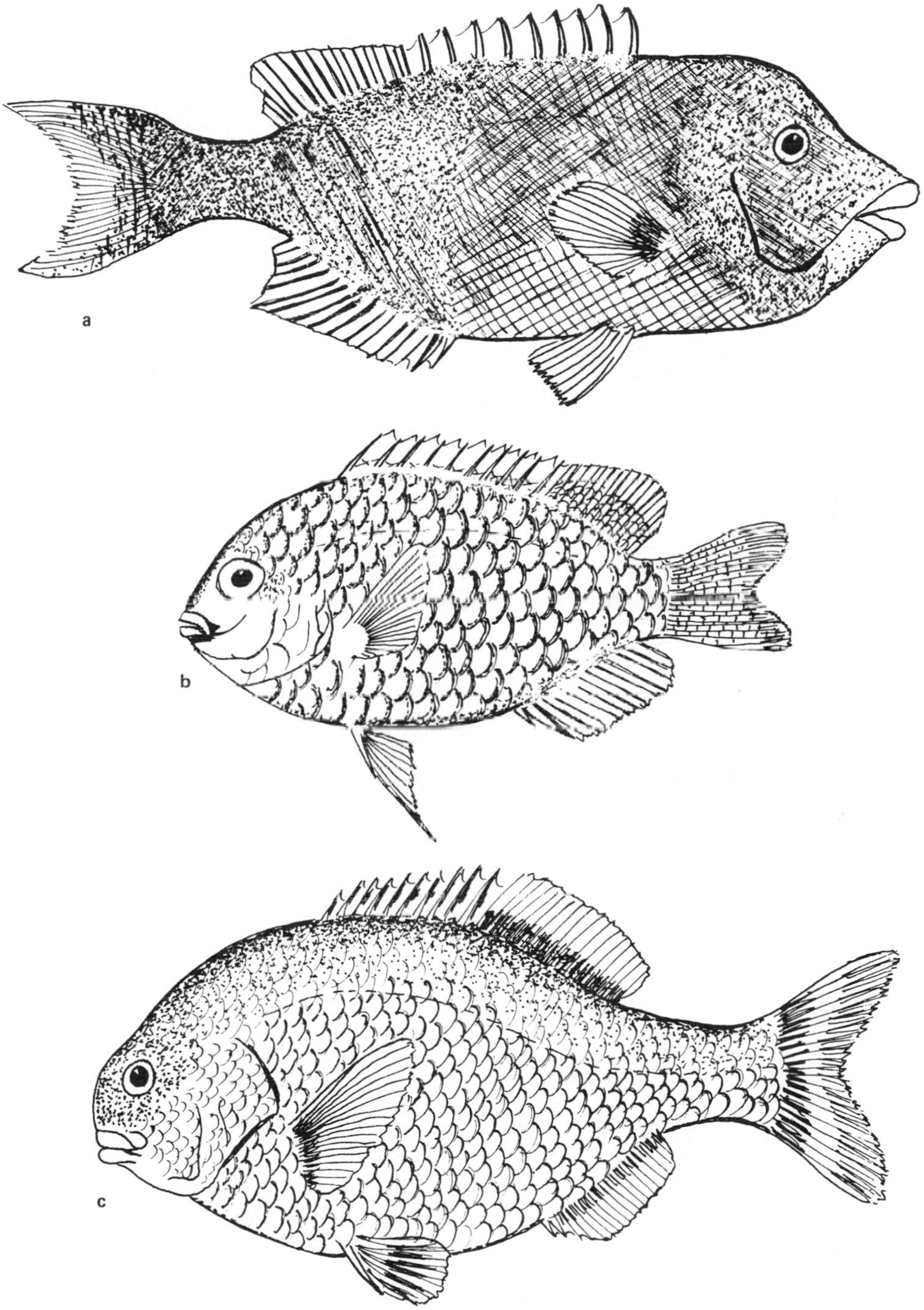

GERALD HESLINGA

Fig. 25.4 North Pacific fishes. (a) Sheephead, *Pimelometopon pulchrum*, 60 cm Labridae, Perciformes. (b) Damselfish, *Pomacentrus inornatus*, 15 cm, Pomacentridae, Perciformes. (c) Black sea perch, *Embiotoca jacksoni*, Embiotocidae, Perciformes.

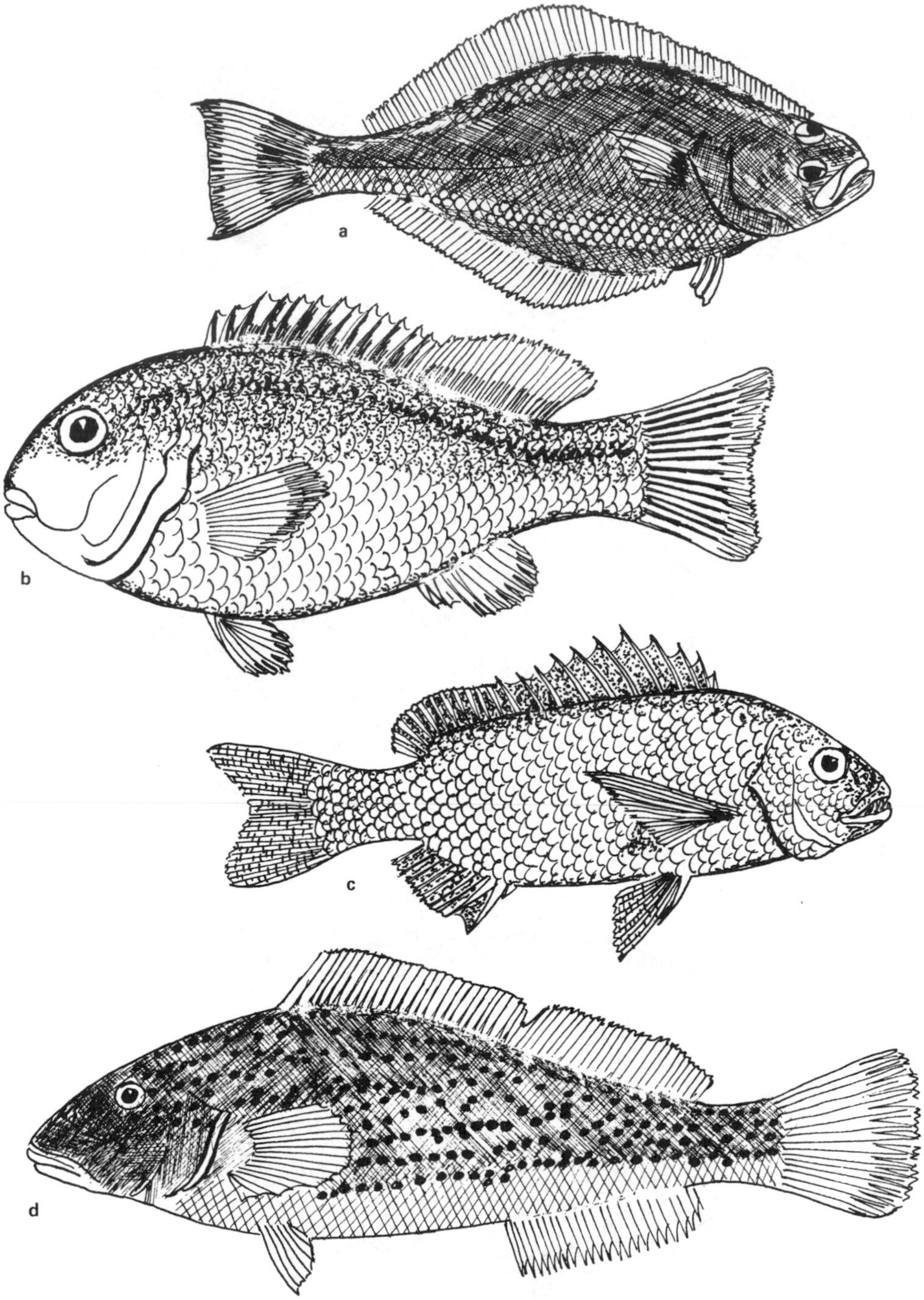

GERALD HESLINGA

Fig. 25.5 (a) California halibut, *Paralichthys californicus*, 90 cm, Paralichthyidae, Pleuronectiformes. (b) Opal-eye, *Girella nigricans*, 40 cm, Girellidae, Perciformes. (c) Porgy, *Chrysophrys cuvieri*, 35 cm, Sparidae, Perciformes (Indo-Pacific). (d) Greenling, *Hexagrammus decagrammus*, 40 cm, Hexagrammidae, Perciformes.

Among the **Mollusca Bivalvia,** the notable Atlantic genus, *Mercenaria,* is lacking from the North Pacific. On the other hand, *Mya,* the soft-shell clam, does occur in the North Pacific, but is naturally represented in western North America by a distinct species *M. truncata;* however, the New England species, *Mya aremaria,* has been naturalized on the west coast.

The crabs of muddy shores include the grapsids *Hemigrapsus* (which also occurs on rocky coasts), and in the mangrove swamps *Aratus* and *Sesarma* can be found.

NEKTON

The fishes of the North Pacific continental shelves belong, for the most part, to the same families as those of the North Atlantic, but there are differences in the degree of representation. These impart a distinctive aspect to the faunas. (See Figs. 25.3 through 25.4.)

The family *Girellidae,* also reported (page 311) under the name of Luedericks as part of the Australian fauna, is represented in the North Pacific. On the Californian coast its members are called Opaleyes or Catalina perch—the name is taken from the large bluish eye of the single coastal species.

Among the distinctive Pacific families is *Hexagrammidae* or greenlings, in which the fins are arranged in much the same manner as in the cod, but there are no barbels. Instead, there are one or two pairs of soft fleshy flaps above the head. They are related to perches, not to cod. Another North Pacific family is the *Clinidae* or kelp fishes, resembling blennies but having the fin spines projecting beyond the fins as a sharp jagged series. The family *Embiotocidae* or Pacific surf perches frequent habitats suggested by the name. They resemble perches but bear their young alive.

The key on page 276 should lead to identifications of North Pacific families, if allowances are made for the additional families noted above.

26

WETLANDS

Salt marsh / Ecotones / Salt-marsh flora / Mangrove sloughs / Distribution of mangroves / Climax formation / Conversion of mangrove slough to dry land

When a bog occurs on low-lying land near the sea it is subject to periodic inundation by saline water, and is then called a **salt marsh** or **fen.** The Fens district of eastern England is an example. Salt marshes have their own special populations of species that are not adversely affected by salinity of the soil. In tropical sands the salt marsh is replaced by a **mangrove.**

Salt marshes can be regarded as transitional regions, or **ecotones,** between the sea, the freshwater, and the land. Since it seems very probable that all the higher plants of the aquatic environment are descended from terrestrial ancestors, it is to be expected that these ecotones would carry vegetation of intermediate types. This proves to be the case. True rushes (*Juncus maritimus*) are represented as well as sedges, such as *Carex litorosa.* One of the bulrushes, *Scirpus lacustris,* occurs both in saline and freshwater areas. The yellow-flowered monkey flower or musk, *Mimulus,* penetrates the salt meadow where freshwater streams meander across an outlet plain. In the Southern Hemisphere a scrubby shrub, *Plagianthus,* occupies salt meadow in Australia and New Zealand, each country having its own series of species; the genus belongs in the mallow family, Malvaceae. A member of the primrose family, Primulaceae, is the genus *Samolus,* which includes a cosmopolitan species, and others that range the salt marshes and estuaries of the Southern Hemisphere. Coastal rocks are also colonized. Plants of this kind are important in preventing subaerial and storm erosion of temperate coasts in both the Northern and Southern Hemispheres, and to this extent they are analogus to the tropical mangroves (otherwise the similarity is slight).

Mangrove Sloughs

On tropical shores where wave action is not severe there develops a fringe forest of trees and shrubs, all of which are tolerant of saline or brackish conditions, that develop stiltlike prop roots or ascending roots (*pneumatophores*) used to prevent the silt from being washed away. These trees belong to various families, but are all spoken of as **mangroves.** One family, the Rhizophoraceae, comprises about 20 genera, including the black mangrove, *Rhizophora mangle,* that ranges the coasts of both the Atlantic and Pacific tropical Americas. Another family, the Combretaceae, includes *Laguncularia,* ranging the tropics of both sides of the Atlantic, and *Conocarpus* with a similar distribution. A third family, the Verbenaceae, is represented by the mangrove genus *Avicennia* in both the New and Old Worlds. Other genera which participate are *Bruguiera, Sonneratia, Carapa,* and *Ceriops.* All show similar adaptation to life in or near salt water and to a tide rising and falling at regular intervals. In Florida the mangroves are zoned such that *Rhizophora,* with its stilt roots, stands next to the sea; then on the landward side is *Avicennia* with pneumatophores; and then *Conocarpus,* which is transitional to the tropical forest itself. The seeds of mangrove trees are adapted for dispersion over saltwater; some species are viviparous, so that the seed has already sprouted a stout anchoring root that penetrates the substrate when the seed falls off the tree. The mangrove belt is somewhat wider than the coral-reef belt because it reaches to northern New Zealand in latitude 35°S, whereas the coral-reef belt ceases at about 30°S. Mangroves may extend along the banks of rivers, where their stilt roots form a kind of picket fence along the river bank, as in New Guinea.

Mangrove sloughs gradually are converted into dry land as the mangrove belt advances into shallow marginal seas. The accumulated silt and organic debris causes a natural reclamation of the seabed into dry land. On many coasts the mangroves are therefore not merely protective, but actively extending the real estate—a matter not appreciated by those who have destroyed the mangrove fringe, with resultant erosion of the Florida coastline. Botanists regard the mangrove association as **climax formation**, because it does not evolve into a forest of different type and regenerates itself with the same species after destruction or injury. This is a rare circumstance in forest, where a **succession** of replacement plant associations usually follows destruction of a forest. Mangroves in the Indo-West Pacific comprise some 10 species of trees, whereas on Atlantic coasts only four species are predominant. There is no equivalent in cool temperate and cold latitudes.

Birds of the Wetlands

Wetlands are important refuge areas for wild fowl. See Chapter 16.

27

MARINE BIOLOGY OF THE SOUTHERN OCEANS

Productivity / Relatively high sea-to-land ratio in Southern Hemisphere / Relative remoteness of southern polar land from southern temperate continental shelves / Fisheries / Dispersal routes / Origin of the Australian and New Zealand shelf biota / Biota shared by Australia, New Zealand, and South America / West-wind-drift dispersal / Biogeographic effects of the Antarctic Convergence / Coastal fishes of Antarctica / Deep-water benthic fishes of Antarctica / Zoogeography of Antarctic fishes / Southern temperate marine algae / Phaeophyta / Chlorophyta / Coelenterata Anthozoa / Sea stars / Regular sea urchins / Heart urchins and sand dollars / Crustacea Decapoda / Southern mollusks / Gastropods / Bivalves / Cyclostomes / Elasmobranchs / Teleosts / Clupeiformes / Beloniformes / Gadiformes / Beryciformes / Zeiformes / Perciformes / Tetraodontoformes / Anguilliformes

Productivity

The productivity of the southern oceans is believed to be much lower than that of northern seas, although considerable uncertainties must exist so long as fishery exploitation of the southern seas remains mainly an offshore and inshore activity with only a few vessels engaged in pelagic fishing. According to Roughley (1966), the Australian fishery yields only one-fortieth of the annual catch on the coasts of the United States, although Australia has a coastline half again as long as that of the United States. The northern oceans at present yield about 95 percent of the world's annual fish catch. Although the number of fishing vessels operating in the southern seas is much less than in northern seas, there appear to be other more fundamental reasons for the low catch. These probably include the following:

1. South of 30° south latitude, only 15 percent of the earth's surface is occupied by land, namely, the southern half of Australia, New Zealand, and the southern extremities of South Africa and South America. The continental shelf regions of the southern oceans are accordingly very limited, and most food chains on which fishes depend are based on marine invertebrate benthos of the continental shelves. In contrast, the earth's surface, north of 30° north latitude is 50 percent land, with a correspondingly higher area of continental shelf bottom. Thus northern productivity is likely to be much greater.

2. Most northern landmasses lie much nearer to the Arctic regions than the corresponding distance of southern lands to the Antarctic. This means that most of the southern coasts are washed by relatively warm ocean currents. Also, the much greater area of sea in the Southern Hemisphere maintains a more even surface temperature, on account of the latent heat of water, whereas in the Northern Hemisphere the surface temperature fluctuates widely during the course of the year, on account of the greater continentality of the Northern Hemisphere. Thus, there is a much more pronounced mixing of shallow and deeper waters in the Northern Hemisphere during the winter months, when the thermocline is broken down, and consequently there is a much greater enrichment of surface waters by nutrients of deep-water origin in the case of the northern seas. This, in turn, yields a much more pronounced spring diatom increase (SDI) with resultant enriched productivity. Measurements by Dakin and Colefax suggest that the plankton production in Australian waters is only about one-third to one-quarter of that of a northern area such as the Irish Sea. Although the plankton production is lower, there is a recognizable annual cycle, with a peak of phytoplankton in October and November (the southern spring), and a peak of zooplankton following this. There is also a secondary plankton peak in the autumn. But the productivity appears to be low, and may bear a direct relationship to the lower fishery production of Australia.

Most Australian fish catch is derived from coastal and estuarine fisheries, using hauled (seine) nets and meshing (gill) nets. The species caught include *Mugil*, *Arripis*, girellids, and others. Trawling and fish traps yield demersal fishes, including snapper, flatheads, tarakihi, and flatfishes. Pelagic fisheries yield tuna, barracouta or snoek, mackerel, anchovy, and others.

Dispersal Routes

Two great continental-shelf highroads lead from the tropical zone into the Southern Hemisphere: one is along the coasts of South America and then, by way of the Scotia arc into the Palmer Peninsula of Antarctica; the other is by way of the Indonesian Archipelago into Australia and Tasmania, with New Zealand receiving immigrants from Australia, apparently by larval or epiplanktonic drift across the Tasman Sea (Fig. 27.1). These two great highways seem to have been in continuous use throughout the Tertiary, to judge by fossil distributions (Fell, 1962).

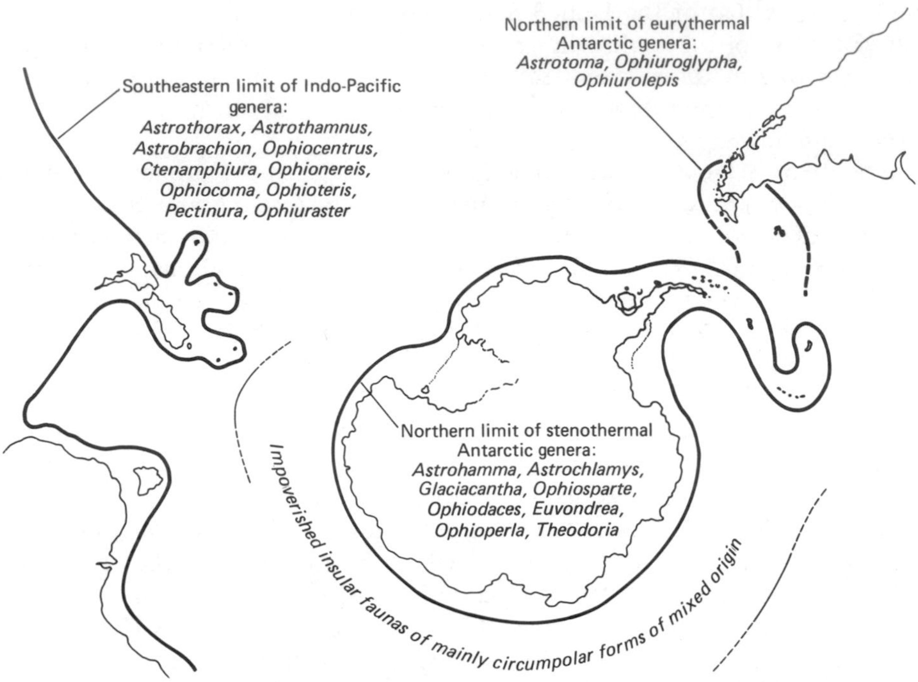

Fig. 27.1 Major dispersal routes available to shelf organisms of the Southern Hemisphere. (Fell, 1962)

Origin of Australian and New Zealand Shelf Biota

Because most mollusks and echinoderms are neritic and these two great groups have left an abundant fossil record, it is possible to reconstruct the origins of much of the continental shelf fauna of the Southern Hemisphere over the past 60 million years. It is clear that Australia and New Zealand have become the southernmost outposts of the Indo-West Pacific fauna, while the Magellanic region, including Antarctica itself, has been a recipient of amphiamerican elements. Doubtless some southern groups have dispersed northward along the same routes, but it is not always easy to determine the direction of dispersion as long as the paleontological record is only partially explored, as is presently the case in most parts of the southern continents.

Biota Shared by South America and the Australian Region

About the middle of the nineteenth century it was becoming clear that some faunal and floral elements of Australia and New Zealand are also shared with South America, although no shallow-water route connects these regions. Early speculation led to the idea of the former existence of land bridges that vanished beneath the sea, but when the bathymetry of the southern oceans was explored, it was realized that the evidence opposed the idea of land bridges in the relatively recent geological past. Then, with the advent of the theory of continental drift, proposed by Wegener in 1910, the similarity of organisms of the southern continents lent support to the theory.

However, steadily accumulating evidence on the actual mechanisms of dispersion now points strongly to the west-wind drift as almost certainly the vector that has transferred, and is still transferring, biota between the southern continents.

West-Wind-Drift Dispersal

In 1953, noting that the genera *Asterodon* (a sea star) and *Pseudechinus* (sea urchin) exhibit greater speciation in New Zealand, their western limit, I suggested that the South American and other scattered southern species of these genera might have reached their present sites as a result of occasional "escape" of elements from the New Zealand plateau. The only mechanism that did not conflict with other data seemed to be the west-wind drift (Fig. 27.2). It was inferred that specimens adhering to brown seaweed might, at rare intervals, complete a trans-Pacific oceanic drift. That asteroids can float long distances on brown seaweed had already been established by Mortensen (1925, 1933); his most notable discovery was that the Indo-Pacific asterinid *Patiriella exigua* had successfully colonized Saint Helena. Mortenson had also observed *Calvasterias* floating on Macrocystis, off the shores of Campbell Island.

During and after the International Geophysical Year (1958–1961), while examining the collections of echinoderms made by the various expeditions, the

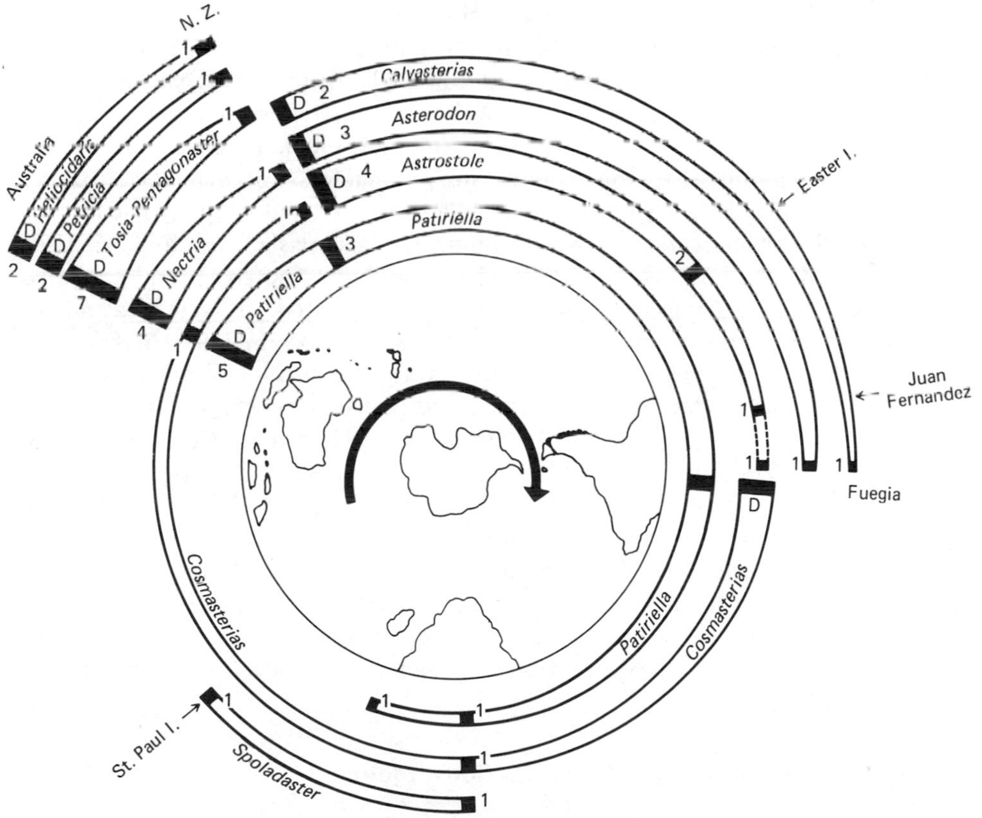

Fig. 27.2 Typical distribution of speciation intensity for organisms having circumpolar west-wind-drift dispersions. The length of the radial coordinate is proportional to the speciation of a genus at the longitude indicated by the angular coordinate. Inferred donor areas marked *D*; the vector is supposedly the west-wind drift itself. (Fell, 1962)

idea of epiplanktonic dispersal of echinoderms naturally exercised my mind; and by 1962 it became feasible to plot the known distribution of many southern, shelf-inhabiting echinoderms, as many more results had become available. It was then obvious that the majority of the genera exhibit a consistent pattern.

Figure 27.2 shows that the histogram containing the species for each genus tends to assume the form of a more or less attenuated triangle, in which the base-line (maximum speciation) occurs at the western end and the apex (minimum speciation) at the eastern end. It may be inferred that, for each of these circumpolar genera, the original source lies at the western end of its range and that the more easterly representatives are derivatives from the western stock. If this is correct then Australia stands as a donor to New Zealand; similarly, New Zealand (including the plateau, and Lord Howe and Norfolk Islands) stands as a donor to Kermadec Islands, Easter Island, Juan Fernandez, and South America. Thus, South America receives from the West Pacific but acts as a donor to South Africa, Australia, and even New Zealand. We are therefore presented with a simple clockwise pattern in which generic stocks tend always to move from west to east. The movement occurs around the Antarctic continent, but not by way of Antarctica. The west-wind drift supplies the only acceptable distributing mechanism, because it exists now and must have existed through much of the Tertiary, if not longer, and it would be plainly absurd to postulate land bridges or continental drift as a vector for genera and species of known recent origin in geological time.

Of course, these patterns can be modified by local factors. Thus, northern Chile can receive or donate by way of the coastal land link with North America as well as by way of the epiplanktonic mechanism, and most elements have evidently come from the Panama route. The warm-water elements of northern Chile are debarred from southern Chile by environmental factors, and here the Antarctic elements enter, but once again, the epiplanktonic route makes its own contribution. *Allostichaster* appears to have its home in Australasia, and to have been transferred to Chile, but has also extended up the Argentine coast, and speciated in southern Brazil. *Pseudechinus* seems to have originated in New Zealand, or Australasia, where more than half of its known species occur, but it has formed at least three subgeneric groups—a green form restricted to New Zealand, a red or red-and-green group shared by New Zealand and South America, and a variegated sparsely spined group shared by the Indian Ocean subantarctic islets and by Australia and New Zealand. When these plots are mapped, one suspects that the genus set out from New Zealand long ago, circumnavigated the globe and, like Magellan's sailors, arrived back whence it started, from the other direction.

Since these theories were first published, McDowall (1970) has shown that the galaxiid fishes (formally held to support the theory of dispersion through continental drift) exhibit in fact a marked west-wind-drift speciation, similar to that of the shelf echinoderms. DeWitt (1971) has observed indications of comparable patterns in some Antarctic fishes.

Biogeographic Effects of the Antarctic Convergence

The constant precipitation of snow, plus the summer melt of sea ice, creates around the coasts of Antarctica a great ring of cold surface water of low salinity that flows radially outward from the continent, driven by the spreading air generated by the Antarctic high. Around south latitudes of 50 to 55° this expanding surface water mass encounters and is overridden by the warmer and more buoyant waters of the west-wind drift. The consequential interface between the two water masses, with its abrupt temperature cline, has now come to be recognized as the effective northern limit of Antarctic faunas, and it carries the name **Antarctic convergence** (see page 84). The biological effects of the convergence may be illustrated by reference to the Antarctic fish fauna.

The earliest collections of Antarctic fishes disclosed the unique character of the fauna, undoubtably reflecting the extreme isolation of the southern lands and their continental shelves, probably reflecting also the long geological duration of such isolation, and perhaps also the long duration of the Antarctic Convergence as a planetary feature. There is, however, a significant peculiarity of the Antarctic shelf, namely, its exceptional depth (500 to 750 m, as opposed to the world average of 132 m). This remarkable fact (now attributed to the downward pressure exerted by the vast ice load of the continent itself) also accentuates the uniqueness of the Antarctic shelf fauna. It means that shelf animals encounter much greater depths and pressures than elsewhere, and so their anatomical characters sometimes resemble those of bathyal animals of the continental slope. DeWitt (1971) has recently examined this matter, and he finds that topographic discontinuities on the seafloor are better features for defining component parts of the fish fauna than the criterion of absolute depth. Nonetheless, he still finds it convenient to distinguish between coastal and deep-water fishes, according to their relationships with corresponding faunas in other parts of the ocean.

Coastal Fishes of Antarctica

These comprise representatives of 14 families. Four of these families, namely, the Antarctic cod, the plunder fishes, the dragon fishes, and the ice fishes, constitute the pièce-de-résistance of the Antarctic endemics, namely, the Notothenioidei, an assemblage of perchlike or blennylike fishes typical of the Antarctic region. Indeed, some 60 percent of all known fishes of Antarctica belong here, and they seem to represent the ancient native fish fauna of Antarctica, which is perhaps 40 million years old or even older.

So odd are these four notothenioid families that some additional comments are justified here. Most peculiar of all seem to be the ice fishes, or family Channichthyidae, which have no respiratory pigment in their blood (which therefore appears whitish!). These fishes are apparently predators upon crustaceans and on other fishes (for information on their respiratory physiology, reference should be made to another text). The Antarctic cod, or Nototheniidae, are a varied but mostly rather sedentary bottom-dwelling stock. They feed on invertebrates and algae and occasionally live near the undersurface of the floating ice shelves. The plunder fishes, or Harpagiferidae, are similar, but lack scales. They have a barbel on the chin, which is of great value in the deep-bottom environment they tend to occupy. The dragon fishes (Fig. 27.3), Bathydraconidae, have an elongate body and lack the anterior spinous dorsal fin.

DeWitt (1971) also distinguishes a second category of coastal fishes that he believes represents a more recent invasion from elsewhere and not, therefore, part of the original Antarctic fauna. Here are included some widely distributed families of fishes, such as the eel pouts (Zoarcidae), the snail fishes (Liparidae), the rays (Rajidae), the moras (Moridae), true cod (Gadidae), flatfishes (Bothidae), and

F. JULIAN FELL

Fig. 27.3 Dragonfish of Antarctica, *Paraclaenichthys charcoti,* Bathydraconidae.

also two other families of Southern Hemisphere origin. These are the codlike Muraenolepidae and the so-called horse or pig fishes (Congiopodidae). There are also in this category two families of jawless fishes or Agnatha, namely, the lampreys and hagfishes. The former are represented by *Geotria australis* (see page 309), long known to be from Australia, New Zealand, and South America, but only recently shown to be part of the Antarctic fauna by Murphy (1964) and Tickell (1964) and to serve as food for albatrosses. The hags are known from two reported species of Myxinidae.

Deep-Water Benthic Fishes

Four families of Antarctic fishes are included under this heading, comprising groups that range bathyal and abyssal zones elsewhere in the ocean. These are the eellike Synaphobranchidae and Halosauridae, the elongate and soft-bodied Brotulidae, and the rattail cod or Macrouridae, well known from some deep-water fisheries, such as that off Portugal. The presence of such fishes in Antarctic deep-water habitats offers no biogeographic problems.

Zoogeography

Analysis of these fish faunas leads to the identification of the following geographic elements: (a) **bathyal and abyssal species** living usually below 2000 m although entering the shelf on occasion; (b) **endemic insular species** living on the shelves of particular islands; (c) **endemic Antarctica species** of west and east Antarctica; and (d) **circum-Antarctic species.**

The last category presents distribution patterns similar to those given for echinoderms and related to the influence of the west-wind drift, as DeWitt notes, although he was unable to recognize distinct decay series. The Scotia Arc is identified as the main portal of entry into Antarctica for shallow-water fishes, again matching the conclusions drawn on the basis of echinoderm distribution.

SOUTHERN MARINE ALGAE

In the southern oceans, as elsewhere, the floating diatoms and other minute phytoplankton constitute the main basis of the trophic pyramid. For a discussion of primary productivity of the southern oceans, see page 292. Closer inshore, on the upper part of the continental shelf and in the intertidal regions, large algae occur, and these are important because they provide physical shelter for other denizens of the shelf ecosystem and they are food of the primary heterotrophs. These plants, which form a great part of the autotrophic elements of the shelf ecosystem, include some of the largest seaweeds known anywhere in the world. The following are some representative examples.

Phaeophyta or Brown Kelps. *Hormosira banksii* more or less replaces the northern *Fucus* on Australasian coasts. Its appearance is distinctive because the thallus is composed of strings of small spherical hollow floats, each about 1 cm across and strung in linear series up to 20 cm long or so, dividing at frequent intervals. This olive-brown alga occurs at the level of low neap tide, or just below the average midtidal level; here it forms very precise horizontal curtains on rocky shores, around the more sheltered coasts of southern Australia and New Zealand.

Macrocystis Pyrifera or Bladder Kelp. This is a kelp of moderately exposed rocky coasts. It grows in long ropelike masses up to 60 m long. The central axis is anchored to a rock holdfast on the bottom of the shelf by means of a robust rootlike lower termination. The free part of the axis slopes at an angle upward, rather like a fisherman's line, with the angle determined by the direction and strength of the current. At intervals along the axis elliptical bladders occur, one at the base of each straplike branch of the huge thallus. The combined flotation power of all the hundreds of cysts serves to hold the whole plant in its erect posture. If a violent storm tears the plant away from the seabed, then the thallus floats to the surface to constitute a raft, capable of drifting thousands of miles to sea and of carrying miniature ecosystems of benthic and epiphytic organisms, plus a cloud of kelp fishes underswimming the whole. Sometimes the alga becomes entangled with a log or logs and with other species of kelp, in which case great rafts up to 15 m in diameter are produced.

The giants of the world's brown algae are the species of the great brown bull kelp, *Durvillea,* with *D. antarctica* and others. In these kelps the actual thallus incorporates the floats as honeycomblike cells forming a sort of pith inside the leathery exterior layers of the giant straps of the thallus. The straps may be up to 30 cm wide, about 1 cm thick, and 100 m long. The Chlorophyta include *Ulva* and generally resemble Northern Hemisphere forms.

COELENTERATA—ANTHOZOA

The general absence of corals from southern waters is partly compensated by the great variety of southern sea anemones. *Calliactis conchicola* is commonly found living attached to the exterior of the shell of gastropods, whether occupied by the snail or by a hermit crab. Recent studies by Ross (1971) have shown that octopuses, which prey upon hermits and other crabs, are extremely sensitive to *Calliactis* toxin and will not attack any host carrying the anemone. A giant inshore anemone *Isovradactic magna* occurs in the low-tide zone of rocks in the surf region. *Cricophorus* is a small olive-colored anemone that lives epiphytically on the olive-colored kelp *Cystophora;* this anemone has a simple marsupial fold of the body wall in which the young are carried. Another viviparous anemone is the dark red *Actinia tenebrosa* that lives on rocks up to high-tide level (consequently exposed to the air for hours, at which time the tentacles are folded inside the columnar part of the body). *Metridium* lives partly imbedded in sand, and has, therefore, a very elongated tubular body. *Corynactis* is an anemone apparently more closely related to corals, for the tips of the tentacles are expanded as beadlike globules carrying concentrations of stinging cells, as in some corals. However, it is a true anemone and does not secrete a calcareous skeleton. *Edwardsia* is a wormlike burrowing anemone, living in sand; it does not have a sucker at the lower end of the tubular body, which instead is pointed and adapted to burrowing.

SEA STARS

Following are the more common genera of starfishes found in shallow water in the Southern Hemisphere, south of about latitude 30° south: Of the lingthorns (family Luidiidae, page 187), *Luidia* is widespread on soft bottom in the subtropical and warm temperate parts of the region. Among combstars (family Astropectinidae, page 187), *Astropecten* is restricted to warmer coastal soft bottom. Two deeper water genera, *Psilaster* and *Persephonaster*, occasionally come up onto the shelf and may be found on soft bottom at about 50 fathoms, though they properly belong to the archibenthic fauna. Apparently these forms ascend to higher levels where suitable submarine canyons intersect the continental slope, providing a migration route from deeper water. A rarer deep-water genus, also ascending to 50 fathoms on occasion, is *Plutonaster*.

Tooth Stars (Family Odontasteridae). The genera of this family (discussed on page 187) are highly characteristic of cool and cold seas of the Southern Hemisphere, especially of southern New Zealand, the subantarctic islands, and southern Chile and Argentina, as well as on the continental shelf of the Antarctic continent itself. In Antarctica in summer they ascend to the most shallow parts of the seabed, retreating into deeper water at the onset of winter, when the shallow seawaters freeze solid. The best known genera are *Asterodon* (Fig. 27.4), with species in New Zealand, the subantarctic, and southern South America, and *Odontaster*, with the same range plus Antarctica. *Asterodon* has two oral spines on each jaw angle, whereas *Odontaster* has only one.

Goniasters (Family Goniasteridae, page 189). Conspicuous southern representatives are *Pseudarchaster, Mediaster, Hippasteria,* and *Pentagonaster*. Of these genera, the most common in shallow water is *Pentagonaster*, a bright orange species of unusually symmetrical appearance, having the penultimate marginals enlarged on each arm. Meristic variation in *Pentagonaster* occasionally occurs, leading to four-armed cross-shaped forms, and six- and seven-armed forms.

Oreasters (Family Oreasteridae). Several tropical genera of this family have already been seen (page 188), and some of these extend into the subtropical parts of the region now under consideration. However, the only conspicuous oreasterid that enters the cooler parts of the Southern Ocean is *Asterodiscus*, a massive form with rounded spines on the upper surface and of a deep brick-red color.

Serpent Stars (Family Ophidiasteridae, page 190). The only genus that ranges south of the tropics is *Ophidiaster*, and this is restricted to the northern subtropical part of the Southern Ocean, from ebb-tide level down to about 10 fathoms. It occurs in New Zealand only in the northern part of the archipelago.

Sun Stars (Family Solasteridae). The family falls in the order Spinulosida (see page 190), and has bipolar distribution. The common Southern Hemisphere species fall in the genus *Crossaster*, but a southern *Solaster* has been seen, although not yet formally named. The New Zealand species seems to be the same as the North Pacific species *Crossaster japonicus*. A similar species occurs off New England.

GEORGE PUTNAM

Fig. 27.4 Southern Hemisphere sea stars. (a) *Odontaster*, New Zealand, South America, Antarctica. (b) *Pentagonaster*, New Zealand. (c) *Heliaster*, Chile. (d) *Allostichaster*, New Zealand, Chile, Argentina, Brazil. (e) *Asterodon*, New Zealand, subantarctic islands, Chile, Patagonia.

Asterinas (Family Asterinidae, page 190). *Patiriella*, sometimes included in *Asterina*, is a common shallow-water genus of Asterinidae. The Australian species commonly have seven arms, the New Zealand and subantarctic forms generally are five-armed. *Stegnaster*, a more flattened form, is remarkable for its distribution: One species occurs in shallow water in New Zealand and a second species is found in deep water in the Caribbean. No other species of the genus are known, so the discontinuous distribution is a puzzle.

Henricias or Blood Stars (Family Echinasteridae, page 191). The genus *Henricia* represents the Echinasteridae in the cooler parts of the range of this widespread family. The Henricias are bipolar and are as common in boreal seas as in the far south of the Southern Ocean. Some species brood their young. They range from sea level down to about 300 fathoms and vary widely, so that the species are very hard to determine.

Order Forcipulatida. The most widespread shallow-water starfishes of the cool temperate and polar regions of both hemispheres, lacking from the shallow waters of the tropics. The ordinal character is the possession of crossed pedicellariae (see page 192). The more conspicuous Southern Ocean genera are as follows. *Calvasterias*, or brood-stars, has five flattened, leathery arms. It is greenish-brown and adheres to large brown algae in the subantarctic region. The eggs are hatched out by the female in a cluster covering her mouth; there is no larval stage. The female does not feed for 6 months or so during the breeding season. The genus has dispersed between New Zealand and South America, apparently by epiplanktonic rafting. *Allostichaster*, or dividing-stars, are usually six-armed or eight-armed species. The arms commonly arranged in two sets, one set larger than the other, resulting from the habit of asexual reproduction by transverse fission. The madreporite is multiple, at least just before a division occurs. This animal is found in New Zealand and in South America. *Stichaster*, or reef-star, usually with 12 arms, the spines reduced to rounded granules, and the whole animal able to cling with great strength to rocks far out in the heavy wave-zone. It occurs in New Zealand. *Astrostole*, is a large seven-armed starfish of the littoral region, occurring on islands to the east of Australia, in New Zealand, and with a related genus on the coast of Chile, probably derived from *Astrostole (Meyenaster)*. It is the largest predatory starfish in the southern seas, reaching up to half a meter across. *Coscinasterias*, *Asterias*-like starfish, but usually has 11 arms, one or more of them likely to be regenerating. Bluish grey with orange tube feet, it is commonly found on oyster beds, where it is a predator.

REGULAR SEA URCHINS

The commoner genera of sea urchins of the Southern Hemisphere, are found south of about 30° south latitude (Fig. 27.5): *Goniocidaris* (page 303) is found in Australia and New Zealand, and is represented also in deep water in Indonesia. *Centrostephanus* (page 202) occurs in eastern Australia and northern New Zealand, resembling *Diadema*, but having spines on the buccal plates and frequenting rock reefs south of the coral-reef zone. Some Northern Hemisphere species are also known. The arbacioids (page 207) are highly characteristic of the Americas, but also occur in Europe. They are totally absent from Australia and New Zealand. *Tetrapygus*, with polyporous amb plates, occurs in Chile and Peru. Of the temnopleurids (page 207), two very delicate shallow-water genera occur in Australia, *Holopneustes* and *Amphipneustes*, the former at least also in northern New Zealand. These are reddish forms with high, egg-shaped tests. A third genus, *Pseude-*

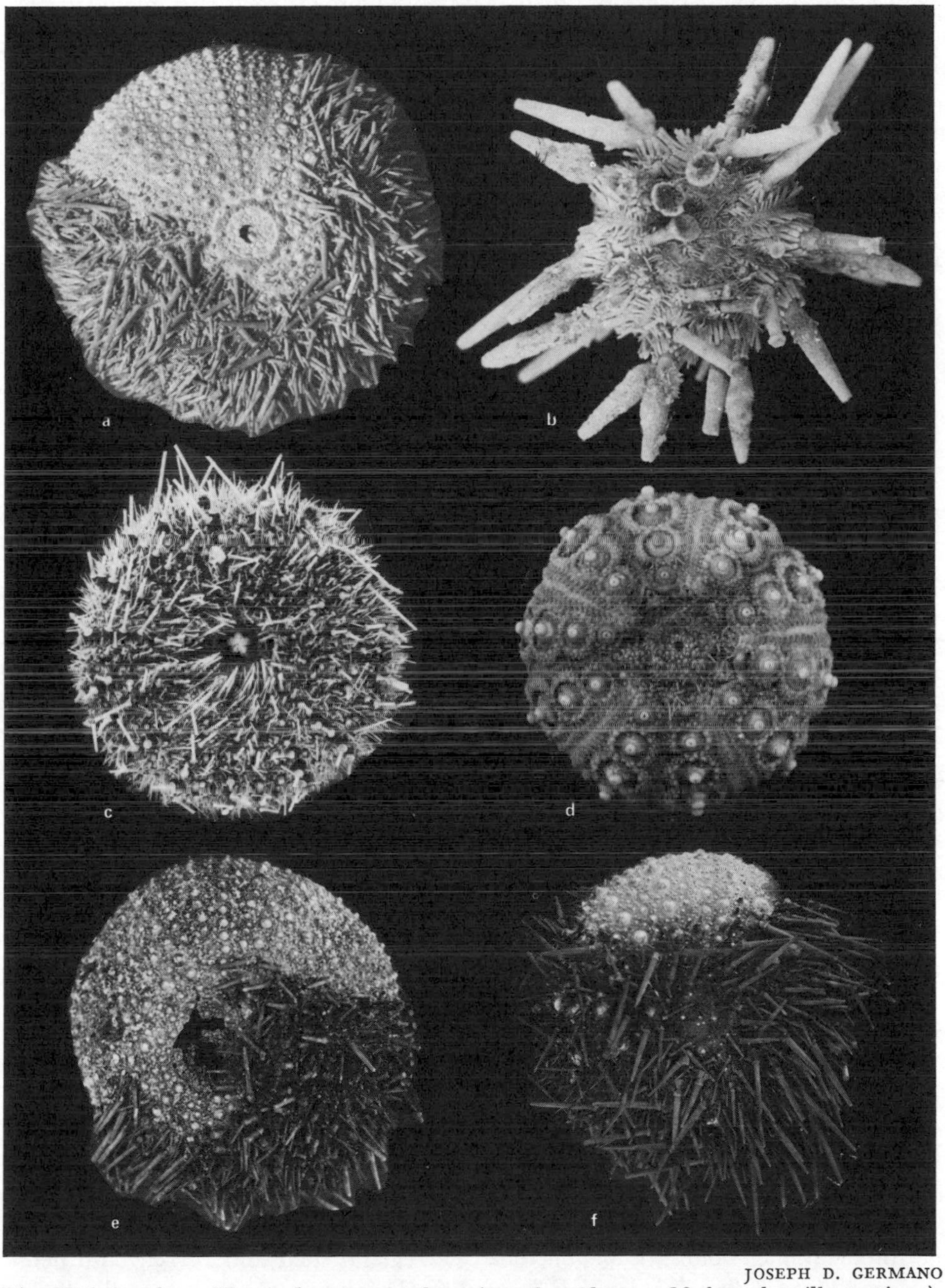

JOSEPH D. GERMANO

Fig. 27.5 Southern Hemisphere sea urchins (see also Chapter 20 for other illustrations). (a) *Loxechinus,* Chile. (b, d) *Goniocidaris,* New Zealand and Australia. (c) *Sterechinus,* Antarctic Ocean. (e) *Caenocentrotus,* southern Australia. (f) *Heliocidaris,* eastern Australia and New Zealand.

chinus, resembles small species of *Echinus*, with three pore pairs per amb plate. It ranges the entire Southern Hemisphere with representatives at various isolated islets in mid-ocean, as well as on the southern coasts of Australia, New Zealand, and South America. There are three species groups, red forms, green and red forms, and green forms. The genus does not occur on Antarctica.

Of the echinoids with more than three pore pairs on each amb plate the following are important Southern Hemisphere members. *Caenocentrotus* has four pore pairs. Its longest spines are shorter than horizontal radius of test. The test is almost always deformed by the presence of a parasitic crab, *Fabia chilensis*, in the rectum. Galapagos, Peru, and Chile. *Heliocidaris*, has seven to ten pore pairs. Its longest spines are as long as horizontal radius of test. Brownish littoral brown-seaweed-eating rocky-reef dwellers found in the surf zone. Common in southeastern Australia and more rare in New Zealand. Absent from other regions. *Loxechinus* has seven to ten pore pairs and short spines. It is greenish or whitish. *Loxechinus* can be found in coastal seas of Chile, where it is the object of an important fishery. Absent from other southern regions. *Evechinus*, has nine pore pairs in 3 vertical rows. Green and edible. New Zealand only. Object of fishery by Maoris, who call the animal *kina*. (The name *kina ariki*, meaning chief urchin, is applied to *Centrostephanus*.)

HEART URCHINS AND SAND DOLLARS

Cassiduloids (Order Cassiduloida, page 210). The genus *Apatopygus* is ovoid, and concave below and vaulted above. Its periproct is supramarginal and located in a deep sulcus. It is found in Australia and New Zealand. This genus resembles the Jurassic and Eocene cassiduloids and is sometimes known as a living fossil.

Clypeasteroids (Order Clypeasteroida, page 211). The genus *Arachnoides* comprises rounded, flattened sand dollars that lack lunules. Their periproct is supramarginal and has so-called combed areas occupying entire ambulacral plates on oral surface. It is found in Australia. A related and similar genus from New Zealand, differing in having several interamb plates on the lower surface instead of one only, was referred by Durham to a separate genus *Fellaster*. In South America, the genera *Encope* and *Mellita* occur, emphasizing the American character of the Patagonian clypeasteroids. *Iheringell* was a related fossil genus of Patagonia in the early Tertiary, and, Hotchkiss and Fell (1972) report it from contemporaneous East Antarctic (i.e., Australasian sector). This new find points to a milder climate and a former American connection with Antarctica.

Heart Urchins (Order Spatangoida, page 212). A characteristic Antarctic genus *Abatus* has marsupia formed from the lateral petals. The species occur on soft bottom around southernmost South America and Antarctica. The genus *Spatangus*, wide ranging in northern seas, has several species on the continental shelf of New Zealand, although little else seems to have been recorded of its distribution in far southern waters. Another wide-ranging genus is *Echinocardium*, which, like the forementioned genus, occurs on the New Zealand shelf and also in both hemispheres (*E. cordatum*, synonym *E. australis*). These two genera are easily distinguished: *Spatangus* has a heart-shaped test with only the subanal fasciole developed, whereas *Echinocardium* has an exceedingly fragile thin shell, with an internal fasciole. The cause of these bipolar distributions is presently unknown.

CRUSTACEA DECAPODA

Lobsters. The Southern Hemisphere lobsters are species of the genus *Jasus*, spiny lobsters without large chelipeds, belonging to the family Palinuridae (page 235) already studied under tropical faunas. Some authorities refer all the common southern lobsters to one species, *Jasus lalandii*, others separate them into different species, one each in New Zealand, South Amerca, South Africa, and Australia, with other specimens known also from various southern islands. Whether all are united as one species or not, they are clearly closely related forms. The larval planktonic stages last for one year, so transoceanic drift of larvae on the west-wind drift seems probable. A larger species *Jasus hugelii* occurs in both Australia and New Zealand. The northern genus *Homarus*, with giant chelipeds, is unknown in southern waters.

Crabs. Of the families of crabs already studied under tropical faunas, the following also have representatives in southern seas: (1) Diogenidae (page 235) with *Diogenes* on sandy shores in Australia. (2) Grapsidae (page 235) with *Leptograpsus, Plagusia,* and *Pachygrapsus* on rocky bottom. On soft substrates and also on estuarine habitats *Sesarma* occurs (see page 235). *Leptograpsus* is the dominant Southern Hemisphere genus, with species on rocky substrates in Australia and New Zealand. In Chile occurs *Cyclograpsus*, with a range extending to California along the west coast of the Americas. Some species are shared by New Zealand and Chile, such as *Hemigrapsus crenulatus* and *Leptograpsus variegatus*. One genus, *Plagusia*, occurring in floating logs, has a range extending through South Africa, Australia, New Zealand, and Chile. (3) Ocypodidae (page 235) are represented on sandy shores by *Ocypode* and *Scopimera*. (4) Majidae (page 238) are represented on rocky coasts by *Naxia*, and on soft substrates by *Hyastenus*.

Of families already noted under northern faunas the following southern representatives may briefly be noted here: (1) Cancridae, with *Cancer* as its main genus. This genus is represented on the Pacific coast of South America by four species, but only one species occurs in Australia and New Zealand, namely, *Cancer novaezelandiae*. (2) Portunidae, with genera *Portunus* and *Ovalipes*, are swimming or surf crabs. *Ovalipes* has three southern species, one each in Australia and New Zealand, the third being shared by South America and South Africa. (3) Xanthidae, with genus *Ozius*, are mud crabs. (4) Pinnotheridae, with genus *Pinnotheres*, are commensal mussels and oyster crabs.

The Southern Hemisphere crabs primarily serve a predatory and scavenger role in the ecosystem, as in the tropical and northern seas.

SOUTHERN MOLLUSKS

The southern mollusks belong for the most part to families that are also represented in northern seas. Notable differences include the absence from southern seas of the family Buccinidae, which is conspicuous in the northern oceans, and the presence of the Struthiolariidae in the New Zealand–South American region, a family entirely absent from other parts of the world. The tropical families Neritidae, Cypraeidae, Cymatidae, and Conidae are rather feebly represented south of 30° south latitude, with a few inconspicuous or rare examples. Entirely lacking are the large strombs, the showy cowries, the miters, the harp shells, the cameo shells, the pearl oysters, and giant clams. The other families already studied under tropical faunas are not restricted to the tropics, and so have for the most part a strong representation in both northern and southern seas. Following is a short review of representative southern genera, arranged by family and by ecological role, in the same sequence as was followed for coral reefs.

GASTROPODS

Primary Heterotrophs of Rocky or Other Hard Bottom

Abalones (Family Haliotidae, page 222). *Haliotis* is represented by much larger species than in the tropics, one of the best-known southern examples being *Haliotis iris*, the Paua of New Zealand, in which the nacre is blue (see Fig. 27.6). Similar but less showy species occur in Australia (Fig. 27.7). Other smaller species include *H. australis* with pink corrugated nacre and the inch-long, delicate *H. virginea*. Because of its use in jewelry, the export of paua shell is prohibited under New Zealand law, except for scientific study. Overkill by scuba divers is causing grave havoc to the inshore populations, and wasteful use of young stages as fishing bait has largely eradicated this beautiful mollusk from the more frequented rocky beaches. The Australian abalone was formerly exceedingly common in Sydney Harbor, but overkill by early Chinese settlers rapidly depleted the stocks. The status is now more or less that of the Californian abalone, and the same ecological association with the brown alga *Macrocystis* occurs in both hemispheres. Sea otters, of course, are absent from the southern food web, but the urchin *Heliocidaris* in eastern Australia seems to play a similar role to *Strongylocentrotus* in cutting the brown alga adrift.

Limpets (page 222). Fissurellidae are represented by *Megathura* on the Pacific coast of South America, by *Diodora* in widespread localities, and by *Scutum* in New Zealand. The family Patellidae is widely distributed, with genera *Patella* and *Cellana*.

Zethalia, *Cantharidus*, and *Calliostoma* (*Maurea*) are southern examples of the top shells. The turbans occur, with *Phasianella* conspicuous in southern Australia and *Cookia* in New Zealand.

Secondary Heterotrophs of Rocky Shore and Hard Bottom (page 224)

Slipper Shells (Family Crepidulidae). Limpetlike shells with an inner shelf, gregarious, widely distributed in both northern and southern seas. The common genera are *Zeacrypta*, *Maoricrypta*.

Round Slipper Shells (Family Calyptraeidae). Limpetlike shells with a shelf or cup within. They range all seas. *Sigapatella*.

Purple Snails (Family Thaididae). Robust shelled rock-frequenting forms, body whorl enlarged and usually ribbed, spire short, able to withstand surf. Predators. *Neothais*.

Volutes (Family Volutidae, page 226). Some warm temperate species of this otherwise tropical family occur in the Indo-Pacific. Examples are *Alcithoe* and *Pachymelon* in New Zealand littoral waters.

Infaunal Mollusks of Sandy and Silty Substrates

Family Turritellidae (page 307). *Maoricolpus* represents the family in New Zealand.

Fig. 27.6 New Zealand gastropods. (a) *Cirsotrema*. (b) *Cantharidus*. (c) Maoricolpus. (d, h) *Haminoea* (Atyidae, related to Bullidae). (e) *Serpulorbis* (Vermetidae). (f) *Cellana* (Patellidae). (g) *Alcithoe* (Volutidae).

GEORGE PUTNAM

Fig. 27.7 Australasian gastropods. (a) *Phasianella* (Turbinidae). (b, d) *Maurea* (*Calliostoma*). (c) *Cerithium*. (e) *Cabestana* (Cymatiidae). (f) *Modelia* (Turbinidae).

Struthiolarias (Family Struthiolariidae). Infaunal detritus feeders. Shell trumpet-shaped, up to 3 inches long, thick, with no notches in the aperture, the outer lip is thickened in the adult. This family is presently typical of New Zealand seas, where it has been present since Eocene times, but Tertiary South American examples are known (*Struthiolariella*). One of these was recently reported from the same Antarctic beds as are mentioned on page 304, under Clypeasteroida. *Struthiolaria*, very common.

Naticas (Family Naticidae, page 230). *Tanea* is a southern example, barely 1 inch across, a cool-water representative of the thick-shelled, brightly colored tropical members of this widely distributed family.

Cameo and Helmet Shells (Family Cassididae, page 230). The Australian and New Zealand representatives, south of 30°, are mainly small forms 1 or 2 inches long. The Australian forms include some with heavy, thick shells, as *Casmaria*; the New Zealand members are thin, delicate-shelled forms, as *Xenophalium*.

Olive Shells (Family Olividae, page 230). The southern forms lack the showy patterns of the tropical olives and also have shorter, thicker shells, with a more pointed spire. *Baryspira* is a New Zealand example, with the same habits as other members of the family.

BIVALVES

In general, the southern bivalves present broad similarities to those of the northern seas. Following are some of the commoner southern examples:

Dog Cockles (Family Glycymeridae). Thick shells of rounded outline, with taxodont dentition. *Glycymeris* occurs on coarse pebble bottoms on coasts with heavy surf.

Oysters (Family Ostreidae). Monomyarian, rounded to elongate shells, usually cemented by one valve to the substrate. Southern genera include *Saxostrea*, on rocks, and *Ostrea*.

Mussels (Family Mytilidae). Shell bluish to brownish, pear-shaped, teeth weak or absent, attached by a byssus, common in intertidal areas. *Mytilus*.

Horse Mussels (Family Pinnidae, page 230). *Atrina* lives in great colonies on mud banks under a fathom of water. Some 20 square miles of seabed, much of it occupied by *Atrina*, was elevated to become dry land by an earthquake in 1931 in Napier, New Zealand. The old shells were still standing, half emergent from the substrate, several years later when grassland had taken over the old seafloor.

Scallops (Family Pectinidae, page 228). *Notovola* and *Chlamys* are southern examples.

CYCLOSTOMES

The lampreys of the Southern Hemisphere all belong to the genus *Geotria*, in which only one species is now recognized as valid, *Geotria australis*, occurring in Australia, New Zealand, and South America. The eggs are deposited far upstream by the 18-inch-long mature parents, which migrate from the sea into rivers at the breeding season. The immature stages are spent in freshwater. Later the body takes on a bright silvery appearance and the lamprey migrates downstream and out to sea. It is now considered likely that the circumpolar distribution has been achieved in the marine phase, utilizing the west-wind drift as the vector between continents.

ELASMOBRANCHS

The sharks and rays of the temperate southern seas are generally similar to those of the tropics, although the sharks tend to be restricted to warm temperate seas and do not penetrate further south than about 44° south latitude, and then only in summer. The cool and cold-water elasmobranchs of the Southern Hemisphere are generally similar to those of the northern oceans and do not call for particular comment.

TELEOSTS

Table 27.1 gives a key to the main families of neritic bone fish of the temperate southern oceans.

Order Clupeiformes

Family Galaxiidae. This family seems to represent the Northern Hemisphere Salmonidae. The various species are distributed on the southern continents and intervening islands, and some species are present on two or more widely separated landmasses. The most widely distributed species is *Galaxias maculatus*, which occurs in Australia, New Zealand, and Chile in freshwater. This family was one of the groups originally cited as clear evidence of the former union of the southern continents. However, it is now known that species like *G. maculatus* spend the first six months of their life in the sea and then migrate inland to reach maturity during the next six months in a freshwater habitat. At maturity, when the fish is about 10 to 16 cm long, the adults return to the sea to spawn. It is probable that such distributions as that cited have been brought about in very recent geological time by transoceanic dispersion.

Order Beloniformes

Family Belonidae. These are called needle fishes in America and Long Toms in Australia. *Lewinchthys ciconia* frequents shallow water, including estuaries, and reaches a length of about a meter. This Australian species lays its eggs on seaweed and on river grass.

Family Hemirhamphidae. The family is represented in the Northern Hemisphere; in America they are called halfbeaks. In Australia where no freshwater garpike occur, they are known as garfish (Fig. 27.8), and in New Zealand they are called pipers. The genus *Hemirhamphis* occurs on all southern coasts, some in rivers, some on the continental shelf. The species illustrated, *H. welshi*, occurs in estuaries and the open sea of Australia.

Order Gadiformes

Family Gadidae. True codfish, members of this family, are rare in the Southern Hemisphere, although various unrelated fishes incorrectly are called cod. One of

Table 27.1 Key to the Main Families of Southern Hemisphere Temperate Teleosts Subclass Teleostei: operculum present, single gill opening, bone present

No finspines	
Pelvic fins abdominal	
Jaws not extended into a beak (order Clupeiformes)	
Dorsal fin set far back, near tail fin. Marine and freshwater	Galaxiidae
Dorsal fin about midway along body	
Belly with spinous scales (herrings)	Clupeidae
Belly without spinous scales	
Mouth large, inferior, extending behind eye (anchovies)	Engraulidae
Mouth small, terminal, not extending behind eye	Dussumieriidae
Jaws extended into a beak (order Beloniformes)	
Both upper and lower jaws form beak (needlefishes)	Belonidae
Only lower jaw entering beak (halfbeaks or pipers)	Hemirhamphidae
Pelvic fins jugular, barbel usually present under jaw (ling)	Gadidae
Finspines present	
Pelvic fins thoracic or jugular	
Pelvic fin with 5 or more soft rays, eyes large (order Beryciformes)	
Dorsal fin continuous, not subdivided into 2 parts (nannygai, etc.)	Berycidae
Pelvic fins with fewer than 5 soft rays	
Dorsal fin divided into 2 or more separate parts	
Dorsal finrays produced as free filaments (John Dories)	Zeidae
Dorsal finrays not produced into free filaments (order Perciformes)	
Pectoral fins with 3 free spines for walking (gurnard)	Triglidae
Pectoral fin normally formed	
Soft dorsal fin followed by finlets or fringe	
First dorsal fin half as long as body (barracoutta)	Gempyllidae
First dorsal fin shorter than half body length	
2 preanal free spines (jacks)	Carangidae
No preanal free spines (mackerel)	Scombridae
Soft dorsal fin never followed by finlets or fringe	
Second dorsal fin and anal fin serrated (flatheads)	Platycephalidae
Second dorsal and anal not serrated (mullet)	Mugilidae
Dorsal fin continuous, not divided into 2 parts	
Pelvic fins thoracic, behind or below pectoral fin base	
Spiny cheek ridge below eye, occipital spines (scorpion fish)	Scorpaenidae
No spiny cheek ridge or occipital spines	
An eye on either side of head	
1 of pectoral finspines enlarged (morwong, tarakihi)	Cheilodactylidae
No elongated pectoral finray	
Caudal fin strongly diphycercal, body slim (kahawai)	Arripididae
Caudal fin homocercal, body not slender	
11–12 caudal rays, cycloid scales (wrasses)	Labridae
15 or more caudal rays, scales ctenoid	
Base of soft dorsal and anal covered by scales	
14 dorsal spines (luedericks)	Girellidae
11–12 dorsal spines (drummers, rudderfishes)	Kyphosidae
Base of soft dorsal and anal not covered by scales	

Table 27.1 *(Continued)*

Maxillary not sliding under cheek (hapuku)	Serranidae
Maxillary sling up under cheek when mouth closes	
Snout convex, palate toothless (tamure)	Sparidae
Snout flattened, palate toothed (snapper)	Lutjanidae
Pelvic fins jugular eyes both on one side of head (order Pleuronectiformes)	
Both eyes on right side of head (right-eye flounders)	Pleuronectidae
Both eyes on left side of head (left-eye flounders)	Bothidae
No pelvic fins. Pectoral fins present	
Mouth tubular, skin containing bony plates (order Syngnathiformes)	Syngnathidae
Eels (order Anguilliformes)—(congers)	Congridae
Body more or less fishlike (order Tetraodontiformes)	
Skin granular, with erectile anterior dorsal spine (leather jacket)	Monacanthidae
Skin spinous, dorsal finrays all soft (porcupine fishes)	Diodontidae

the few southern members is the ling, *Lotella callarias*, which frequents the rocky shelf off Australia.

Order Beryciformes

Family Berycidae. These somewhat resemble the tropical squirrelfishes. Members of the family are widely distributed; *Beryx* and *Actinoberyx* are found off South Africa. The well-known nannygai or redfish, *Centroberyx affinis* (Fig. 27.9), is an Australian food fish, which is bottom dwelling across the whole width of the continental shelf.

Order Zeiformes

Family Zeidae. John Dories very similar to the northern *Zeus faber* (Fig. 27.10) occur in the Southern Hemisphere, although opinions differ as to their identity. The common South African species is referred to *Z. japonicus*, and the best-known of several Australian species is *Zeus australis* with a similar or identical species in New Zealand waters.

Orders Pleuronectiformes and Syngnathiformes

Members of these two orders in the southern oceans closely resemble those of the northern seas. See Fig. 27.11.

Order Perciformes

Family Mugilidae. The sea mullet, *Mugil cephalus*, is very widely distributed in warm temperate seas of both hemispheres. The species occur off all southern continents. They frequent shelf waters during their 3-year growth period, then assemble in schools off estuaries to await favorable conditions to migrate to sea for spawning. In Australia these conditions are met when the west wind blows.

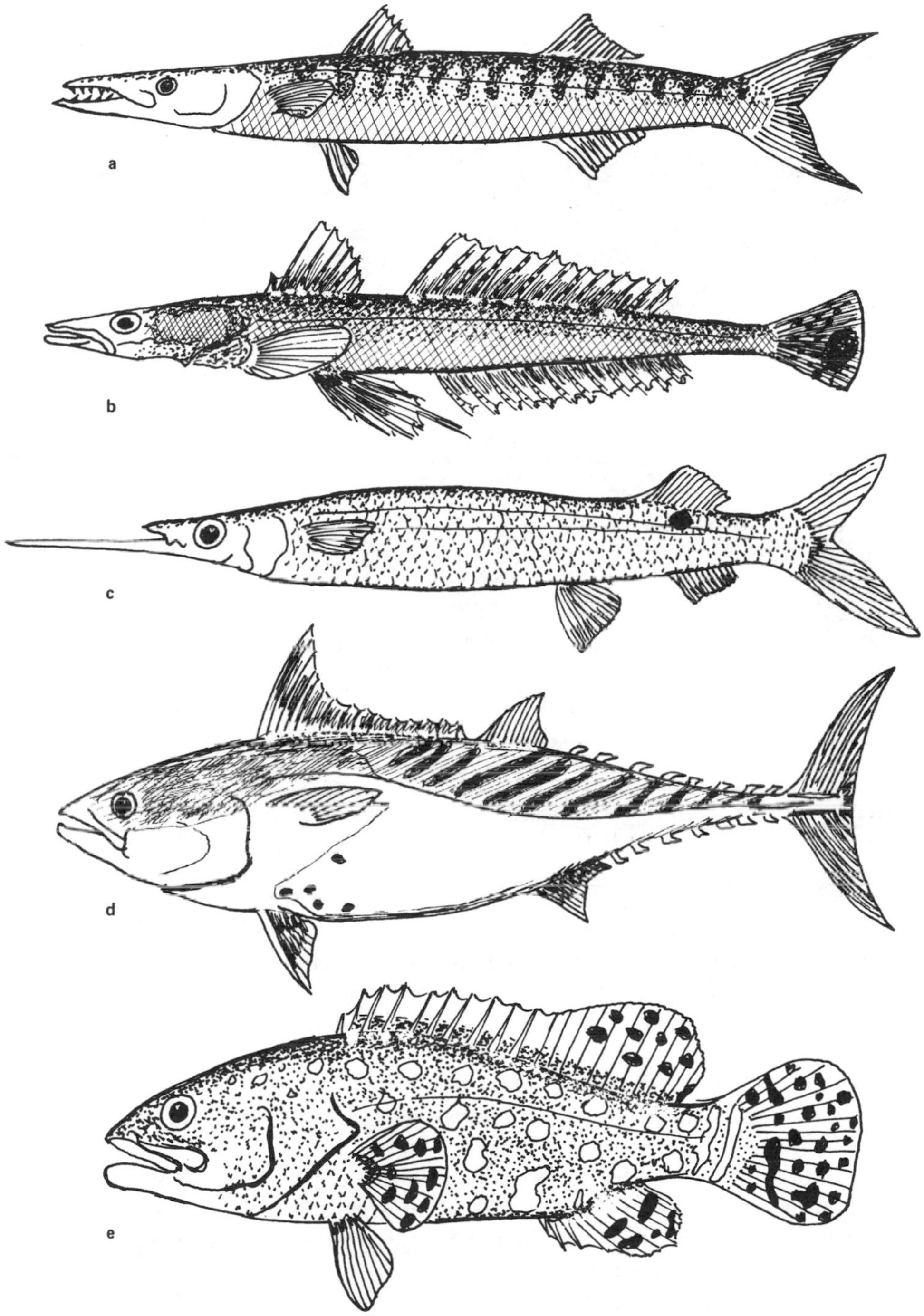

GERALD HESLINGA

Fig. 27.8 Australian marine fishes. (a) Barracuda, *Sphyraena jello*, Sphyraenidae, Perciformes. (b) Flathead, *Platycephalus fuscus*, 80 cm, Platycephalidae, Perciformes. (c) Garfish (called halfbeak in America), *Hemirhamphus welshi*, 30 cm, Hemirhamphidae, Beloniformes. (d) Tuna, *Euthynnus alliteratus*, 1 m, Scombridae, Perciformes. (e) Groper (grouper in American usage), *Epinephelus lanceolatus*, 50 cm, Serranidae, Perciformes.

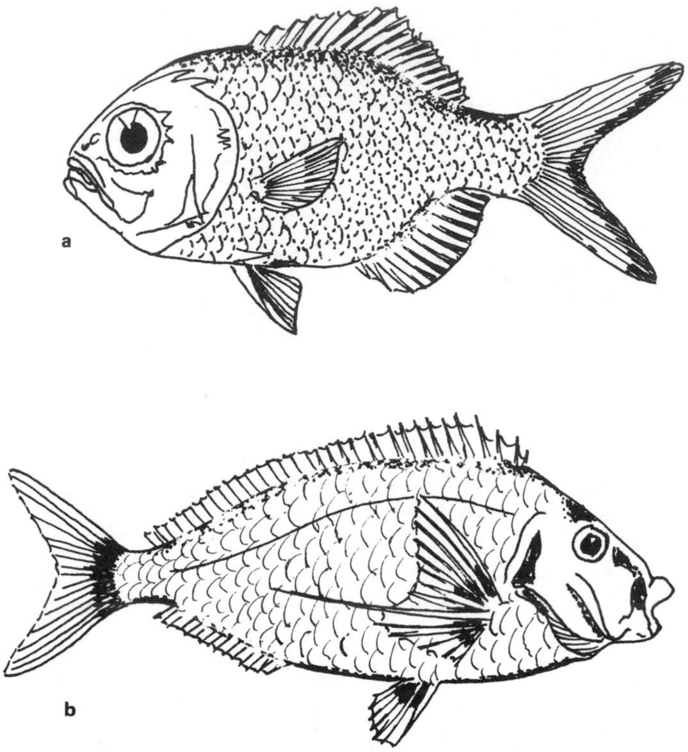

GERALD HESLINGA

Fig. 27.9 Australian fishes. (a) Nannygai (Redfish), *Centroberyx affinis,* Berycidae, Beryciformes. (b) Morwong, *Nemadactylus morwong* (the New Zealand tarakihi is similar), to 60 cm, Cheilodactylidae, Perciformes.

Several of the Perciform families have already been studied under tropical-reef faunas. These will briefly be noted here by reference to southern representatives. The vernacular names used in Australia and New Zealand are given, the New Zealand name generally being the Polynesian (Maori) name. Where these names differ from the vernacular name used in the United States, the American equivalent is given in parentheses.

Family Serranidae. Groper or bass (Grouper), *Epinephelus lanceolatus* (Fig. 27.8) is an Australian example. *Promicrops lanceolatus* may reach 800 pounds. Found in Australia and South Africa on the continental shelf.

Family Carangidae. Silver trevally, *Usacaranx georgianus,* found on continental shelves with surf in Australia. The yellowtail kingfish, *Seriola grandis,* may be pelagic or a shelf dweller. Found in the Southern Hemisphere. Other species of *Seriola* occur elsewhere in the southern seas.

Family Lutjanidae. Sea perch (Snapper of American usage), *Lutjanus superbus,* found on Australian reefs; other species occur on other southern continental shelves. See Fig. 27.12.

Family Sparidae. Snapper or tamure (Porgies of American usage), *Chrysophrys gutulatus,* frequent Australian shelves; young stage also called a bream in Australia; the same genus occurs in South Africa but different species.

a

b

GERALD HESLINGA

Fig. 27.10 Southern Hemisphere fishes. (a) John Dory, *Zeus faber*, 50 cm, Zeidae, Perciformes, cosmopolitan. (b) Topsail drummer, *Kyphosus cinerascens*, 25 cm, Kyphosidae, Perciformes.

Family Labridae. Blue groper (wrasse of American usage), *Achoerodus gouldii*, grows to 60 pounds, found on reefs in southern Australia. The Australian parrotfish (wrasse of American usage), *Pseudolabrus gymnogenis*, also frequents reefs in Australia and Lord Howe Island; these genera are absent from South African waters.

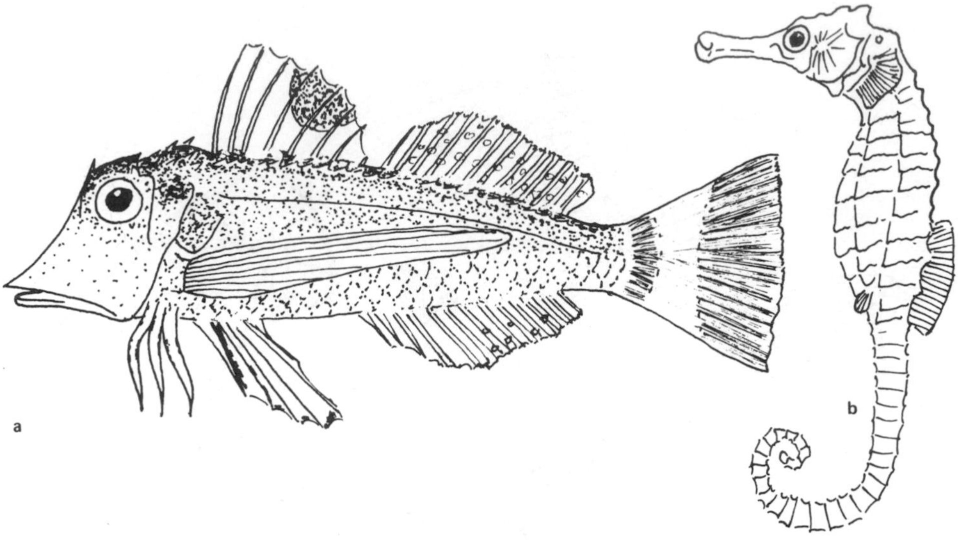

Fig. 27.11 Southern Hemisphere fishes. (a) Flying gurnard, *Dactyloptera papilio*, 20 cm, Triglidae, Perciformes. (b) Sea horse, *Hippocampus planifrons*, 10 cm, Syngnathidae, Syngnathiformes.

Family Scomberomoridae. Spanish mackerel, *Pneumatophorus australasicus*, found in New Zealand and Australia; *Scomberomorus* and other genera occur off South Africa.

Family Scombridae. Bluefin tuna, *Thunnus thynnus*, may reach 1000 pounds. It is both pelagic or shelf dwelling and is found in the southern oceans.

Family Scorpaenidae. Rock cod (scorpion fish of American usage), *Scorpaena cardinalis*, a nonvenomous and edible reef fish of Australia. *Scorpaena natalensis* occurs in South Africa.

The following are Perciform families lacking from the tropics:

Family Triglidae. Sea robins, or gurnards, spinous-armored bottom fishes that walk about on the seafloor with the aid of three long spines arising from each pectoral fin. *Corrupiscis kumu* is wide ranging in the southern seas and also in the North Pacific. The flying gurnard, *Dactyloptera papilio* is another example.

Family Cheilodactylidae. Morwong of Australia and Tarakihi of New Zealand, *Nemadactylus* spp., shelf-food fishes, characterized by the enlarged central pectoral fin ray.

Family Gempyllidae. Barracouta (called snoek in South Africa and sierra in South America) *Leionura atun* (or *Thyrsites atun*), is pelagic and is found in southern oceans. Not to be confused with the aggressive and dangerous barracuda of tropical seas.

Family Arripididae. Kahawai of New Zealand (so-called salmon of Australia) *Arripis trutta*, shelf, food fish.

Family Platycephalidae. Flatheads, *Platycephalus spp.*, found on the South African shelf. Various genera occur in Australian waters (e.g., *Planiprora fusca*, the Australian dusky flathead), generally in estuaries.

Family Kyphosidae. Drummers (Rudderfishes of American usage), *Segutilum sydneyanum*, frequent rocky and weed-grown shelves.

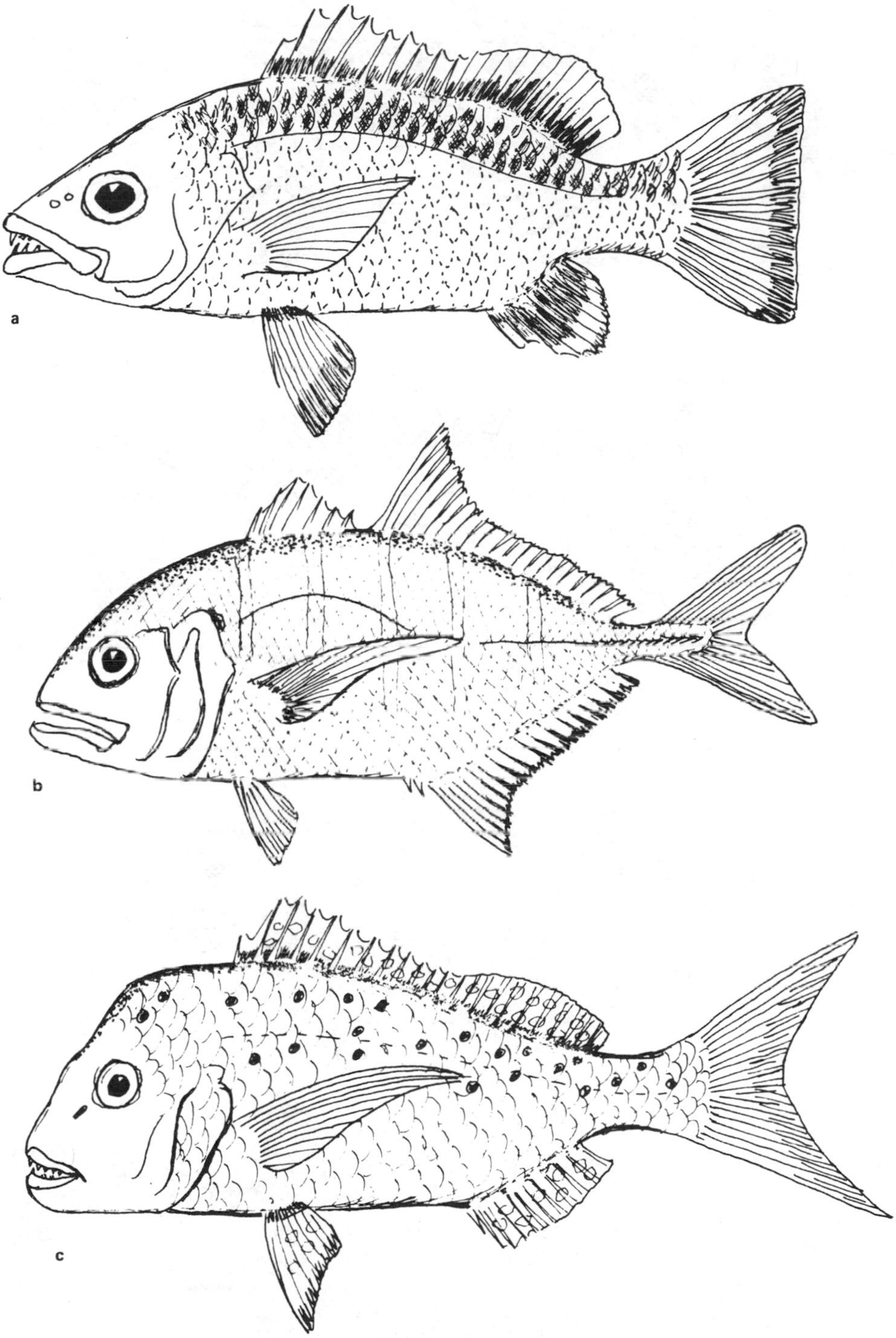

GERALD HESLINGA

Fig. 27.12 (a) Mangrove jack (sea perch in Australian usage; snapper in American usage), *Lutjanus argentimaculatus*, 35 cm, Lutjanidae, Perciformes. (b) Great trevally, *Caranx sexfasciatus*, 30 cm, Carangidae (jacks in American usage), Perciformes. (c) Snapper, *Chrysophrys auratus*, Sparidae (Porgies of American usage), Perciformes.

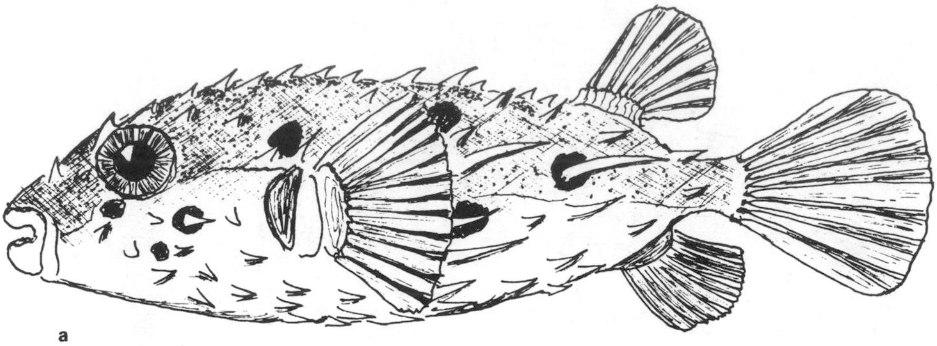

a

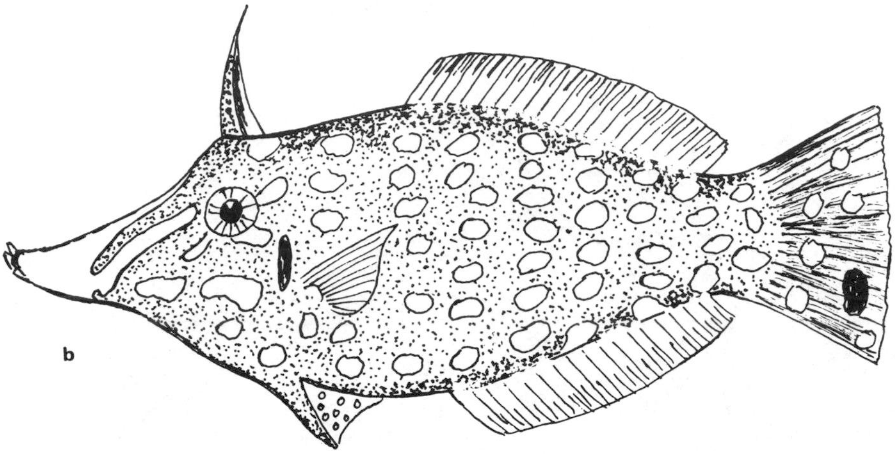

b

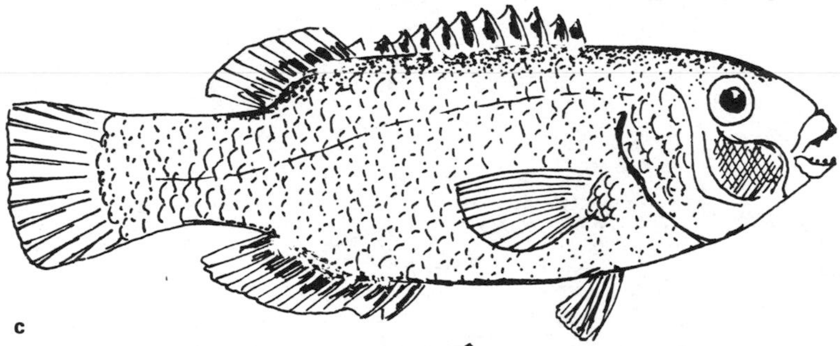

c

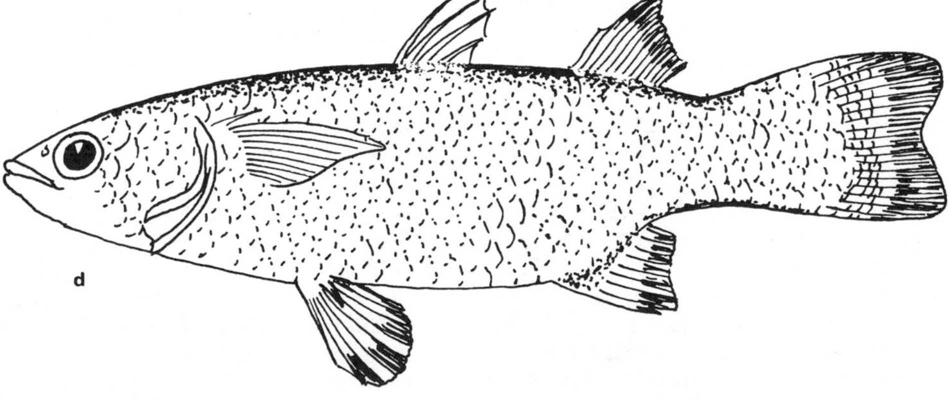

d

Order Tetraodontiformes

The southern representatives of this order generally parallel the tropical examples already studied. The most common are the following families: **Family Monacanthidae** containing the leatherjackets (triggerfishes in American usage) with *Cantherines* and *Nelusetta*; and **Family Diodontidae** containing porcupine fishes, with *Allomysterus*. See Fig. 27.13.

Order Anguilliformes

The eels of the southern seas are similar to those of the northern seas and are referred to the same genera. *Anguilla australis* grows to over 5 feet long and weigh 46 pounds. They are the largest of their kind, believed to breed in deep water off Tonga, otherwise inhabiting freshwater. Congers, family Congridae, are exclusively marine. Eels are called tuna by the Polynesians, whereas other nations call tunny by that name.

GERALD HESLINGA

Fig. 27.13 [Opposite] Australian fishes. (a) Porcupine fish, *Tragulichthys jaculiferus*, 25 cm, Diodontidae, Tetraodontiformes. (b) Leather jacket, *Oxymonacanthus longirostris*, 18 cm, Monacanthidae, Tetradontiformes. (c) Giant pig fish, *Achoerodus gouldii*, 1.5 m, Labridae (wrasse in American usage), Perciformes. (d) Mullet, *Mugil tade*, 10 cm, Mugilidae, Perciformes.

28

THE DEEP-SEA FAUNA

Early attempts to fathom the ocean / The Laplace equation / Can life exist in the abyss? / Pioneer studies of Ossian and Michael Sars / Hypothesis that the deep sea is the refuge of ancient forms of life / Atlantic studies / The *Challenger* Expedition / The *Bulldog* sampler / The *Blake* Expeditions / Alexander Agassiz and Wyville Thomson / Modern sounding procedures / Stratification of benthos / Sampling procedures / Underwater photography / Relationship of organism to substrate / Disturbances of the substrate / Detection of bottom currents / Population densities / Maximum depths of the oceans / Deep-sea food webs / The trenches / Systematic content of Abyssal and Hadal biota

The first attempts to fathom the deep sea were made by Columbus in 1492 in the mid-Atlantic and by Magellan in 1521 in the mid-Pacific. As was inevitable, both attempts failed for neither investigator had the logistic resources to carry a sounding line long enough to reach to the ocean floor, and neither had the remotest idea of the depth of the water column (2000 fathoms) beneath his vessel. So the depth of the ocean floor remained a mystery for another 250 years. Then, in 1775, Pierre Simon, Marquis de Laplace, announced from his study chair at the Ecole Militaire, in Paris, that the average depth of the Atlantic Ocean is approximately 13,000 feet, or 2.5 miles. He performed this remarkable measurement by developing a more perfect theory of tides and then solving an equation.

Accepting Newton's explanation of the tides being due to the attraction of the moon and the sun, he engaged upon a detailed study of the propagation of a tidal wave across an ocean. Ideally, if Newton's explanation is correct, one would except the high-tide crest to lie beneath the moon and sun when these bodies are in the same direction in space, as at the time of the new moon. On the equator, where the surface of the earth rotates at a surface speed of 1000 miles per hour, the tidal wave should sweep across the ocean at the same speed, so as always to lie beneath the heavenly body whose attraction raises it.

But a study of ships' logs disclosed to Laplace that these circumstances by no means hold. He found that ships' officers reported that on the West African equatorial coast, the high tide occurs 2 *hours after* the moon passes the meridian, whereas on the eastern coast of Brazil, on the other side of the Atlantic, the equatorial high tide occurs 6 *hours after* the moon passes the meridian. Evidently, therefore, something delays the tide as it sweeps across the Atlantic from east to west, following the apparent motion of the moon (i.e., the real west-to-east motion of the rotating earth). Since the earth's rotation causes all points on the equator to move at 1000 miles per hour, if there were no delays, the tidal wave would cross the equatorial Atlantic in 3 hours, for the Atlantic has a width of 3000 miles at that point. As there was an observed delay of 4 hours, the total time taken by the tidal wave to cross 3000 miles must be $3 + 4 = 7$ hours.

Laplace now sought an explanation for the observed delay by investigating the way in which a long-period wave is transmitted through the water. He discovered that there is a limitation on the speed at which a long-period wave can travel, the parameters being set by the value of the constant of gravitation, g, and the depth of the fluid transmitting the wave, such that

$$v = (gd)^{1/2}$$

where v is the velocity of propagation, g is the constant 32 feet per sec^2, the acceleration produced by gravity at the earth's surface, and d is the depth of the fluid, in this case the sea. Thus, for a tidal wave to travel at its expected maximum speed, so as always to crest beneath the moon, moving therefore at 1000 miles per hour (1467 feet per second), we have, by rearranging the terms of the equation above,

$$d = \frac{1467^2}{32} = 67{,}000 \text{ feet, or } 13 \text{ miles}$$

Evidently, therefore, the Atlantic Ocean must be shallower than 13 miles. To find its average depth, Laplace inserted into the equation the observed speed of the wave that, as noted above, crosses 3000 miles in 7 hours. This yields a speed of 637 feet per second, whence we have

$$d = \frac{637^2}{32}$$
$$= \text{approx. } 13{,}000 \text{ feet, or } 2.5 \text{ miles}$$

Can Life Exist in the Abyss?

The advances in science during the eighteenth and early nineteenth centuries led some inquiring naturalists to ponder the implications of Laplace's astonishing discovery. It was now realized that if the average depth of the ocean is 2½ miles, then by far the greater part of the seabed must lie in utter darkness, under pressures that had hitherto been considered appropriate only to the interior of a planet. Priestley and his fellow chemists in England, France, and Germany demonstrated that sunlight was essential for the production of sugars and other organic carbohydrates in plants and that animals lack this power to make organic substances from inorganic ingredients. It seemed, then, that the floor of the ocean must be a lifeless desert. Calculations disclosed estimated pressures listed in Table 28.1; and, in face of such estimates, few naturalists or other scientists would dare to predicate the existence of life. These inferences, however, were shown to be false in 1864 by the pioneer work of two Norwegian naturalists, who may be said to have initiated the modern era in deep-sea research.

East of Lofoten Fjord, within the Arctic Circle, the floor of the ocean drops rather steeply to a depth of 1700 fathoms, or nearly 2 miles. Upon these cold, mysterious waters just over a century ago the pioneering techniques of deep-water sampling were first developed by Georg Ossian Sars. His father, Michael Sars, a priest and naturalist, ministered to a small parish on the Norwegian coast. He wrote occasional scientific reports on the marine animals of the district and encouraged his son to make collections.

Ossian, however, did more. He studied the fishing techniques of the local fishermen and then adapted and improved them until he acquired the skill to make bottom collections in the offshore deep water. Then, one day in 1864, his line brought up something unusual—a tiny yellow crinoid with delicately branched arms and a slender stalk about 2 inches long, terminating in a tuft of threads that were evidently rootlike structures to hold it in the sea mud. It was a living sea lily, a type of crinoid differing from the well-known unstalked feather stars by retaining the stalk in the adult stage. In the feather stars the stalk is a temporary embryonic structure that is later discarded. Michael Sars named the discovery *Rhizocrinus lofotensis* and published a detailed account of it 4 years later.

News of the discovery was greeted with astonishment and excitement by zoologists for two reasons. These were:

1. Life had now been proved to exist in deep water, whereas previous researches in the Mediterranean 20 years earlier had led the great naturalist Edward Forbes to declare that no life exists below 300 fathoms.
2. The tiny stalked crinoid resembled similar but much larger fossil sea lilies found imbedded in sedimentary rocks that had formed from the beds of ancient seas of past geological periods. In particular *Rhizocrinus* was considered by the English authority on the group, W. B. Carpenter, to match rather closely certain sea lilies of the Jurassic period.

Not surprisingly the discoveries made by the Sars, taken with the enthusiastic comments of Carpenter, triggered speculation that the hitherto unplumbed depths of the open oceans might prove to be a place of refuge for organisms that had supposedly become extinct millions of years ago. This hypothesis was not entirely justified by events, but the important discovery had been made that the deep

Table 28.1 Ocean Depths and Estimated Pressures

DEPTH	PRESSURE
1000 m (3300 feet)	3/4 ton per square inch (110 atmospheres)
2000 m	1 1/2 tons per square inch (220 atmospheres)
3000 m	2 1/4 tons per square inch (330 atmospheres)
4000 m	3 tons per square inch (440 atmospheres)

ocean is indeed the abode of life and not an empty desert. Other curious and mistaken ideas were dispelled one by one. Some scientists spoke of a mysterious zone in deep midwater where all sunken ships, bodies of drowned sailors, and so on, were supposed to remain suspended in water whose density was so great under pressure as to equal that of the items mentioned. Water is in fact almost incompressible, and objects sink to the floor of the deepest seas. Research ships not uncommonly dredge from great depths such items as beer cans thrown overboard from their own refuse the day before—and less commonly objects thrown or fallen from passing ships 200 years ago or more—all lie upon the seabed, and with them lie the teeth of sharks which shed them millions of years ago, and the hard petrosal bones of whales that lived and died 10 thousand years past.

Atlantic Studies

Norwegians were gratified to discover the fame of their native sons, and Michael Sars was offered a chair at the University of Christiania (where, to be truthful, his students found his lectures unbearably dull); Georg Ossian Sars went on to become a noted marine biologist. But the most far-reaching effect was that on both sides of the Atlantic immediate plans were put into execution to carry out deep-water sampling. Between 1867 and 1869 two American expeditions, the *Corwin* and *Bibb*, were active under the direction of Pourtalès; and from 1868 to 1871 a series of English expeditions was launched by Wyville Thomson, Carpenter, and Jeffreys, namely, the *Lightning*, the *Porcupine*, the *Valorous*, and the *Shearwater*. In New England, Alexander Agassiz of Harvard's Museum of Comparative Zoology offered his services to both Pourtalès and Wyville Thomson, and he participated in the early sorting and studies of the new collections, later assuming responsibility for reporting upon the deep-water sea urchins (Fig. 28.1). The crinoids were assigned to Carpenter.

The *Challenger* Expedition

In Britain public support was now found for those branches of science related to the sea. Queen Victoria's cabinet was induced to act upon the persuasive arguments of Wyville Thomson and others, and a major deep-sea exploring expedition was planned in considerable detail and provided with all the investigatory equipment and scientific staff that the resources of the day could encompass. This expetion of the 1870s, known as the voyage of H.M.S. *Challenger*, resulted in a great series of scientific reports, most of which are still in constant use with new editions appearing even in recent years. The expedition, which sampled the deep seas of most of the world from Arctic to Antarctic, marks the solid foundation of our present understanding of life in deep waters.

There is a certain irony about the meteoric growth of oceanography at this epoch, for, in the light of after-knowledge, we now perceive that the British scientists had apparently overlooked a significant discovery made as early as 1860 by Her Majesty's Ship *Bulldog*. In the course of a cruise performed in that year the assistant engineer on board, named Steil, had invented an improvement in sounding machinery by means of which the vessel was able to recover about one pound of deep-sea mud from a depth of 2000 fathoms. Furthermore, one of the samples

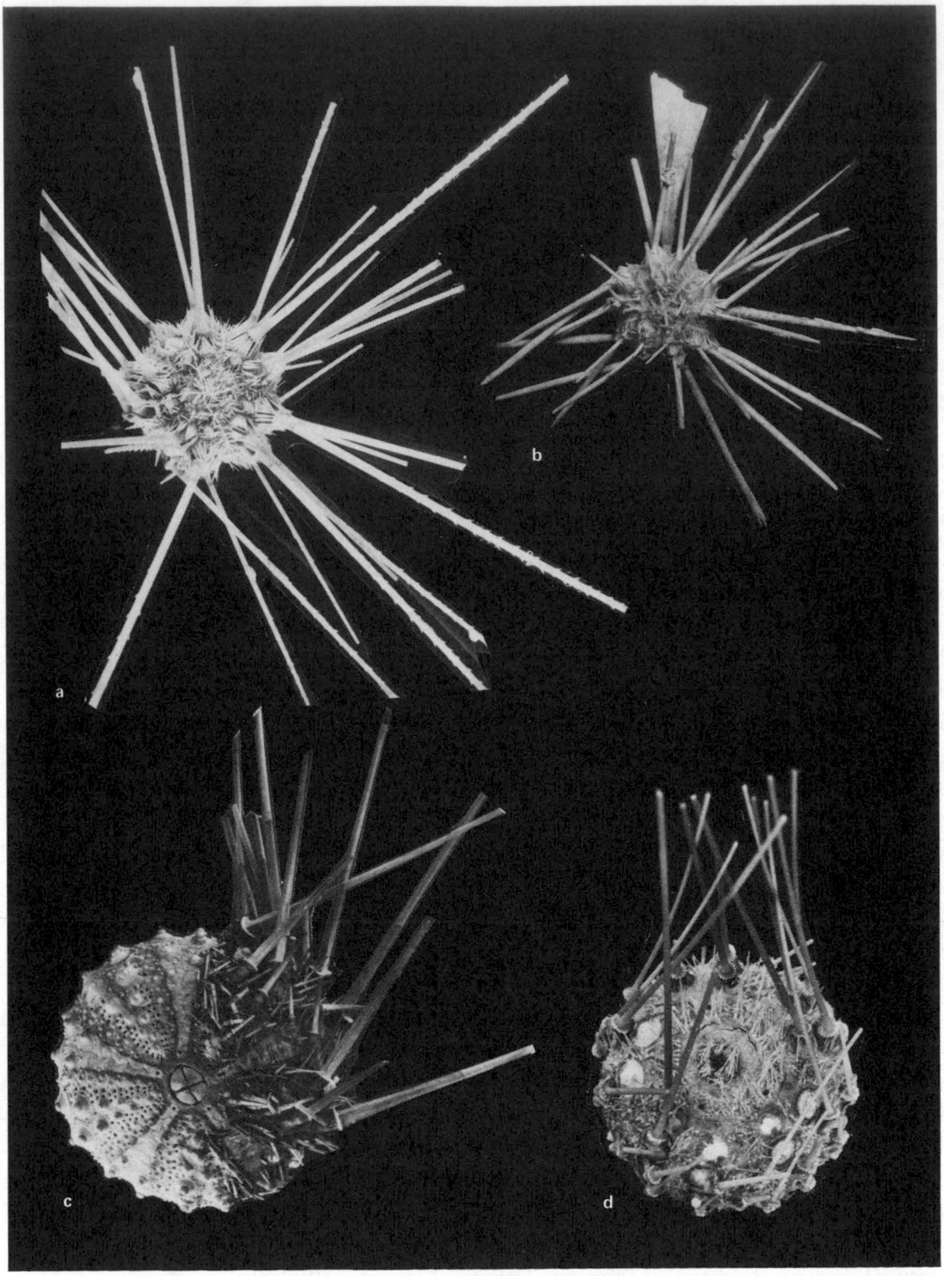

JOSEPH D. GERMANO

Fig. 28.1 Deep-water urchins. (a) *Histocidaris*. (b) *Cidaris blakei, Caribbean*. (c) *Coelopleurus*. (d) *Aspidocidaris*.

brought up from 1260 fathoms had yielded a starfish. Now, as we realize today, starfishes are among the most archaic animals surviving, and their extant members include representatives of forms of life that existed half a billion years ago, twice as ancient as existing crinoids, whose lineage seems to have begun in the Triassic period. Even the oldest extinct crinoids, from the Ordovician period, are no older than the oldest known kinds of starfishes. In other words, the *Bulldog* in 1860 had already disclosed the same kind of information as came from the lines of the Norwegian naturalist, but the scientific world was not then ready to appreciate this fact. Nor is that all. A careful review of the Admiralty archives reveals that Sir John Ross, working in Baffin Bay in the year 1816, had not only invented a clamshell grab capable of bringing to the surface 6 pounds of seafloor mud from a depth of 1000 fathoms, but also, using this device, he had actually recovered a living starfish from 800 fathoms. This event occurred eight years before Professor Forbes announced that no life exists in the sea below a depth of 300 fathoms. Clearly the much talked-of breakdown in communications is no prerogative of our own tumultuous age! These afterthoughts constitute, of course, no criticism of the eager band of nautical-minded naturalists now setting sail to fathom the uttermost parts of the ocean.

By 1872 H.M.S. *Challenger* was ready to sail on what was to be her epoch-making cruise. Everyone was now talking, thinking, dreaming crinoids. Everyone included Robert, the talkative African gray parrot, renowned for his technical repertoire, and whose exploits have been faithfully recorded in the official *Narrative* of the voyage of the *Challenger*. His most notable utterance is recorded to have been: "What! Two thousand fathoms and no bottom—Ah, Dr. Carpenter F.R.S." The *Narrative* mentions also Robert's companion, of the same species but opposite sex, who inadvertently took a bath one day in a beaker of boiling potash and so passed into solution and out of history. Perhaps the most memorable day's dredging was July 14, 1874, off the Kermadec Isles north of New Zealand, when no fewer than 11 species of crinoid were taken from 600 fathoms (Fig. 28.2). This was "probably the richest ground dredged by us at all," wrote Mosely afterward in 1879, when *Challenger* was once more back in British waters. On the outward leg of the voyage, Alexander Agassiz had joined the ship briefly at Halifax, Nova Scotia. Later, Wyville Thomson wrote to him from the Straits of Magellan suggesting that Agassiz rejoin the team in Edinburgh, there to participate in the sorting of the collections and their distribution to the specialists in various countries who had been asked to report upon the results. This work was shared by Thomson, Murray, Agassiz, and also the German biologist Haeckel. From surviving letters of the period we learn that Agassiz had a reputation for quiet gravity whereas Haeckel was remembered for his boisterous merriment during the long busy days. At length the crinoids were all delivered safely to Carpenter, and Agassiz could retire to Massachusetts to gloat over the sea urchins.

The *Blake* Expeditions

Agassiz's respite was brief. By 1877 he was once more at sea, this time invited to take charge of the scientific work on board the U.S.S. *Blake*, under the command of the (then) Lieutenant-Commander Sigsbee. The two men, each gifted with mechanical skills, made an excellent team, introducing several important advances in deep-sea research equipment. The substitution of wire rope for hemp and the introduction of the Blake trawl both date from this period. Off the coast of Cuba the crinoid *Pentacrinus* (*Cenocrinus* in modern parlance) began to appear in the trawl, at first only as broken fragments; but, on a second visit to the area, Sigsbee was able to secure 20 complete specimens. The following year, 1878, *Blake* was placed at Agassiz's disposal for another cruise, on which occasion some crinoids were taken off Havana. By March 11, 1879 we find Agassiz writing to tell Wyville Thomson:

> I am surprised to see how many of the *Challenger* things I have brought up—I have quite a number of stems of *Rhizocrinus rawsoni* but only got ten heads complete. As for the other *Pentacrinus* I have enough now to feel satisfied that there are *two* species, and I shall be able to send you both. What do you think of bringing up in one haul 124 of them. I thought I should jump overboard when the tangles came up loaded with them. This brings me to ask you if you want any more for dissection? Ludwig who wrote the paper on *Rhizocrinus* wanted also *Pentacrinus* just before I left, but as you had them in hand I did not wish to send him any specimens for work till I knew what you were about. . . .

A month later, April 9, 1879, Agassiz wrote to Milne-Edwards in France:

> About the end of June I shall be able to send you the collections [of crustaceans] . . . I have a very interesting collection of hermit crabs that occupy fragments of bamboo and other pieces of wood transported to great depths by the trade winds. The collection of ophiurans is astonishing. The Sponges, Mollusks, Fishes, with the Annelids, Echinoderms and Actiniae make a collection but little inferior to that of the *Challenger*. What do you say to 300 *Pentacrini*! Two kinds!

The following year Agassiz was once more on board the *Blake*, for the third cruise.

The foregoing historical material and extracts from letters and diaries serve to convey something of the excitement and energy of the founders of modern oceanography, and they also illustrate the international character of marine science and good personal relations that can develop between scientists of various nationalities. This aspect, happily, has always characterized the field, and has led to such successful programs as the International Geophysical Year (1957–1959), the International Geophysical Year (1957–1959), the International Indian Ocean Expeditions of the 1960s, and comparable current programs. Some of the results of these researches may now be considered.

Modern Sounding Procedures

During the nineteenth century experimental oceanography developed as a separate discipline, following the need for seafloor surveys prior to the laying of the first transoceanic cables. Long tapered cables and wires were produced in lengths great enough to reach the deepest parts of the ocean. The invention of a meter for measuring the tension applied to a suspended cable permitted direct fathoming, for the touchdown on the seafloor could now be detected by the fact that the tension on the cable became stabilized when once the free end began to pile up on the bottom. A clinometer measured the angle of the cable, permitting solution of the vertical side of the triangle whose three angles were known, and whose hypotenuse was the cable of known length. These methods disclosed depths of 5 miles in certain parts of the sea.

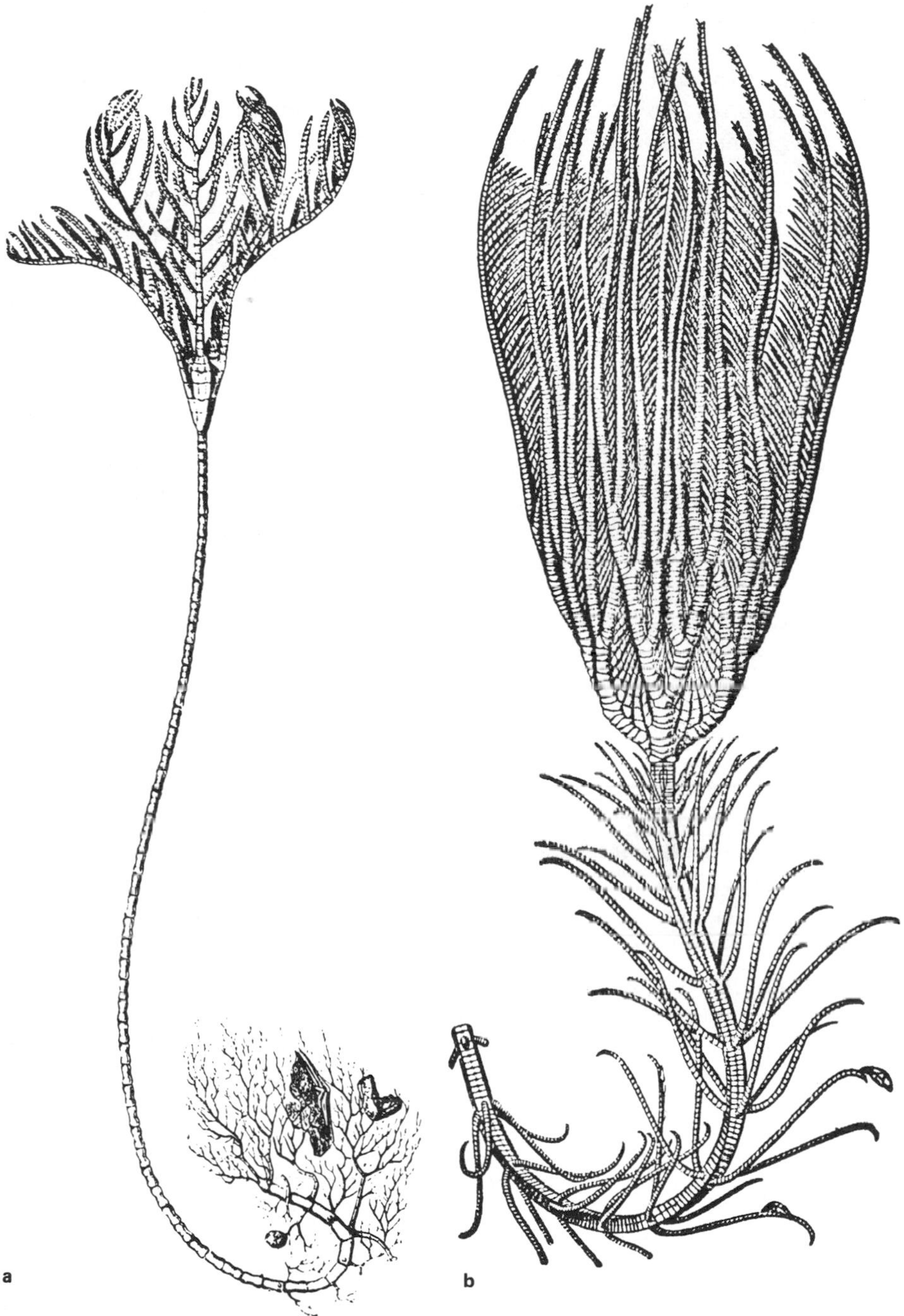

Fig. 28.2 Deep-sea crinoids or sea lilies. (a) *Bathycrinus,* after Sars. (b) *Metacrinus,* taken by H.M.S. *Challenger,* after Carpenter. The discovery of these animals by Sars and the *Challenger* expedition marked the beginning of deep-sea research.

Modern soundings are made with the aid of echo sounders, which give a continuous recording along any transect followed by a ship. Thus very detailed topographic maps of the ocean bed are now becoming available at relatively little cost.

Furthermore, on account of the ability of sound waves of particular frequencies to penetrate the sediments of the seafloor and to be reflected from the various discontinuities that occur between sediments of different consistency, it is possible for the echo sounder to yield charts showing the successive layers below the seafloor. Again, when deep sea cores are obtained, the bit may wear out before the desired penetration is completed. In such cases, if the reflective properties of the penetrated layers have been recorded by the sounding equipment, similar properties may be recognized in the lower layers recorded by the sounding equipment, even though those layers have not actually been penetrated or sampled by the coring devices. Such vessels as the *Glomar Challenger*, for example, presently engaged on coring the ocean bed, are able to extrapolate results into layers presently beyond the reach of the corer.

Stratification of Benthos

The *Challenger* and later expeditions established that beyond the continental shelf the seabed slopes more steeply; this steeper part, known as the **continental slope,** plunges to about the average depth computed by Laplace, 2½ miles or 2000 fathoms, and then becomes more nearly level. Actually there are numerous submarine ridges and depressions. In some restricted regions, usually near land, the earth's crust seems to be thrown into great vertical folds, with deep **trenches** extending to depths of 5 and 6 miles below sea level. The benthos lying upon, or imbedded in, the floor of the ocean is classified by depth in Table 28.2.

Sampling Procedures

Samples of deep-water benthos are obtained by what are essentially the same methods as those for the continental shelf. The collecting instruments are dredges, beam trawls, and grabs, as well as sledges and corers (Fig. 28.3)—especially the Pfleger corer (Fig. 3.8). The principles are the same, but the manner of operation demands more sophisticated controls and more experienced or more ingenious operators.

Underwater photography has become an indispensable technique both in qualitative and quantitative investigations. Besides showing life in its natural state, photography enables more random sampling of the epifauna than do methods such as dredge or trawl sampling. To obtain maximum information, photography and

Table 28.2 Vertical Stratification of Benthos

DESCRIPTIVE TERM	DEPTH AT WHICH BENTHOS OCCURS (IN METERS)	EXAMPLES
Littoral	0–5	Biota of beaches, reefs, mangrove sloughs
Sublittoral	5–20	Region commonly entered by scuba divers
Neritic	0–200	Benthos of the continental shelf
Archibenthal	200–1,000	Benthos of upper continental slope
Bathyal	1,000–2,000	Benthos of lower continental slope
Abyssal	2,000–4,000	Fauna of the general oceanic floor
Hadal	4,000–10,000	Fauna of the deep trenches

NOTE: For *hadal,* the term *ultrabyssal* (Russian, *ultraabyssal*) is also used.

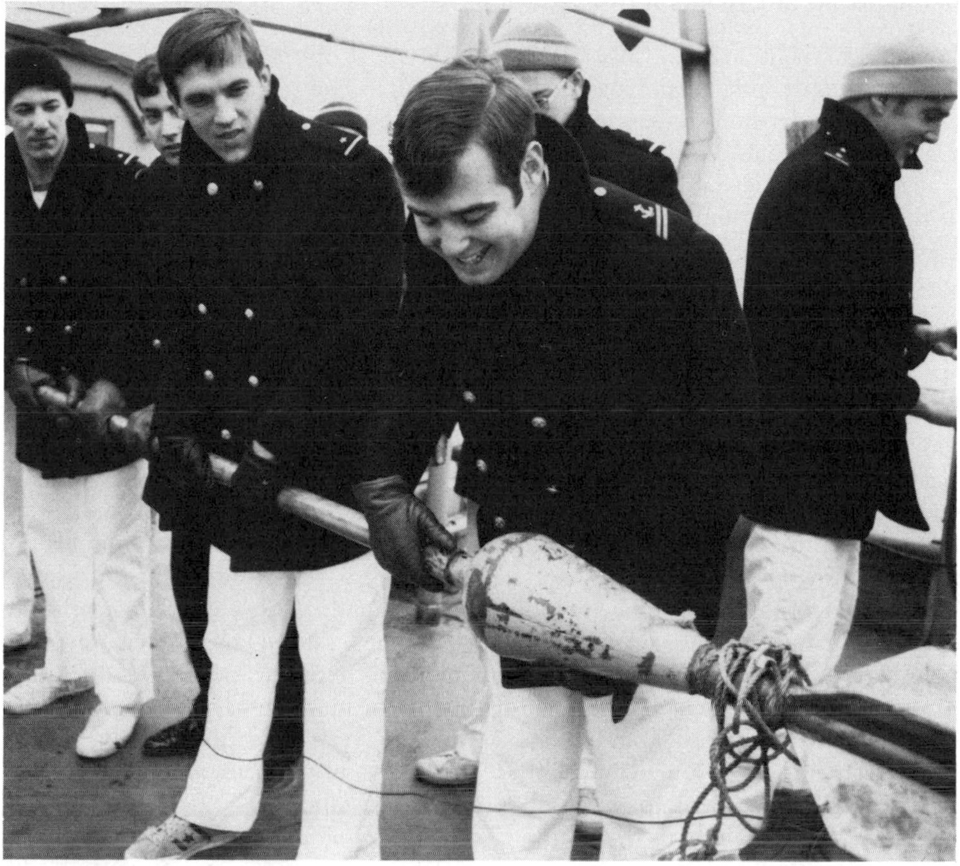

JAMES F. CLARK

Fig. 28.3 Piston corer for sampling sediments of the seafloor.

more traditional collecting methods should be used together, each supplementing the other. In this way much can be learned about the ecology of a region, including feeding habits, modes of locomotion on abyssal substrates, reaction to light and temperature change, community formation, and population densities of underwater life. Seafloor photographs, if stereo pairs are obtained, provide valuable information on the topography and topographic arrangement of biota. It does not reveal much, if anything, about the infauna, however. On the other hand, traditional dredging sampling often causes the epifauna to take flight before the dredge passes over, and the method therefore tends toward selective sampling of the infauna. A combination of techniques gives a more nearly balanced and randomized sample.

To be most effective, seafloor photographs should be taken at frequent intervals along the track of any ship that is trawling or dredging for biological specimens. If possible, photographs should not be taken without also securing actual biological samples, for identification of images in photos is always difficult, often impossible, without comparative material in the form of specimens of biota from the same or nearby station. Neither should biological samples be taken without seafloor photographs, unless it is unavoidable; photographs nearly always throw extra light upon details of the environment and the ecology of the animals collected.

Relationship of Organism to Substrate

While some benthic organisms tolerate a wide range of seafloor conditions, many are restricted to particular types of bottom. This is not always evident from the collections taken from trawls and dredges, because these instruments may traverse a variety of bottoms during a single transect. Here a useful adjunct to the collecting procedure is provided by taking periodic bottom photographs during the progress of the transect. Bottom may vary from naked or thinly silted rock to coarse or fine gravel, or soft silt or ooze. On the harder substrates many of the organisms will be anchored to the hard objects, either permanently, as in the case of sponges, or corals, and similar sessile forms, or facultatively mobile as in the cases of crinoids, with clasping organs such as cirri. The epifauna on a soft substrate may itself provide a hard substrate for such mobile elements as crinoids or ophiuroids.

Soft substrate may form flat submarine plain or the floor of a submarine canyon carrying a characteristic population of organisms so constructed that they either do not sink into the soft material or else have means of extricating themselves from it. A typical population of this type in deep water will comprise in large part echinoderms, especially ophiuroids and holothurians. The ophiuroids do not sink into the soft substrate because their weight is distributed by their long, horizontally disposed arms. The holothurians are usually very delicately constructed consisting mainly of watery fluids in fine membranes. Photographs taken from the U.S. Navy bathyscaphe *Trieste* show that they can actually walk across the sediment on the tips of their lateral rows of tube feet (Fig. 28.4). Some benthic animals such as free crinoids and echinoids develop very elongated cirri or spines. These evidently provide a stable support by sinking just so far into the sediment as is needed to reach a state of equilibrium without submerging vital organs in the substrate. Sea urchins of the family Aspidodiadematidae possess exceptionally long, hollow curved spines that are very delicately constructed and so do not add appreciably to the weight, yet are very effective in supporting it in a soft environment. These echinoids are almost invariably shattered in the process of bringing them to the surface, photography alone provides an undistorted record of their ecology.

Disturbances of the Substrate

Photographs in submarine canyons sometimes disclose relatively steep profiles, resembling hillsides. If the substrate is soft, then crevasse-like openings are sometimes visible, suggesting that periodic slumping must occur. Doubtless many fossil assemblages, in which many animals of the same species and age are found preserved in the same bedding plane, may have been entombed through sudden slumping of soft substrate of this type. (See Fig. 28.5.)

Feeding habits are related to substrate. Probably the main source of nutriment in the substrate itself in the deep-sea environment is the bacterial content, together with its by-products. To secure enough nutritives from such a source, holothurians swallow relatively immense quantities of mud, thus continuously disturbing the upper layers and preventing any very precise stratification of matter falling from above. Hence considerable care is needed in distinguishing the evidence of living bottom communities (biocoenoses) from that of secondarily associated evidence of materials fortuitously brought together after death (thanatocoenoses). This warning particularly applies to grab or core samples, where quantitative treatment is often

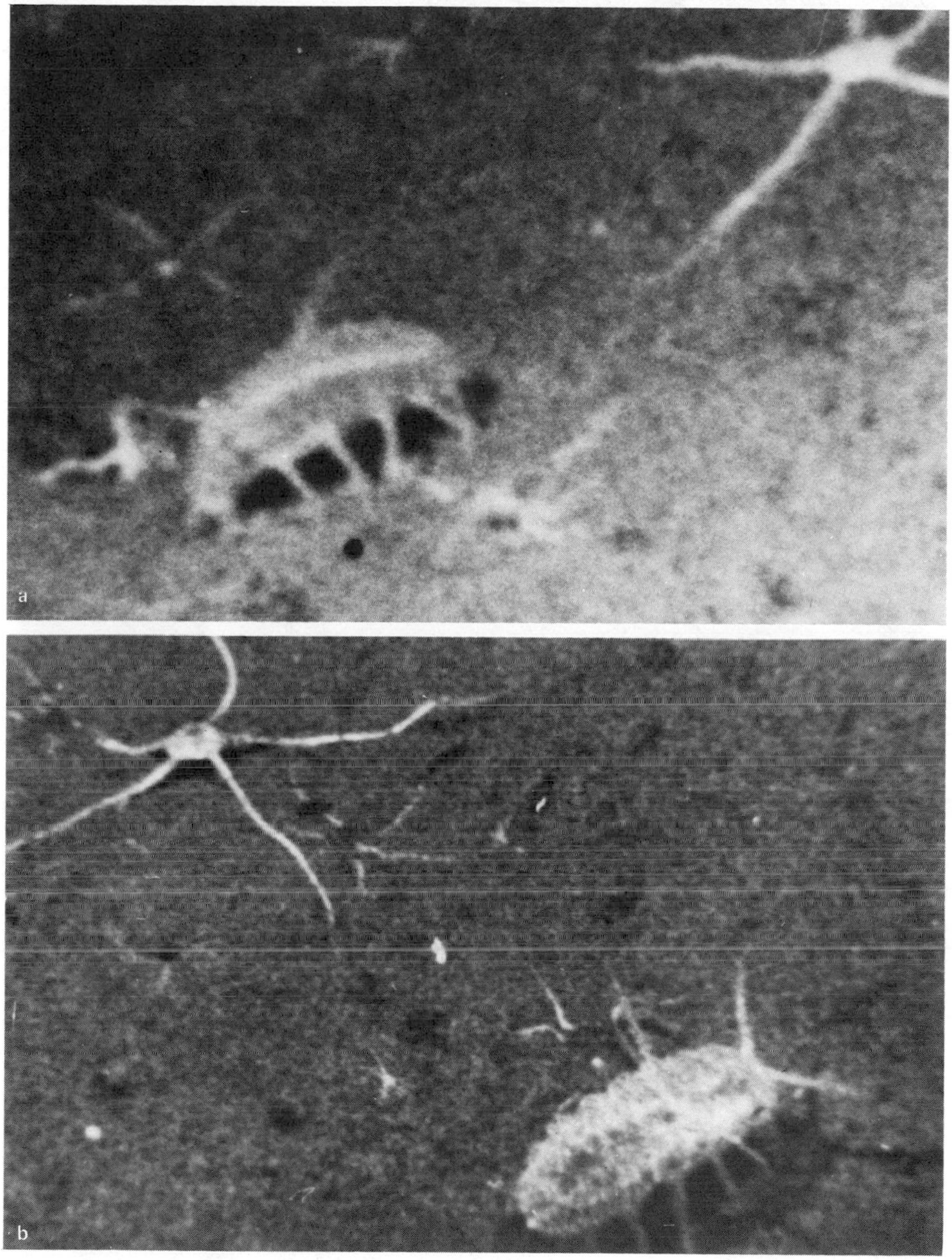

U.S. NAVY, ERIC BARHAM

Fig. 28.4 Details of North Pacific seafloor at a depth of 750 fathoms. Holothurians (*Scotoplanes*) walking on enlarged tube feet, other enlarged tube feet are held out as "tentacles." In view (a), the holothurian has the mouth directed toward the mud as if sucking in the substrate. The ophiuroids (*Ophiomusium lymani*) are feeding on detritus. Photos taken from the U.S. Navy bathyscaphe *Trieste*.

accorded the samples. Seafloor photography warns us that it is highly improbable that annual increments of sediment and microfossils can survive for long in undisturbed horizontal layers on the bottom of the sea, except under unusual abiotic

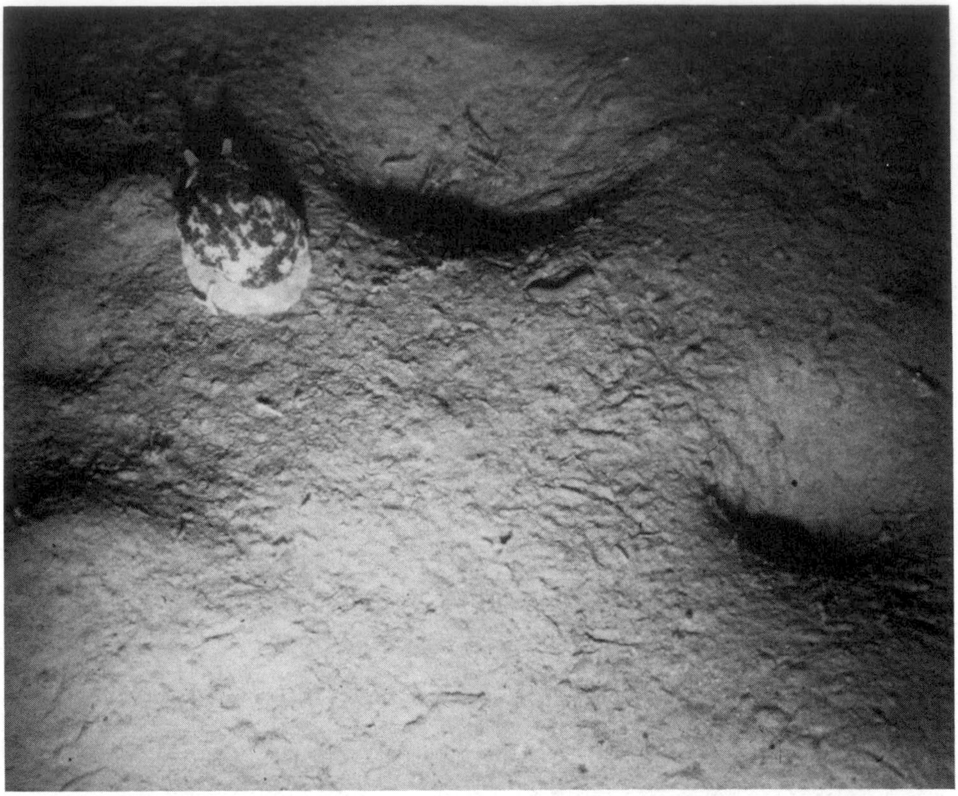

NEW ZEALAND OCEANOGRAPHIC INSTITUTE, JOHN BULLIVANT

Fig. 28.5 Barren mud of the Antarctic seafloor, photographed by remote control, on the continental slope off McMurdo Sound. An unidentified sea slug is creeping past depressions apparently produced by unseen infaunal organisms.

conditions. Indeed, the castings of holothurians are among the commonest objects on the seabed disclosed by photography, and these imply the constant dislocation of sediments. Castings fall to pieces at a touch, so they are never taken in a recognizable state in sampling devices operated by remote control; here then is an area of study where seafloor photography has proved critical.

Detection of Bottom Currents

Bottom photographs often disclose a nonrandom orientation pattern bottom communities, more particularly the orientation of sessile members of the community. Fixed animals, and also those errant forms that adopt a sedentary habit when environmental conditions permit, may sometimes be seen directing the mouth or feeding organs upward. Presumably this means that they are collecting fragments of organic materials, or small organisms, that are falling from above. This would imply relatively little movement in the water mass. Under such conditions normally mobile wandering holothurians may adopt the posture of a sea-anemone, standing upright with the body anchored by the posterior tube feet and the oral tentacles at the upper, free end, held out in a ring. A random orientation of such erect organisms would suggest the presence of micronekton, or swimming plankton, capable of being captured from various directions if the water mass remains still.

Photographs of the Antarctic seafloor frequently show a definite orientation of the polychaetes such that all will have the free tentacle-bearing end of the body directed the same way (Fig. 28.6). It is inferred in these cases that a gentle bottom current is flowing toward the animals from the direction in which they are pointing. Sponges, which can create their own feeding currents by the vigorous activity of the flagellated cells lining their gastral chambers, generally show quite random orientations, and mobile benthos, such as ophiuroids, asteroids, and mollusks, appear in photographs to be creeping randomly. In general, mobile animals disregard the current.

NEW ZEALAND OCEANOGRAPHIC INSTITUTE, JOHN BULLIVANT

Fig. 28.6 Mixed benthic community on the Antarctic seafloor. The bottom current evident from the biassed orientation of the tubicolous polychetes. The current flowing from the upper right corner toward the lower left corner of field.

Population Densities

Transects utilizing photographic techniques can provide good data on epifaunal elements. In most areas, however, there are also likely to be infaunal elements, such as burrowing sea urchins and burrowing asteroids. These are likely to be missed by any technique wholly dependent upon photography. Consequently the photographic method cannot replace traditional dredging procedures. Clearly, the biological samples are needed, not only to provide accurate determinations of the fauna visible in photographs, but also to yield data on the infauna that is not visible in photographs. Conversely, without photography, small but numerous members of the fauna may be lost through inappropriate mesh size, or at least reduced numerically in the samples, thus confounding qualitative analysis. Highly mobile species escape from the mechanical sampler and can only be taken by lucky accident, whereas the camera registers their existence without difficulty. The camera also will record peculiar attitudes, for example, fishing postures and plankton-trapping postures, that throw light on the bottom ecology. So neither camera nor dredge alone is sufficient, for neither one can replace the other. Population densities cannot accurately be studied without cameras. The high incidence of feather-stars in Antarctic seafloor trawl and dredge samples has been interpreted in the past as indicating high population densities of crinoids in the Antarctic. Photographs do not confirm this. It is more probable that feather-stars are disturbed by the passage of the trawl, swim slowly above the seafloor, and become immediately entangled in the net, whereas other smaller or more mobile elements escape the net (Fell, 1967). Thus, populations should be studied from photographs, or better, from photographs plus dredge or trawl samples.

Maximum Depths of the Oceans

The greatest depths of the oceans occur in relatively narrow elongated **trenches,** usually not wider than about 100 km and up to about 2000 km long. Table 28.3 shows the deepest trenches of the Pacific, Atlantic, and Indian oceans.

Deep-Sea Food Webs

Since there is no sunlight reaching the deep-sea floor, there can be no autotrophic organisms there. The postures of microphagous benthos shows that food particles arrive either as plankton shower from above, as inflowing suspended materials, or as organisms in bottom current (Fig. 28.7). Thus there is a materials input and therefore also energy input from elsewhere. The deep-sea communities must ultimately be dependent upon other communities for their continued existence. Carnivorous members of the deep-sea nekton doubtless feed upon the overlying nekton of the mid-water masses (Fig. 28.8). The latter through the diurnal mechanism of the rise and fall of the plankton and the accompanying vertical migrations of the associated nekton at successive levels, probably set in motion a continuous vertical transfer of energy and biomass, from above downward. Thus, the deep-sea biota must be viewed as an extension of the biota of the shallower parts of the oceans and not as an independent ecosystem.

The deep-sea environment plays a critical part in the rejuvenation of the planktonic ecosystem by way of the ascending currents (upwelling) that carry

Table 28.3 Deepest Trenches of the Oceans

OCEAN	REGION	TRENCH	MAXIMUM DEPTH *Kilometers*	*Miles*	*Fathoms*
Northwest Pacific	Marianas	Mariana	11.02	6.85	6028
Southwest Pacific	Tonga	Tonga	10.88	6.76	5950
Northwest Pacific	Philippines	Philippines	10.50	6.52	5740
Southwest Pacific	New Zealand	Kermadec	10.05	6.24	5494
Northwest Pacific	Japan	Idzu-Bonin	9.99	6.21	5460
West Atlantic	Caribbean	Puerto Rico	8.39	5.21	4585
East Indian	Indonesia	Java	7.45	4.63	4074

dissolved nutrients to the surface from the bottom where the bacterial degraders are active. One is tempted to speculate upon the possibility that the rather sudden rise to eminence of diatoms in the epiplankton in the Cretaceous period, 100 million years ago, may perhaps have been related to an invasion of the deepest parts of the ocean by significant new biota at that time. For example, most of the groups of deep-sea fishes belong to orders whose earliest known members are of Cretaceous or Eocene age.

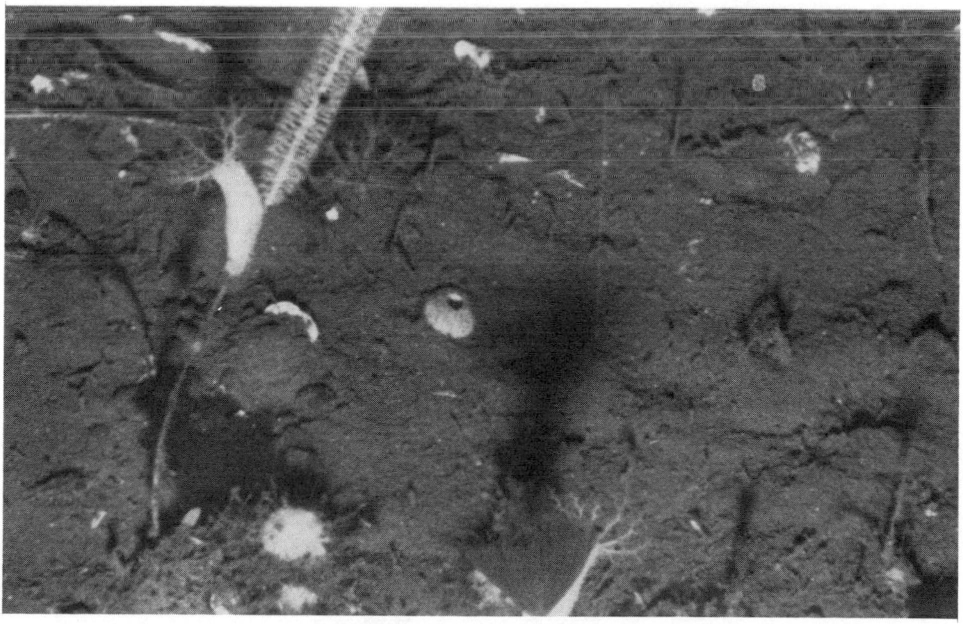

NEW ZEALAND OCEANOGRAPHIC INSTITUTE, JOHN BULLIVANT

Fig. 28.7 Plankton fall from above suggested by the upwardly directed tentacles of the holothurians, which here assume the posture of sea anemones.

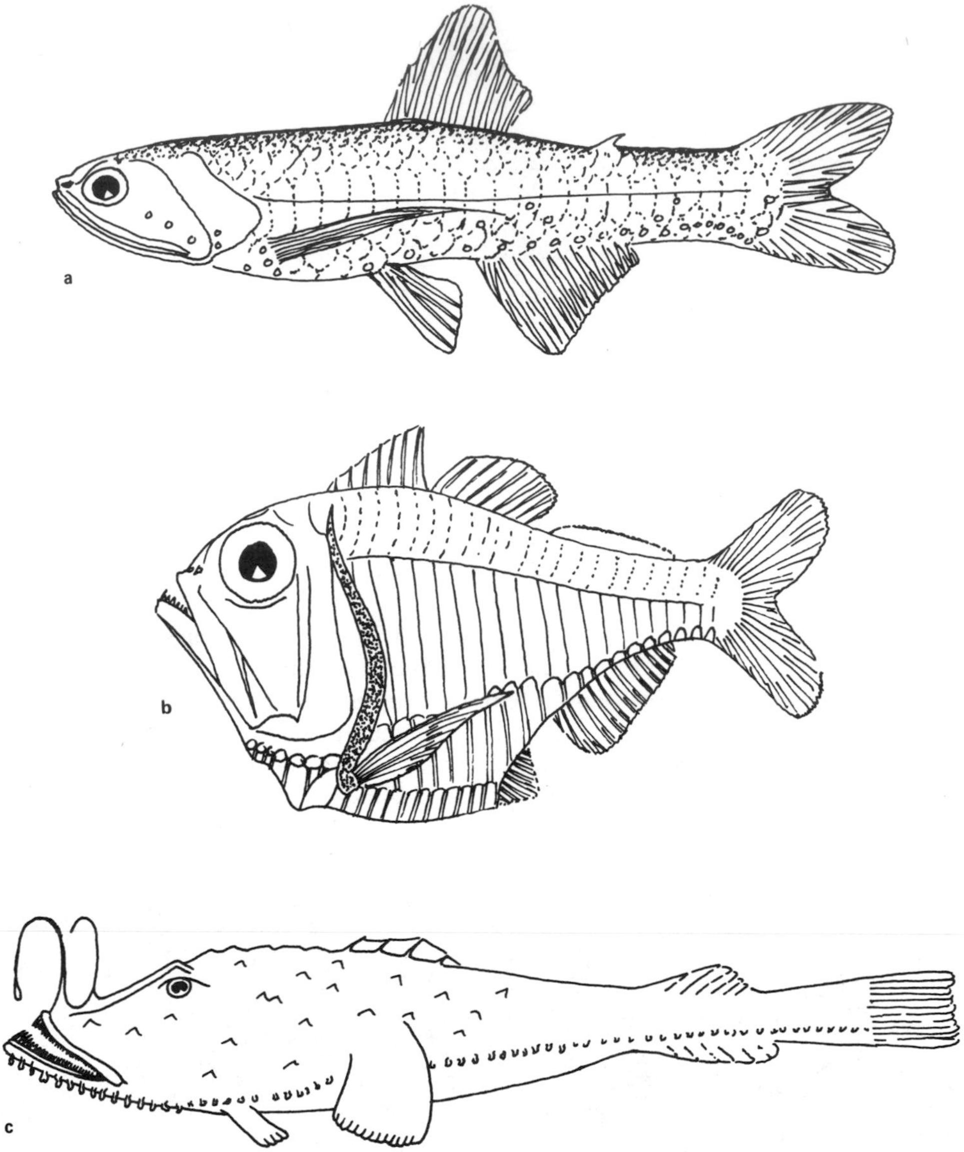

GERALD HESLINGA

Fig. 28.8 Deep-water fishes (a) Lanternfish, *Lampanyctus crocodilus,* 12 cm, Myctophidae. (b) Hatchetfish, *Argyropelecus olfersi,* 5 cm, Sternoptychidae, cosmopolitan. (c) Angler *Lophius* sp., Lophiidae, Lophiiformes.

Systematic Content of Abyssal and Hadal Biota

The phyla represented in the benthos and nekton of the deepest parts of the oceans prove to be the same as those encountered on the continental shelf and in the overlying waters. (See Table 28.4.)

Of groups not previously discussed, the crinoids or sea lilies are members of the phylum Echinodermata and have a structure similar to a starfish that is inverted and usually attached to the bottom by a stalk (the feather stars have no stalk. The Pycnogonida are spiderlike marine animals, classified in the Arthropoda, and also occur in shallow polar seas. Deep-water fishes usually have a whiplike tail if they frequent the bottom. Some have chemical luminescent organs, as for example the lantern fishes (family Myctophidae). Some of these rise to the surface at night, when they may even be caught and eaten by oceanic birds such as petrels.

Table 28.4 Representative Genera of Animals Conspicuous in Deep-Water Environments

GENERIC NAME	PHYLUM OR OTHER GROUP NAME	DEPTH AT WHICH TAKEN (IN KM)	TRENCH WHERE TAKEN
Hyalonema	Sponge (Porifera)	6.86	Kurile
Nereis	Bristle worm (Annelida)	7.29	Banda
Scalpellum	Barnacle (Crustacea)	7.00	Kermadec
Serolis	Isopod (Crustacea)	3.70	(abyssal)
Nucula	Clam (Mollusca)	6.77	Kermadec
Xylophaga	Log-boring mollusk	6.29	Banda
Elpidia	Sea cucumber (Holothuroidea)	9.74	Bonin
Urechinus	Sea urchin (Echinoidea)	4.48	cosmopolitan
Bathycrinus	Sea lily (Crinoidea)	9.05	Kurile
Amphiophiura	Brittlestar (Ophiuroidea)	6.04	Atlantic
Porcellanaster	Sea star (Asteroidea)	7.58	Mariana
Eremicaster	Sea star (Asteroidea)	7.24	Aleutian
Freyella	Sea star (Asteroidea)	6.18	Kermadec
Macrourus	Deep-sea cod (Chordata)	1.00	bathypelagic

29

SOME PROBLEMS OF MARINE BIOGEOGRAPHY

The Tethyan legacy / Where was India during the Jurassic? / Transatlantic immigration / Can shelf benthos cross an ocean?

THE TETHYAN LEGACY

Peninsular India has been an elevated landmass for the past 150 million years—seemingly an island continent. There are marine sediments still surviving to mark the former existence of a northern coastline, which ran from Cutch in the northwest eastward along the present course of the Narbada valley to Baroda. The sea that washed this ancient coast is termed Tethys, and it extended as a shallow epicontinental body of water across southern Asia and North Africa to the Mediterranean region. A relatively uniform marine fauna occupied Tethys from Europe in the west to the Himalayan region in the east, although in the Jurassic period we have Indian sediments only in the Cutch region.

The southern coastline of the Indian island continent is not presently known from Jurassic sediments, but from the Cretaceous onward we have surviving deposits from the Tiruchirapalli district of southern India. They too are richly fossiliferous and yield information on the nature of the fauna that inhabited the sea to the south of India, presumably the forerunner of the existing Indian Ocean.

During the latter part of the Jurassic period, the coral-reef belt of the world was about 60° of latitude in breadth, about the same as today, but it was tilted about 25° toward the north on the Eurasian side of the earth. For lack of Australasian fossils we cannot say if there was a similar tilt to the south on the opposite side of the planet. If there was, then it would imply a difference in the position of the poles, as indeed is postulated by Bain (1963), Fell (1967, 1968), and other writers. Putting aside the question of polar wandering, all data are in agreement that the entire marine faunal assemblages of India for the Jurassic, Cretaceous, and Tertiary periods comprise elements of the coral-reef zone. The corals themselves are richly represented and they, together with the associated invertebrate phyla, including echinoderms, belong to typical Tethyan genera that ranged from Europe across southern and central Asia, and also in many cases occurred in tropical and subtropical America. It would appear, then, that the subcontinent of peninsular India straddled the coral-reef belt and carried reefs along both its southern and northern courses.

Further, and this is particularly true of the Jurassic of Cutch, and the Cretaceous beds of the Narbada valley, the marine invertebrate fauna of the northern coastline was extremely similar to that of the European Tethys. No author had doubted that northern India at these epochs formed the southern margin of Tethys. During the Cretaceous period the coral-reef belt continued to girdle the earth, although not as tilted as before, for its outliers extended to about 37° south latitude and to 50° north latitude in Europe. The Tertiary beds that succeed the Cretaceous continue to register tropical forms, with the reef belt gradually assuming its present direction, symmetrical with respect to the existing equator and poles. Echinoid genera of modern aspect begin to appear in the beds from Eocene onward, at first shared with Europe, later becoming restricted to the present tropical belt. It would appear, then, that these modern genera were original inhabitants of Tethys, but died out one by one in the northern regions of that sea as the earth's climate changed or as patterns of circulation in the ocean were altered. This topic has been discussed elsewhere (Chapter 10).

The relevant point here seems to be that the modern Indian echinoid fauna is the lineal descendant of the original Tethyan fauna, and it owes this heritage to the fact that India has always formed the southern margin of the Tethyan region. After the elevation of the Himalayas at the end of the Tertiary, the realm of Tethys became restricted to the present northern portion of the Indian Ocean. According to Furon (1963), the coast of the Indian Ocean was already evident in East Africa by Permo-Triassic times. Lack of sediments presumably explains our inability to detect an Indian Ocean shoreline on the Indian Ocean subcontinent until the Jurassic. Furon reports that the later Jurassic faunas of Madagascar are very closely correlated with those of Cutch, so this would seem to imply intercommunication between Tethys and the Indian Ocean by late Jurassic times.

The paleontological picture, then, suggests that India has occupied the northern Indian Ocean continuously since at least late Jurassic times, and that its northern peninsular region has lain near the boundaries of both Tethys and the Indian Ocean. The paleontology also implies that throughout the 150 million years since the mid-Jurassic, the coral-reef belt of the earth has had approximately the same breadth as it has today, and India has straddled the reef belt continuously. And it is precisely in these conclusions that paleontologists now find themselves at odds with geophysicists.

Geophysical Evidence

On the basis of paleomagnetic observations, Adie (1965) and Creer (1966) deduce that India formed part of a supercontinent, Gondwanaland, located over or near the geographic South Pole in the early Mesozoic. The supercontinent is inferred to have suffered dismemberment in mid-Jurassic time. In Adie's version, the process terminated in the Cretaceous. In Creer's view the disruption started about 150 to 200 million years ago (Triassic to mid-Jurassic). Adie (1965) states that "India drifted at least 60° of latitude northward in post-Jurassic time." Bullard (1969) says "Paleomagnetic work shows that India has been moving northward for the past 100 million years." McElhinny (1969) deduces that an India–Madagascar–Antarctica block broke away from Africa between 155 and 100 million years ago, thus opening up the Indian Ocean for the first time. An India–Madagascar block then separated from Antarctica and at first drifted southward, before reversing its course to move northward.

There is some mutual disagreement between these various versions, but all agree that India once lay far to the south of its present position, and that this southern origin is to be dated to Jurassic time. The various versions differ as to the time of arrival of India in its present situation, ranging from "post-Jurassic" or Cretaceous to some unspecified time during the last 100 million years. The same authors attribute similar translocations to Australia, except that McElhinny suggests that eastern Australia may have separated from the rest of Australia, and then subsequently rejoined it. On the other hand, Audley-Charles (1966) finds "no evidence to support the contention of continental drift between the continents of Asia and Australia during the Mesozoic, on the contrary all the indications are that the spatial relationships between Asia and Australia have not altered since the beginning of the upper Triassic at least."

Discussion

I believe that the evidence is rather compelling that the breadth of the coral-reef belt has not significantly varied during geological time. If, therefore, India lay some 60° of the latitude to the south of its present position during Jurassic time, it is highly improbable that the Cutch Jurassic marine beds and their included corals could have been formed on the Gondwana basement rocks where we now find them. Yet the geological evidence appears to indicate that the Cutch Jurassic beds rest on their original basement (and have not, that is, been overthrust on to Gondwana basement from elsewhere). Similarly, the evidence appears secure that the reef-bearing beds of the succession of Cretaceous sediments at Tiruchirapalli did in fact form on the basement rock where we now find them. This is also the case with the Cretaceous beds of the Narbada. Further, it seems inexplicable that the Cretaceous Narbada so closely resembles its European Tethyan counterpart, if in fact the Narbada beds were deposited in some remote and unspecified location in the far southern ocean. There is also inconsistency between the African date for the origin of the Indian Ocean as given by Furon (Permo-Triassic) and the inferences of McElhinny cited above. Lastly, it is peculiar, to say the least, that the Tethyan-derived forms that inhabit modern Indian seas should have been preceded by closely related congenera or identical genera in the Mesozoic Indian sediments laid down in some far southern ocean. It is as if the early Indian corals and echinoids had been forewarned that one day India would be attached to Asia, and so they had better take care to adopt tropical Asian characteristics well in advance of their projected arrival date 150 million years in the future.

If we accept the paleontological and paleomagnetic data as both equally well founded, then we would have to conclude that the coral-reef belt and the Tethyan fauna extended through a breadth of 100° of latitude in Jurassic times, although this strange distension of the presumed tropical fauna would apply only to the Indian Ocean segment of it. My own view is that the paleomagnetic data may have been misinterpreted; it is probable that Jurassic India and its contemporary reefs occupied the same geographical position as does modern India. On that view there is no mystery about the Tethyan heritage of a Tethyan land.

TRANSATLANTIC IMMIGRATION

Two principal dispersal mechanisms affect shelf benthos: shallow seafloor connections and surface currents. The latter affect also pelagic organisms. The effect of these factors, including the west-wind drift, has been studied in the Southern Hemisphere (Chapter 27). Just as in extant faunas, so we can also analyze fossil faunas at specified Tertiary horizons and thereby ascertain (in a general sense) the relative importance of former shallow seafloors and former ocean currents, distinguishing their effects from those that might result from former land bridges or from supposed continental drift. In seeking the origin of the Cenozoic shelf faunas of the Atlantic margins of America, the rich Caribbean faunal region proves to be of critical importance.

A review of the generic content of 30 families of shelf echinoids that have had representatives on the Caribbean and Atlantic coasts of the Americas discloses that:

1. Of the 28 families with representative genera in both American and Tethyan seas, the Tethyan representatives appear earlier in the geological record, before the American occurrences.
2. The closest external relationship of the Caribbean echinoid faunas has always been with the west Tethyan fauna (that is, the fauna of that part of Tethys now corresponding to the Mediterranean and West African seas).
3. The American Atlantic and Caribbean faunas have always been relatively impoverished as compared with west Tethyan faunas. Of 28 families represented in both regions, 314 genera are recorded from west Tethys as against only 137 genera from the Atlantic and Caribbean coasts of the Americas.
4. Whereas the west Tethys contributed numerous genera to the Americas, those families that originated in the Americas, such as the Neolaganidae, Echinarachniidae, Monophorasteridae, and Mellitidae, have not contributed a single representative to European, Mediterranean, or West Africa seas, neither have they any representative in any west Tethyan Teriary fauna.
5. The geological record discloses that successive increments of west Tethyan genera have entered the Atlantic and Caribbean faunas of America throughout the Tertiary, apparently as a continuing process. The immigration is evidently still taking place, the latest genera to arrive in America would seem to be those that still have **sibling species** on either side of the Atlantic or that are presently represented by **identical species** in West Africa and in the Caribbean.
6. Severe losses of genera have occurred in the Caribbean at intervals during the Tertiary, thereby depriving its fauna of genera that still flourish in the east Tethyan region (Indo-West Pacific). Among echinoids now surviving in the east Tethyan region, which formerly lived as west Tethyan immigrants in the Caribbean, we may note *Phyllacanthus* and *Prionocidaris*, both represented by

Caribbean fossils. A review of published data on corals (Vaughan, 1919; Wells, 1956) shows that comparable conclusions would be drawn for these reef organisms, with which the echinoids would be ecologically related.

These facts are clearly of considerable biogeographic importance and, since they throw new light on the oceanography of the Atlantic during the Tertiary era, they deserve careful study. It is obvious that a profound relationship exists between the Caribbean and west Tethyan faunas. The fundamental basis of the observed similarities must still exist today, since the extant faunas on either side of the Atlantic have the same relationship to each other as extinct faunas on either side of the Atlantic, despite the fact that the genera involved in the earlier similarities are not the same ones as are involved in the present-day similarities. Thus, it is not a case of two faunas resembling each other because each descends from the same ancestry. Rather, it is a case of repeated immigrations of generic stocks, so that whatever changes occur on the eastern side of the Atlantic are eventually reflected on the western side, because the eastern populations send fresh colonists to the west. Our problem, then, is to ascertain the means by which this sequence of events has come to pass.

We have to isolate a vector that has successfully transferred shallow-water benthic organisms across a minimum of 5000 km of deep oceanic waters. Furthermore, the vector must be one which has operated for the last hundred million years, is still operative today, and acts in such a manner as to favor crossings from east to west, while discouraging or prohibiting traffic in the reverse direction.

It is immediately evident that a continent drift hypothesis fails on all counts. Seafloor cores taken by the *Glomar Challenger* have shown that the Atlantic has been continuously a deep-water basin at least so far back as the Tithonian stage of the Jurassic (at which horizon the drill-bit broke); obviously no serious attention need now be given to any continental drift hypothesis requiring drift onset in the Cretaceous or later.

A land-bridge hypothesis, such as that of Termier and Termier (1952, Cartes 29 and 32), postulates transatlantic bridges in the Paleocene and again in the Oligocene but does not have them in the Miocene or any later epoch. To meet our requirements a land bridge would have to exist throughout the entire Tertiary, would have to be selective in permitting access to some groups while denying it to others and would have to be equipped with one-way turnstiles to prohibit return crossings.

The only vector possessing all the required properties—long persistence through geological time, unidirectional east-to-west transfers, and unavailability to large terrestrial organisms—must surely be an ocean current.

CAN SHELF BENTHOS CROSS AN OCEAN?

The ability of echinoderms to cross ocean gaps of up to 8000 km has been inferred in Chapter 27 on the basis of speciation intensity. Genera drift in a clockwise direction around Antarctica, with maximum speciation at their oldest inhabited base, which lies always to the west, and minimum speciation in their newest colony, which lies always to the east. The vector in this instance is identified as the west-wind drift.

The mode of transoceanic drift can either be flotation as a pelagic larval stage, if this lasts long enough to complete a crossing (Fell, 1953; Scheltema, 1966), always provided the animal in question has a larval stage; or by epiplanktonic drift, in the course of which mature stages are carried across the ocean surface adhering to floating material such as large brown algae. The latter method appears to be more significant in the Southern Hemisphere, where brown algae reach very great size and have been shown to drift distances up to 11,000 km. The seastar *Calvasterias* was observed by Mortensen (1925) drifting in the Southern Ocean, adhering to floating *Macrocytis*; *Calvasterias* is one of the genera supposed by Fell to have colonized Patagonia from New Zealand, where it has its maximum speciation. The alga *Macrocystis* itself seems to have dispersed by drifting, having encircled the Southern Hemisphere and also followed the course of the Humboldt current northward along the west coast of South America to Galapagos; the North Pacific occurrences of *Macrocystis* may well have originated from drifted examples that survived the equatorial crossing in a cool glacial phase of the Pleistocene (when equatorial surface temperatures dropped to about 19°C).

Fragments of the sea urchin *Heliocidaris*, which inhabits brown algae at the intertidal level in eastern Australia have been recovered from the top of the Aotea seamount, 1200 miles to the east, at a depth of 1000 m (Fell, 1962). Carbon dating showed that all the fragments were too young to yield a measurable carbon age, although some were clearly older than others; the newest fragments retained their original pigmentation. The genus is believed to have recently recolonized New Zealand from Australia. It was formerly present in New Zealand, but apparently became extinct during the glacial periods, though surviving in Australia and recolonizing New Zealand in interglacial warm periods. It apparently drifts across the Tasman Sea on floating brown seaweed carried by the East Australian current. The fragments on the seamount were presumably derived from individuals that fell off floating alga. The top of the seamount is current swept, hence fine silt was removed and larger objects, such as shells, remain exposed. On the rest of the floor of the Tasman Sea, averaging 2000 fathoms, the slowly accumulating silt and ooze effectively conceals the remains of other voyagers, such as *Heliocidaris*, the remains of which must fall continuously, eventually to undergo solution (as normally occurs at great depths and pressures).

So far as concerns the Atlantic, some relevant data can be cited. Mortensen (1933) discovered that the common Indo-Pacific asteroid *Patiriella exigua* has colonized the mid-Atlantic isle of Saint Helena. The species ranges in the Indian Ocean so far south and west as to reach the Cape of Good Hope, where it has locally adopted the habit of living on the large brown laminarian alga *Ecklonia*, a genus common in all parts of the Southern Ocean. Mortensen found abundant evidence on Saint Helena to show that quantities of *Ecklonia* drift to the island on the Benguela current, which sweeps past the Cape and then strikes northward toward the mid-Atlantic islands. The sea star has adapted to its new Atlantic habitat deprived of its *Ecklonia*, for this alga cannot survive the warmth of the tropics and lies withered and dead on the beaches where it is cast ashore. Here also we have very clear evidence that the voyage was epiplanktonic, because *Patiriella exigua* is known to have direct development, without any larval stage.

A similar case is exhibited by the echinoid *Rotula orbiculus*, a member of the characteristic West African family, Rotulidae. The family originated in western Tethys in the late Tertiary, and during the Plicene it achieved a distribution along the West African coast. During the Pleistocene coolings, its range contracted to equatorial West Africa. The family has not yet entered the American region, but the species named above has apparently reached the mid-Atlantic islet Ascension.

In these cases we are doubtless witnessing the early stages of what could eventually be a successful transatlantic passage. The remaining part of the crossing, if it is completed, will presumably utilize the South Equatorial current, and the portal of entry into the Americas will be near the easternmost extremity of Brazil.

Taking into account the four genera of echinoids that still have identical species in the Caribbean and West Africa (namely, *Diadema, Tripneustes, Lytechinus,* and *Echinometra*) and that, therefore, probably performed the Atlantic voyage during the latter part of the Tertiary (since the geological life span of species of regular echinoids is of the order of 10 million years), we might be tempted to infer that these genera only colonized the Americas quite late in the Tertiary. This, however, is proved incorrect by the geological record.

The case of *Tripneustes* is apposite. The distribution of the Recent species *Tripneustes ventricosus* is as follows: West Africa from the Gulf of Guinea to Walfisch Bay, Ascension Island, Fernando Noronha Islands (off Brazil), Trinidad, Caribbean north to Florida and Bermuda Island. Here we have laid out in a most unmistakable manner the whole course of the transportation route from West Africa to Bermuda, namely the Benguela current, the South Equatorial current, and the Gulf Stream, with colonies established on the various islands en route. We see in this case what may be expected to become the future distribution pattern of *Rotula orbiculus.*

Nevertheless, it would be wrong to suppose that we have here the documentation of the westward expansion of *Tripneustes* as a genus; all we have is the record of one particular species, apparently the last one to move westward. Reference to the earlier geological record shows that the genus *Tripneustes* ranges back in time to the Miocene, some 25 million years ago, and that before the end of the Miocene (about 12 million years ago) at least one species had already entered the Caribbean region. During the Pliocene another species became established on the Californian coast, *Tripneustes californicus,* with an arrival date perhaps around 5 or 10 million years ago; this species is considered by Mortensen to stand in the lineage of the extant west American species *Tripneustes depressus.* Thus we see that a single genus has colonized the central American region on at least three occasions, two of which occurred at times when the Central American Seaway was still open, giving access to the Pacific coast. It would seem probable that over the long time span during which a species of regular echinoid can exist without evident morphological evolution, it must have happened that the same species crossed the Atlantic on many different occasions. The later migrants interbred with their earlier precursors as they and their descendants became reunited in the west.

Thus we cannot simply count the number of transatlantic crossings by counting the number of immigrant species that we find in the fossil record. The actual number of crossings must be legion, and all that we can recognize are the number of species that were involved.

The remarks on *Tripneustes* apply equally well to *Echinometra,* another regular echinoid with identical species, namely *Echinometra lucunter,* living on either side of the Atlantic. The distribution of *E. lucunter* is as follows: West Africa (Dakar to Angola), Saint Helena, Ascension Island, Brazil (southward to Desterro), Caribbean (northward to Florida and Bermuda). This indicates a late Tertiary or

Pleistocene immigration extending over an indefinite period of time, probably not longer than about 5 million years (fossil specimens have not yet been recognized). Evidence of earlier immigrations of *Echinometra* is provided by the fossil record in the cases of other species. The oldest known species appears in the Paleocene of the Indian Tethys (about 60 million years ago), but a species had already reached Cuba by the Oligocene (about 30 million years ago). In the west Tethyan region *Echinometra* lingered on through the Miocene in France, after which it vanished from the Mediterranean, whose waters now began to cool following the elevation of the Middle East and the consequent isolation of the west Tethyan seas from the Indian Ocean. *Echinometra* survived, however, on the African coast; elsewhere the modern representatives of the genus still occupy eastern Tethys (now the Indo-West Pacific) and also the Central American region; the earlier Caribbean immigrants passed through the Panama Seaway to reach the Pacific coast.

INDEX

75 76 77 9 8 7 6 5 4 3 2 1